From Shipbreaking to Sustainable Ship Recycling

Legal Aspects of Sustainable Development

General Editor

David Freestone

This series will publish work on all aspects of the international legal dimensions of the concept of sustainable development. Its aim is to publish important works of scholarship on a range of relevant issues including conservation of natural resources, climate change, biodiversity loss and the role of international agreements, international organizations and state practice.

VOLUME 5

From Shipbreaking to Sustainable Ship Recycling

Evolution of a Legal Regime

By

Tony George Puthucherril

MARTINUS
NIJHOFF
PUBLISHERS

LEIDEN • BOSTON
2010

This book is printed on acid-free paper.

Library of Congress Cataloging-in-Publication Data

Puthucherril, Tony George.
From shipbreaking to sustainable ship recycling : evolution of a legal regime / by Tony George Puthucherril.
p. cm. -- (Legal aspects of sustainable development ; 5)
Includes bibliographical references and index.
ISBN 978-90-04-17491-7 (hardback : alk. paper) 1. Ships--Scrapping. 2. Ships--Recycling--Environmental aspects. 3. Salvage. 4. Marine pollution--Law and legislation. I. Title.
K1188.A8P88 2010
344.04'62--dc22

2010003713

ISSN 1875-0923
ISBN 978 90 04 17491 7

PRINTED IN THE NETHERLANDS

Contents

Series Editor's Preface

This is the fifth volume in the Martinus Nijhoff monograph series on *Legal Aspects of Sustainable Development* published under my general editorship. The aim of this series is to publish works at the cutting edge of legal scholarship that address both the practical and the theoretical aspects of this important concept.

I am pleased to include this major study by Tony George Puthucherril in the series. Shipbreaking poses a number of key sustainable development issues, yet it has not to date attracted much academic attention. Recycling the materials used to build huge oceangoing vessels must be a sound idea. However, when these vessels are exported from the developed world to be broken up in some of the world's poorest countries, including India, China and Bangladesh, the process poses basic north-south environmental justice issues. These issues are even more pronounced when this work is done in yards and on beaches in developing countries with minimal, if any, basic safety precautions for the workers or protection of the environment from the toxic and dangerous substances that may be released in the process.

This study was started at Dalhousie Law School well before the negotiation of the International Convention for the Safe and Environmentally Sound Recycling of Ships was finalized in Hong Kong in May 2009. Tony George Puthucherril was therefore ideally placed to produce what is probably the first authoritative assessment of the Convention. His study however is more than just this. He looks at the global significance of shipbreaking from a multidisciplinary point of view; he looks in detail at the attempts by the Indian authorities to regulate this growing industry including a consideration of the role of the Indian Supreme Court – a now familiar actor in the field of Indian environmental law after a number of seminal cases on environmental issues. Then he considers the patchwork of pre-existing international rules and guidelines, which derive from the work of the International Maritime Organization, the International Labour Organization as well as the Basel Convention Secretariat. Indeed it was these three bodies that collaborated in the development of the 2009 Convention. After a critical – in both senses of the word – assessment of the Convention itself, he poses a number of issues for future consideration and action. It is clear that the Convention is to be welcomed as a major step forward. However, as Professor Chircop points out, it could take some years more before it enters into force. In the meantime further actions can be taken to implement and to supplement the conventional regime.

In this light I trust that this timely work will provide a well informed and important stimulus for forward progress on this important agenda.

David Freestone
Washington DC

Foreword

This book is one of the first, if not the very first, substantial work on the international law of shipbreaking and ship recycling. The subject of shipbreaking is a complex problem with many difficult and interrelated issues. It has long been a virtually unregulated industry at the international level. It has stood out in sharp contrast to the otherwise substantial body of international regulation for maritime safety, environmental protection and security, mostly under the auspices of the International Maritime Organization (IMO), and work safety regulation of the International Labour Organization (ILO). Shipbreaking somehow fell between the cracks of international regulation.

And yet the need for regulation has been cogent for a long time. The industry has had a generally poor occupational health and safety and marine environmental protection record. Every year hundreds of workers in the yards of Asia, among others, suffer injuries and tragic deaths at work in shipbreaking yards. Workers are frequently not properly equipped and trained, including when they are expected to handle material such as asbestos, a toxic material used for insulation on board ships. In the yards of South Asia, ships are beached after their final voyage, and during the process of breaking they release a variety of toxic wastes directly into the coastal and marine environment. Shipbreaking also concerns the recycling of scrap metal, a highly valued commodity in the countries where the breaking occurs. The industry employs thousands of people in those countries and is a significant contributor to local and regional economies, mostly in regions which are under-developed. The bulk of the industry is located in Asia, mostly in a few places that include China, Bangladesh and India, and even though the ships to be broken were owned and operated mostly from outside of this region. This reality has produced social equity undertones, juxtaposing values and interests of rich and poor countries.

Recently, the international community has come to the realization that this problem cannot continue to be unregulated. In the space of a few years the International Convention for the Safe and Environmentally Sound Recycling of Ships was developed and in May 2009 it was formally adopted at a conference in Hong Kong. This initiative has its origins in collaboration between the IMO, ILO and the Secretariat for the Basel Convention on the Control of Transboundary Movements of Hazardous Wastes and their Disposal. It will be several years before the Convention will receive the required number of ratifications to bring the instrument into force. The hope is that the main source and shipbreaking States will become parties. It is possible to criticise the Convention for what it has not been able to achieve, or address effectively. However, few would not concede that it is an important achievement and addition to the international regulation of the industry.

Against this backdrop, Tony George Puthucherril's book breaks new ground in the fields of international maritime law and international environmental law. The book emerged from seminal graduate work the author completed while at Dalhousie Law School. He undertakes a massive, brilliant and passionate exercise in setting out the problem, tracing its history and economic significance. He deconstructs the problem as it is manifested in some of the world's major shipbreaking States and highlights the many domestic challenges within each of those. He also ably explains a complex relationship between existing international regimes, notably for shipping, labour and transboundary movement of hazardous wastes, and explains how each of these addressed issues related to the problem, but without effectively addressing the problem, hence the need for a new international instrument. He provides numerous insights into economic, social, political and institutional processes, frequently interwoven. At the end of the day, any international effort in addressing the problem had to consider issues of equity in international relations and international law. The author provides the reader with a guided tour of the Convention, providing in-depth analysis and numerous insights along the way. This book promises to be "the work" on the subject for years to come. It will serve to enlighten those that wish to seek an in-depth understanding of how the international community embarked on an exercise in policy and legal development to respond to a major contemporary problem.

Professor Aldo Chircop
Dalhousie Law School, Halifax
NS, Canada

September 2009

Acknowledgements

This work is the product of the encouragement and support that I received from several quarters, without which it would not have come into being.

First and foremost, I express my deepest debt of gratitude to Honourable Judge Sandra E. Oxner O.C., Chairperson of the Commonwealth Judicial Education Institute, Canada. I am grateful to Judge Oxner who was instrumental in bringing me to Canada and for supporting my scholarship. Without her steadfast support and magnanimity, this study would never have materialised.

I would like to thank my professors and colleagues at the Marine & Environmental Law Institute, Dalhousie Law School, Halifax, Canada. In particular, I am indebted to Professor Aldo Chircop for his intellectual inspiration, invaluable guidance and encouragement; Professor David VanderZwaag for offering illuminating comments and suggestions, which has in a great measure enhanced the quality of this work; and Professor Moira L. McConnell for reading and responding to this work and for providing me with materials and advice at all crucial stages. My sincere thanks are also due to Professor Richard Devlin (Dalhousie Law School) who inspired in me critical legal thinking.

I express my immense sense of gratitude to my law teachers in India, particularly to Professor (Dr.) N.R. Madhava Menon, (Founding Vice-Chancellor of the National Law School of India University and the National University of Juridical Sciences; Founding Director of the National Judicial Academy, India; and Member, Commission on Centre-State Relations, India) my mentor, who has always encouraged me to think and work beyond traditional legal confines. I am also grateful to Professor V. Vijayabalan (former principal of the Government Law Colleges at Trivandrum and Ernakulam and former Dean of Law, Mahatma Gandhi and Calicut Universities) who has been a pillar of strength and has provided me with guidance and support at every stage of my career. At this juncture, I fondly remember the late Professor B. Dharmarajan (former principal of the Kerala Law Academy Law College, Trivandrum) who introduced me to the first lessons in law.

Honourable Madam Justice Roshan Dalvi of the Bombay High Court and Honourable Mr. Justice Madan B. Lokur of the Delhi High Court have always encouraged me in my academic endeavours and I am obliged to both of them. I am also grateful to Dr. K.P. Kylasanatha Pillay (Senior Advocate, Supreme Court of India) for his academic support.

I am also grateful to the series editor, Professor David Freestone, for his encouragement, as well as Mr. Peter Buschman, and Ms. Ingeborg van der Laan at Brill for their assistance in making this book a reality. I would also like to thank

Ms. Susan Rolston of Seawinds Consulting Services, Halifax, Canada, who provided invaluable language and editorial assistance.

I am fortunate to be blessed by a large circle of friends many of whom have gone out of their way to help me overcome several obstacles during the writing of this work. I warmly thank David Dzidzornu (J.S.D. candidate, Dalhousie Law School) for his constructive comments and continued motivation. My utmost gratitude is due to Mr. Mathews Abraham and to Mrs. Thankamma Mathews of Halifax, who provided me with the affection and care that I needed being so far away from home. I am grateful to Mr. Charles Nellari (Vice President, Intlex Inc., USA) who went out of his way to ensure that some of my personal issues did not affect my academic work. I am appreciative of the efforts of Mr. Jessel Vinohar Rodricks (Advocate, High Court of Kerala), and Ms. Geeta Oberoi (Ph.D. candidate, Delhi University) both of whom worked to ensure that I have access to the necessary background materials.

Last, but certainly not least, I convey my gratefulness to my 90 year old grandmother, Mrs. Lily George, for her prayers and blessings. I would also like to thank my 'Acha' and 'Amma', and 'Achachan' and 'Mummy' for their love and encouragement. Ms. Lekshmi Vijayabalan, my wife needs special mention for the love and care she showered on me while I was busy writing this book.

Tony George Puthucherril
Halifax, Nova Scotia

September 2009

List of Abbreviations

ACM	asbestos containing materials
AERB	Atomic Energy Regulatory Board
AFC	International Convention on the Control of Harmful Anti-Fouling Systems on Ships, 2001
AIDS	acquired immune deficiency syndrome
BIMCO	Baltic and International Maritime Council
BWC	International Convention for the Control and Management of Ships' Ballast Water and Sediments, 2004
CEC	Central Empowered Committee
COP	conference of parties
CPCB	Central Pollution Control Board
CRZN	Coastal Regulation Zone Notification
dB	decibel
DOC	document of compliance
DWT	deadweight tonnage
EA	environmental assessment
EFTA	European Free Trade Association
ENGO	environmental non-governmental organisation
EPA	Environmental Protection Agency (United States)
ESM	environmentally sound management
EU	European Union
FOC	flag of convenience
GMB	Gujarat Maritime Board
GT	gross tonnage
HPC	High Powered Committee
ICS	International Chamber of Shipping
IHM	inventory of hazardous materials
ILO	International Labour Organization
IMO	International Maritime Organization
IMOGSR	IMO Guidelines on Ship Recycling
INR	Indian rupee
IRRC	International Ready for Recycling Certificate
ISRA	International Ship Recycling Association
IWPSR	Industry Working Party on Ship Recycling
JWG	Joint Working Group
LDT	light displacement tonne
LOS Convention	United Nations Convention on the Law of the Sea, 1982

MARAD	Department of Transportation, Maritime Administration (United States)
MARPOL 73/78	International Convention for the Prevention of Pollution from Ships, 1973 and its Protocol of 1978
MEPC	Marine Environment Protection Committee (IMO)
NDRF	National Defence Reserve Fleet
NEPA	National Environmental Policy Act
NGO	non-governmental organisation
NIMBY	not in my back yard syndrome
NMHA	National Maritime Heritage Act
OECD	Organisation for Economic Co-operation and Development
OEWG	Open-ended Working Group
OPA	Oil Pollution Act
PAHs	polycyclic aromatic hydrocarbons
PCBs	polychlorinated biphenyls
PVC	polyvinyl chloride
RRF	Ready Reserve Force
SCMC	Supreme Court Monitoring Committee
SDIC	Ship Decommissioning Industries Company
SGAV	Specific Guidelines for Assessment of Vessels
SMB	State Maritime Board
SOLAS	International Convention for the Safety of Life at Sea, 1974
SPCB	State Pollution Control Board
SRFP	Ship Recycling Facility Plan
SRP	Ship Recycling Plan
TBT	tributyltin
TEC	Committee of Technical Experts
TGSD	Technical Guidelines for the Environmentally Sound Management of the Full and Partial Dismantling of Ships
TRO	Temporary Restraining Order
TSCA	Toxic Substances Control Act
UK	United Kingdom
UNEP	United Nations Environment Programme
UNESCO	United Nations Educational, Scientific and Cultural Organization
US	United States
USD	United States dollar
VLCC	very large crude carrier
WRC	Nairobi International Convention on the Removal of Wrecks, 2007
WSR	Waste Shipment Regulation
WTO	World Trade Organization

Chapter 1

Introduction

I. Background

The dumping of hazardous wastes in the Third World has long been an intractable issue, despite the international legal regime[1] put in place to protect the victims of this trade. Popularly known as "garbage imperialism",[2] one of its more sinister forms is the shipbreaking industry wherein ships that have outlived their economic utility are dismantled for scrap steel and other equipment which are recycled and reused. Even though ship recycling is performed in nearly 79 countries,[3] most of the heavy dismantling takes place on the tidal beaches of the South. It is here that there is a huge market for end-of-life ships from the North.

Among the leading shipbreaking nations, the countries on the Indian subcontinent, namely, Bangladesh, India and Pakistan, have been the undisputed leaders.[4] Whether Alang in India, Chittagong in Bangladesh, or Gadani in Pakistan, mention shipbreaking and the first picture that it conjures is that of geriatric ships resting peacefully on these once pristine beaches, awaiting their last rites. However, beneath this deceptive calmness lies a storm. Ever since the industry started functioning in these locations, the sandy beaches have turned

---

[1] *Basel Convention on the Control of Transboundary Movements of Hazardous Wastes and their Disposal*, 22 March 1989, 1673 U.N.T.S. 126, 28 I.L.M. 657 (entered into force 5 May 1992) [*Basel Convention*].

[2] See Petra K. Kelly, "The Need for Eco-Justice" (1990) 14 Fletcher F. World Aff. 327 (HeinOnline) (accusing the West of indulging in garbage imperialism). See also Rozelia S. Park, "An Examination of International Environmental Racism through the Lens of Transboundary Movement of Hazardous Wastes", Note, (1998) 5 Ind. J. Global Legal Stud. 659 at 660 (WLeC) (pointing out that the shipment of hazardous wastes from developed to developing countries is environmental racism). See Nisha Thakker, "India's Toxic Landfills: A Dumping Ground for the World's Electronic Waste" (2006) 6 S. D. L. & Pol'y 58 (HeinOnline).

[3] Ron Hess *et al.*, *Disposal Options for Ships*, RAND Monograph Report (RAND Corporation, 2001) at 41, online: RAND Corporation <http://www.rand.org/pubs/monograph_reports/MR1377/>. Even though performed in several countries, shipbreaking in the majorioty of these countries is restricted mainly to the dismantling of government ships, fishing boats or pleasure craft. Most of the business is handled on the Indian subcontinent (almost 90 percent). France, Interdepartmental Committee on the Dismantling of Civilian and Military End-of-Life Ships, *Le rapport de la Mission Interministérielle portant sur le Démantèlement des Navires civils et militaires en fin de vie*, annex II at 2–3 (27 March 2007, Chair: Xavier de la Gorce), online: SGMer <http://www.sgmer.gouv.fr/IMG/pdf/Annex_2_Dismantling_countries.pdf>.

[4] Amy Yee, "Wreckers in deep water" *Financial Times [London]* (18 April 2007) 4, online: FT .com <http://www.ft.com/home/us>.

gooey and black, and the waters of the Arabian Sea and the Bay of Bengal that wash their shores are covered with floating oil globules. The environment in these regions has been scarred and marine biodiversity has been systematically ruined. These yards resemble huge battlefields with metal pieces, asbestos sheets, thermocol, glass bits, equipment from the ship, oil, and other substances strewn all over.[5] Almost everything, from the steel to the fixtures and furniture, to pieces of sanitary ware, kitchen utensils and other items salvaged from the ship, end up in local markets.[6] At certain places, discarded wastes are burnt, sending huge columns of smoke into the atmosphere.[7]

The ship scrapping industry on the Indian subcontinent generates a whole range of economic activities, but it comes at an enormous cost. The job of breaking a ship is extremely hazardous and lethal. The shipbreakers who do these jobs, often without protective gear, are exposed to a series of life threatening substances like asbestos, polychlorinated biphenyls, residual oil, and situations such as explosions and falling steel.[8] Resembling ants that work diligently on carcasses, the undertakers in these yards use their raw power, sledgehammers and torch cutters to rip apart the huge bellies of ships.

From the point of view of safety and environmental protection, shipbreaking happens under totally unacceptable conditions. Consequently, many of the workers in these shipbreaking yards contract lethal diseases and the water, the soil and the coastal habitats are heavily polluted. This industry raises fundamental questions of human rights, environmental justice, and equity.[9]

Theoretically, the ship scrapping industry should be a sustainable enterprise as it removes redundant tonnage by providing a new lease of life to rusty steel and to the fixtures on board a vessel. However, shipbreaking operations contain elements that undermine its sustainability. Why is this situation persisting?

Shipbreaking is symbolic of what emerging and developing economies will continue to encounter in the context of the new economic opportunities afforded by globalisation. The central question is whether these seemingly beneficial

---

[5] David Dodds, "Breaking up is Hard to Do: Environmental Effects of Shipwrecking and Possible Solutions under India's Environmental Regime", Comment, (2007) 20 Pac. McGeorge Global Bus. & D.L.J. 207.

[6] The steel re-appears as reinforcing rods which are used in the construction industry. The generators, light fittings and batteries are also reused. Even the hydrocarbons on board are reclaimed to be used as fuel in steel rolling mills and brick kilns. *Recycling of Ships: Report of the Correspondence Group*, Submitted by the Co-ordinator of the Correspondence Group, IMO/MEPC 46/7, 18 January 2001, ¶ 3.1.2 (KR-CON) [*Report of the Correspondence Group*].

[7] Dodds, *supra* note 5 at 217–20.

[8] Dwarik Nath Rath, "In the face of death" *PUCL Bulletin* (June 2003), online: What's new on the site <http://www.pucl.org/by-month-2003.htm>.

[9] John F. Sawyer, "Shipbreaking and the North-South Debate: Economic Development or Environmental and Labor Catastrophe?", Comment, (2001) 20 Penn St. Int'l L. Rev. 535 (WLeC).

economic prospects are ephemeral, since operationalisation of many of them entails suffering on the part of a vast majority of the population (who are already economically, socially and politically marginalised and have very limited bargaining power) to benefit a few.[10] Distributive justice is a far cry in most of these inequitable societies, and rarely do economic benefits trickle down to the poor.[11] However, governments and other stakeholders, while highlighting the economic benefits which the scrapping industry provides, prefer not to look at these disturbing facts in the name of ensuring the ideal climate for foreign investment and development.[12] This raises the question, what does 'sustainable development' mean for the Third World? Even though literature on sustainable development abounds,[13] the concept of sustainable development has been reduced to an oxymoron, particularly, in some of the Third World countries. In the guise of striking a balance between environment and development, development often receives precedence at the expense of genuine environmental and human right considerations.[14] This is particularly true in the ship demolition industry and warrants a detailed legal analysis.

---

[10] B.S. Chimni, "A Just World under Law: A View from the South" (2007) 22 Am. U. Int'l L. Rev. 199 at 211 (WLeC) (pointing out to the existence of two India's: one for the rich and the other for the poor).

[11] The preamble to the *Constitution of Bangladesh 1972* reads as follows, "Further pledging that it shall be a fundamental aim of the State to realise through the democratic process to socialist society, free from exploitation-a society in which the rule of law, fundamental human rights and freedom, equality and justice, political, economic and social, will be secured for all citizens." The preamble to the *Constitution of India 1950* reads, "WE, THE PEOPLE OF INDIA, having solemnly resolved to ... secure to all its citizens: JUSTICE, social, economic and political." The preamble to the *Constitution of Pakistan 1973* states, "Wherein the principles of democracy, freedom, equality, tolerance and social justice, as enunciated by Islam, shall be fully observed."

[12] *Ibid.* Surya Deva, "The Sangam of Foreign Investment, Multinational Corporations and Human Rights: An Indian Perspective for a Developing Asia" [2004] Sing. J.L.S. 305 at 316 (HeinOnline).

[13] See generally World Commission on Environment & Development, *Our Common Future* (Oxford: Oxford University Press, 1987). See also David VanderZwaag, "The Concept and Principles of Sustainable Development: Rio-Formulating Common Law Doctrines and Environmental Laws" (1993) 13 Windsor Y.B. Access Just. 39 (HeinOnline); James P. Karp, "Sustainable Development: Toward a New Vision", Essay, (1994) 13 Va. Envtl. L.J. 239 (WLeC); Emily Fisher, "Sustainable Development and Environmental Justice: Same Planet, Different Worlds?" (2003) 26 Environs Envtl. L. & Pol'y J. 201 (WLeC); Molly Harriss Olson, "Accepting the Sustainable Development Challenge" (1995) 31 Willamette L. Rev. 253 (WLeC).

[14] For instance see *Dahanu Taluka Environment Protection Group v. Bombay Suburban Electricity Supply Co. Ltd.* (1991), [1991] 2 S.C.C. 539 (India S.C.). The Indian judiciary has been a "Lion" in its environmental activism in relation to small businesses, small-scale developmental projects, local, state and federal state regulatory agencies whenever there has been a clear dereliction of duty. It has been a "Lamb" in respect of large-scale developmental projects, large business enterprises, and multinational corporations, and in cases where the state, private or foreign entities are closely allied with the government. Ayesha Dias, "Judicial Activism in the Development and Enforcement of Environmental Law: Some Comparative Insights from the Indian Experience" (1994) 6 J. Envtl. L. 243 at 261–62 (HeinOnline).

As a result of the timetable laid down by the International Maritime Organization (IMO) on the phasing out of single-hull ships,[15] and given the present global economic crisis which has severely impacted the shipping industry, it can be expected that the beaches in the developing world will soon be full of rotting ships awaiting decent burials.[16] Shipowners and recyclers stand to reap a windfall from this activity. Unfortunately, this situation will also translate into greater environmental degradation and more human casualties, unless the workings of this industry are re-engineered to allow only socially and environmentally responsible scrapping.[17]

An important normative response at the international level to deal with some of the intractable issues that have plagued the ship recycling industry for the last several decades is the IMO-sponsored *Hong Kong International Convention for*

---

[15] Several of the single-hull very large crude carriers are being converted into iron ore carriers rather than having them scrapped. For further details, see Cowan Thant Zin, "Single-hull VLCC owners opt for conversion" (6 August 2008), online: tankerworld <http://www.tankerworld.com/news/2008/08/72785>. This development may affect present predictions regarding the number of ships that will be available for scrapping in 2010. See also Recycling of Ships: Recycling Capacity for Entry into Force Criteria for the International Convention for the Safe and Environmentally Sound Recycling of Ships, Submitted by India, IMO/MEPC 58/3/14, 13 August 2008, ¶ 13 (KR-CON) (pointing out that in 2001 when MEPC adopted regulation 13G of annex I and later in 2003 when this was amended there were several predictions that the rate of demolition of single-hull tankers would sky-rocket. However due to changed market conditions this did not happen since many single hull tankers were converted into double hull tankers and then into bulk carriers and FPSOs).

[16] See generally Greenpeace, "Destination Unknown: European single hull oiltankers: no place to go" *Greenpeace International*, (December 2004), online: Greenpeace International <http://www.greenpeace.org/raw/content/india/press/reports/destination-unknown.pdf>. In 1989, the environmental catastrophe occasioned by the sinking of the 'Exxon Valdez' in Alaska led to growing concerns regarding the suitability of single-hull oil tankers to prevent maritime accidents. This led to a phase-out scheme for single-hull oil tankers in the United States with the *US Oil Pollution Act of 1990* (OPA 90). The IMO also introduced international requirements for double hulling of oil tankers in 1992 by amending annex I of MARPOL73/78. However, the phase-out timetables were not identical to the scheme under OPA 90. In 2000, the single-hull tanker 'Erika' spilled some 20,000 tonnes of oil off the coast of Brittany, France. Subsequently, the IMO introduced a revised phase-out scheme which was still less strict than OPA 90. The European Union also adopted *Regulation 417/2002*. In 2003, another single-hull oil tanker, the 'Prestige', was lost off the Galicia coast, Spain, leading to an amendment to European Union *Regulation 417/2002* by means of *Regulation 1726/2003*, which aligned the European law with the OPA 90 scheme. Subsequently, upon a proposal of the European Union member states, the IMO decided to introduce a similar acceleration of the phase-out of single-hull oil tankers at a worldwide level. EC, Directorate-General Energy & Transport, *Oil Tanker Phase Out and the Ship Scrapping Industry: A Study on the Implications of the Accelerated Phase out Scheme of Single Hull Tankers Proposed by the EU for the World Ship Scrapping and Recycling Industry* (EC, 2004) at 22–29, online: European Commission <http://ec.europa.eu/transport/maritime/safety/doc/prestige/2004_06_scrapping_studyen.pdf>.

[17] More than a thousand oil tankers will leave service creating an insatiable demand for more recycling facilities. This will increase the danger to developing countries. For instance, the Government of India is planning a new shipbreaking facility at *Kakinada* beach in the State of Andhra Pradesh, despite protests from local communities. Already oil tankers are being abandoned along the West African coastline. Recycling of Ships: The Need to Develop a Definitive and Consolidated List of Single Hull Oil Tankers Subject to Phase out Regulations, Submitted by Greenpeace International, IMO/MEPC 53/3/3, 12 May 2005 (KR-CON).

the Safe and Environmentally Sound Recycling of Ships, 2009 (Ship Recycling Convention) (a copy of the text of the Ship Recycling Convention is enclosed as an appendix to this book).[18] The convention was recently adopted at the International Conference on the Safe and Environmentally Sound Recycling of Ships, held from 11 to 15 May 2009 at Hong Kong. The Ship Recycling Convention is expected to significantly reshape both the international and domestic regimes to manage all aspects of this industry. However, even before the ink could dry, the convention has been criticised by environmental non-governmental organisations (ENGOs) as "a legal wreck" for being inadequate.[19]

The international ship recycling regime is thus at a crucial stage in its evolution. It is therefore useful to evaluate whether the ongoing efforts at the international level can lead to the development of a sustainable model for ship recycling. This study probes the following questions to determine the prospects of developing an international legal regime for ensuring the sustainable development of the ship recycling industry:

1. Is the ship recycling industry a truly global activity?
2. Can a unilateral national approach to regulate the shipbreaking industry be effective, at least in that particular jurisdiction?
3. In what ways can existing international legal instruments be utilised to ensure sustainable development of the shipbreaking enterprise, and is this framework an adequate response?
4. Does the Ship Recycling Convention contribute to sustainable ship recycling?
5. At a minimum, what needs to be done to develop an international legal regime in order that the activity moves away from being mere shipbreaking to sustainable ship recycling?

These questions are addressed in Chapters Two to Six. An outline of what is explored in each is offered as follows:

As a first step to developing a strategy for ensuring the sustainable development of this enterprise, the peculiarities of the shipbreaking industry needs to be understood. Accordingly, Chapter Two, *The Global Business of Shipbreaking*, maps out

---

[18] For the text of the Convention, see International Conference on the Safe and Environmentally Sound Recycling of Ships: Agenda Item 8, Adoption of the Final Act and Any Instruments, Recommendations and Resolutions Resulting from the Work of the Conference: Hong Kong International Convention for the Safe and Environmentally Sound Recycling of Ships, 2009, Text Adopted by the Conference, IMO/SR/CONF/45, 19 May 2009, online: SJÖFARTSVERKET <http://www.sjofartsverket.se/pages/19514/sr-conf-45.pdf> (opened for signature 1 September 2009) [*Ship Recycling Convention*].

[19] Basel Action Network, Media Release, "I.M.O. Ship Recycling Convention Denounced as "Legal Shipwreck": Activists Call for Ban on Toxic Ship Beaching" (11 May 2009), online: Basel Action Network <http://www.ban.org/BAN_NEWS/2009/090511_imo_legal_shipwreck.html>.

some of the characteristics of the transnational ship recycling industry. These include the centrality of the shipbreaking industry in the growth and development of shipping, factors that render this industry dangerous, and the different players and their stake in the ship scrapping business. It also provides a snapshot of shipbreaking practices from around the world with a focus first on the industry in South Asia, including China, followed by a review of the experience in OECD countries.

Chapter Three, *Limitations of a National Response to Regulate the Global Shipbreaking Industry: A Study of the Indian Experience,* uses the Indian regulatory model on shipbreaking to test the hypothesis that unilateral national approaches are insufficient to regulate an industry that is transnational in character. Until recently, India was the primary shipbreaking nation, a position now taken over by Bangladesh. The main reason for this change is that, in comparison to other players on the subcontinent, India has developed a comprehensive law on shipbreaking, stories of human rights violations and environmental degradation are better documented, and its apex court has played a significant role in developing a legal framework for regulating shipbreaking activities. This chapter examines the Indian legislative response to the problem, and reviews the judicial contribution to developing the regulatory framework for sustainable ship recycling. Case studies involving the arrival of certain toxic ships in Indian waters are utilised to examine the interplay of market forces and the efficacy of national regulatory regimes to prove why unilateral national regulatory approaches to shipbreaking are bound to fail.

The subject of ship recycling is highly complex due to the international dimension of the shipping industry. A major reason why national regulatory regimes have failed is that relevant domestic laws are incapable of regulating transnational actors who operate outside its jurisdictional ambit. What aggravates the regulatory dilemma is that that there are very few mandatory standards on shipbreaking at the international level. Accordingly, Chapter Four, *Contemporary International Law and Ship Recycling*, examines the scope of the extant international regulatory regime *via* four steps. First, it analyses the ship recycling industry from the perspective of international hazardous waste management standards, where the focus is on the *Basel Convention on the Control of Transboundary Movements of Hazardous Wastes and their Disposal, 1989* (Basel Convention), the Basel Ban, and the Technical Guidelines for the Environmentally Sound Management of the Full and Partial Dismantling of Ships. The second line of discussion examines shipbreaking from the standpoint of the law of the sea. Thereafter, the role of the International Labour Organization (ILO) in shipbreaking is discussed. Finally, the issue is analysed from the perspective of international maritime law. The concluding argument posits the need for a strong stand-alone binding international legal instrument on ship recycling.

The long wait for an international legal regime on ship recycling may perhaps come to an end once the Ship Recycling Convention comes into force.

Chapter Five, *Deciphering the Ship Recycling Convention*, is an analysis of its terms. The chapter considers the question of whether this convention is an improvement on the Basel Convention and the IMO Guidelines on Ship Recycling, 2003[20] or whether it derogates from the protections afforded by these instruments. The argument that there are serious lacunae in the Ship Recycling Convention necessitating the need for improvements if a sustainable legal regime on ship recycling is to emerge at the international level, is examined.

As will be seen, the activity of breaking ships for steel results in gross human rights violations and environmental damage, primarily, because simple economic development considerations can override respect for the environment and the human rights of workers, and industry leaders fail to recognise the interrelatedness of these three elements. The concluding Chapter Six sets out certain prescriptions for a sustainable model for ship recycling. The argument is that when the suggested guideposts, along with others relevant to the issue-area, are incorporated into the evolving legal regime on the subject, the ship recycling industry would then begin its journey towards sustainability in the aspect of its operations.

II. A Note on Terminology

Often termed as 'shipbreaking', 'ship demolition', 'ship scrapping', 'shipwrecking', 'ship decommissioning' – the idea to be conveyed by each of these terms is the same, namely, the process of tearing down the hull for its steel components and other parts which are, thereafter, recycled. Interestingly, the ILO uses the term "shipbreaking"[21] while the Basel Convention describes the process as "ship dismantling".[22] A new entrant into the debate is "ship recycling" endorsed by the IMO.[23]

---

[20] See generally IMO Guidelines on Ship Recycling: Res. A.962(23) Adopted on 5 December 2003 (Agenda Item 19), A 23/Res.962, 4 March 2004 (KR-CON) [IMOGSR].

[21] See generally *Programme on Safety and Health at Work and the Environment (SafeWork)*, online: International Labour Organization <http://www.ilo.org/global/lang–en/index.htm>. Shipbreaking or ship scrapping has been defined by the United States Occupational Safety and Health Administration as, "any breaking of a vessel's structure for the purpose of scrapping the vessel, including the removal of gear, equipment, or any component of a vessel." U.S., Environment Protection Agency, *A Guide for Ship Scrappers: Tips for Regulatory Compliance* (EPA 315-B-00-001) (Washington, D.C.: Office of Enforcement and Compliance Assurance, 2000) at 1–1, online: U.S. Environment Protection Agency <http://www.epa.gov/oecaerth/resources/publications/civil/federal/shipscrapguide.pdf>.

[22] UNEP, Conference of the Parties to the Basel Convention on the Control of Transboundary Movements of Hazardous Wastes and their Disposal, 6th Mtg., (9–13 December 2002) Technical Guidelines for the Environmentally Sound Management of the Full and Partial Dismantling of Ships, UNEP/CHW.6/23, 8 August 2002, online: Basel Convention <http://www.basel.int/meetings/cop/cop6/cop6_23e.pdf>.

[23] See generally IMOGSR, *supra* note 20.

Amongst these terms, "ship recycling" seems to be the most appropriate. It is much broader as it implies an element of sustainability. The Ship Recycling Convention defines "ship recycling" as

> the activity of complete or partial dismantling of a ship at a Ship Recycling Facility in order to recover components and materials for reprocessing and re-use, whilst taking care of hazardous and other materials, and includes associated operations such as storage and treatment of components and materials on site, but not their further processing or disposal in separate facilities.[24]

However, as can be seen from this definition, the 'sustainability' element is reflected only from a narrow resource utilisation perspective[25] and, to a lesser degree, in the need to take care of the hazardous materials. The costs to human health and the environment and the need for strategies to address these issues have been ignored.[26] In other words, sustainability has to be addressed not only in objectives; it should also be seen in the processes that characterise the functioning of this industry.

Whatever term an individual prefers, each describes this activity more or less in terms of dismantling a ship and recovering materials for re-processing. Accordingly, these terms are used inter-changeably throughout this book. But, as seen from the title to this study, the emphasis is placed on its prescription, which calls for a focused attention on sustainability both in objectives and in the processes of the industry. No matter which term is used, this study seeks to encourage the larger and more important goal of 'sustainable ship recycling'.

A Final Note on Terminology

Even though vessel is a generic term, not every vessel that is employed in daily use can be considered to be a ship.[27] Under the Ship Recycling Convention, a ship is defined as, "a vessel of any type whatsoever operating or having operated in

---

[24] *Ship Recycling Convention, supra* note 18 art. 2(10).

[25] Approximately, in terms of weight, more than 99 percent of the material of most ships can be economically reused (most of the recycled material is scrap steel). See Recycling of Ships: Comments on the Draft Convention and the Inventory of Hazardous Materials, Submitted by the Community of European Shipyards' Associations, IMO/MEPC 55/3/15, 18 August 2006, ¶ 3 (KR-CON). *Report of the Correspondence Group, supra* note 6 ¶ 3.1.3 (noting that to produce one tonne of finished steel from scrap requires only 11gJ of energy as against 25gJ to produce one tonne from iron ore).

[26] France, Interdepartmental Committee on the Dismantling of Civilian and Military End-of-Life Ships, *Le rapport de la Mission Interministérielle portant sur le Démantèlement des Navires civils et militaires en fin de vie*, at 12 (27 March 2007, Chair: Xavier de la Gorce), online: SGMer <http://www.sgm er.gouv.fr/article.php3?id_article=52>.

[27] Edgar Gold, Aldo Chircop & Hugh Kindred, *Maritime Law* (Toronto: Irwin Law, 2003) at 143.

the marine environment and includes submersibles, floating craft, floating platforms, self elevating platforms, Floating Storage Units …, and Floating Production Storage and Offloading Units … including a vessel stripped of equipment or being towed."[28] Accordingly, the terms ship and vessel will also be used interchangeably throughout this book.

[28] *Ship Recycling Convention*, *supra* note 18 art. 2(7).

Chapter 2

The Global Business of Shipbreaking

I. Introduction

Ships often evoke strong emotions about new economic opportunities, conquests and adventures.[1] A major reason for the present-day economic prosperity that many Western countries enjoy in comparison to their Asian counterparts can be traced historically to their strong navies[2] and merchant fleets enabling their writ over the seas and, in due course, the land.[3] Like newborn babies, ships are welcomed into our midst – "champagne and christening galas, but what of their final hours?"[4] Some ships are "fortunate to die nobly in battle or ram icebergs on their maiden voyage".[5] However, the overwhelming majority of ships that have attained superannuation after having put in some 25 to 30 years of service on unforgiving seas end up on the once pristine beaches of Asia. Here, they are torn down for their steel and other parts, which are put to more productive use.

An activity as old as shipbuilding, shipbreaking is an essential and integral component of the shipping business.[6] Once they have outlived their use, ships have been subjected to the axe to recover useable materials.[7] However, with the

---

[1] Edgar Gold, *Maritime Transport: The Evolution of International Marine Policy and Shipping Law* (Massachusetts: Lexington Books, 1981) at 1–24 (providing an overview of the origins of shipping).

[2] However, this does not mean that Asian civilisations did not have navies. For further details, see R.P. Anand, *Origin and Development of the Law of the Sea: History of International Law Revisited* (The Hague: Martinus Nijhoff Publishers, 1982) at 12–34.

[3] *Ibid.* at 72–115.

[4] Julie McElroy-Brown, "Shipbreaking at Alang, India "What is the Right Thing for this Place?"" (2006) 3 [unpublished, archived at University of Washington Libraries Digital Initiatives Unit], online: University of Washington Libraries <https://digital.lib.washington.edu/dspace/bitstream/1773/2630/1/McElroyBrown_project.pdf>.

[5] John F. Sawyer, "Shipbreaking and the North-South Debate: Economic Development or Environmental and Labor Catastrophe?", Comment, (2001) 20 Penn St. Int'l L. Rev. 535 (WLeC).

[6] Ron Hess *et al.*, *Disposal Options for Ships*, RAND Monograph Report (RAND Corporation, 2001) at 41, online: RAND Corporation <http://www.rand.org/pubs/monograph_reports/MR1377/>.

[7] In the 18th century, ships were sold to the breaker for recovering spars, firewood, iron and brass parts to be used in new ships or to be re-melted. Old lines were shredded to make oakum for caulking the seams of new ships. *Ibid.* at 42. See also Recycling of Ships: Report of the Correspondence Group, Submitted by the Co-ordinator of the Correspondence Group, IMO/MEPC 46/7, 18 January 2001, ¶ 1.1 (KR-CON) (noting that in the past, serviceable material from wooden ships were also used to build wood-framed houses and structures like jetties or beach quoins) [Report of the Correspondence Group].

advent of large modern ships with diverse structural complexity, the ship recycling industry became complicated and dangerous. Till the 1970s, large shipyards in the United States and in Europe performed shipbreaking. With mounting economic costs for safe demolition of ships, industrialised countries found a convenient way to dispose of obsolete ships – pack them off to the developing world where there is a great hunger for steel for development activities.[8]

Initially, the industry relocated to Taiwan[9] and Korea.[10] By the early 1990s, realising the dangers which the activity entailed, these countries followed the West in discouraging shipbreaking.[11] Thereafter, business gravitated primarily to the beachheads of Third World countries, namely, India, Bangladesh, and Pakistan.[12]

---

[8] Sawyer, *supra* note 5 at 536.

[9] After the Second World War, the industry was established in Taiwan. It expanded rapidly and Taiwan became a world leader in shipbreaking. Demolition took place in two state-owned sites at the deepwater port of Kaohsiung using specially built berths and dockside cranes. However, by the early 1990s, Taiwan closed the demolition yards and replaced them with a container terminal. Martin Stopford, *Maritime Economics* 2d ed. (London: Routledge, 1997) at 486.

[10] In the 1980s, South Korea was the third largest shipbreaker in the world with a market share of about 17 percent. The activities were carried out in two demolition yards owned by Hyundai. By late 1980s, the yards were closed down due to rising wages. *Ibid.*

[11] See Chung-Te Fu & Shian-Chee Wu, "Bioaccumulation of Polychlorinated Biphenyls in Mullet Fish in a Former Ship Dismantling Harbour, a Contaminated Estuary, and Nearby Coastal Fish Farms (2005) 51 Mar. Pollution Bull. 932 (ScienceDirect). This study investigated the severe bioaccumulation of PCBs in fish from Ann-Ping harbour in Taiwan, which operated a ship dismantling business in the 1980s. It notes that even though PCB contamination stopped for over a decade, the residual contaminants in the soil and sediments contribute to the body burden of fish residing in the estuary and in the harbour. *Ibid.*

[12] The top ten dismantling countries in terms of the number of dismantled merchant ships ≥ 100 GT in 2006–2007: Bangladesh (ships in 2006–161; ships in 2007–118; total ships in 2006–2007 – 279; percentage of total – 28.5); India (ships in 2006–113; ships in 2007–155; total ships in 2006–2007–268; percentage of total – 27.3); Turkey (ships in 2006–60; ships in 2007–50; total ships in 2006–2007–110; percentage of total – 11.2); Pakistan (ships in 2006–22; ships in 2007–38; total ships in 2006–2007–60; percentage of total – 6.1); Denmark (ships in 2006–32; ships in 2007–18; total ships in 2006–2007–50; percentage of total – 5.1); China (ships in 2006–30; ships in 2007–14; total ships in 2006–2007–44; percentage of total – 4.5); Spain (ships in 2006–5; ships in 2007–23; total ships in 2006–2007–28; percentage of total – 2.9); United Kingdom (ships in 2006–8; ships in 2007–13; total ships in 2006–2007–21; percentage of total – 2.1);United States of America (ships in 2006–10; ships in 2007–7; total ships in 2006–2007–17; percentage of total – 1.7); Norway (ships in 2006–8; ships in 2007–5; total ships in 2006–2007–13; percentage of total – 1.3). EC, Commission, *Impact Assessment for an EU Strategy for Better Ship Dismantling* (Brussels: SEC (2008) 2846), at 11, online: European Commission <http://ec.europa.eu/environment/waste/ships/pdf/impact_assessment.pdf> [Commission, *Impact Assessment*]. The top ten dismantling countries in terms of tonnage of dismantled merchant ships ≥100 GT in 2006–2007: Bangladesh (total tonnage 2006–2007(1,000 GT) – 5,025; percentage of total – 51.7); India (total tonnage 2006–2007(1,000 GT) – 2,413; percentage of total – 24.8); China (total tonnage 2006–2007(1,000 GT) – 5,025; percentage of total – 51.7); Pakistan (total tonnage 2006–2007(1,000 GT) – 632; percentage of total – 6.5); Turkey(total tonnage 2006–2007 (1,000 GT) – 410; percentage of total – 4.2); Unknown (total tonnage 2006–2007(1,000 GT) – 201; percentage of total – 2.1); USA (total tonnage 2006–2007 (1,000 GT) – 165; percentage of total – 1.7); Canada (total tonnage 2006–2007 (1,000 GT) – 41; percentage of total – 0.4); Denmark (total tonnage 2006-2007 (1,000 GT) – 26; percentage of total – 0.3);

This chapter reviews the existing patterns and practices in the shipbreaking industry to understand its global character and some of its intractable issues from two perspectives. First, shipbreaking on the Indian subcontinent, using Alang as representative of shipbreaking yards in Bangladesh and Pakistan, is examined. Second, the in-house dismantling capabilities of OECD countries are reviewed with Turkey, the undisputed leader in shipbreaking in the OECD, the focus. In addition, a critique on the European Union's Waste Shipment Regulations (WSR)[13] and the controversy over the export of ghost ships from the United States' National Defence Reserve Fleet is offered. The major features of this global industry and these case studies arguably lead to the conclusion that international standards for ship recycling, along with greater North-South cooperation, are necessary to ensure sustainable ship recycling.

II. The Shipbreaking Industry: An Overview

The shipping industry is comprised of four basic markets: the new building market which trades in new ships; the freight market which provides sea transport services; the sales and purchase market which trades in second hand ships; and finally, the demolition market where obsolete vessels are sold and purchased.[14] In this scheme, the demolition market, an integral part of the shipping industry, plays an important "buffer role"[15] to balance the shipping industry. At any given time, only limited numbers of new ships are built.[16] Since most of

---

Norway (total tonnage 2006–2007 (1,000 GT) – 25; percentage of total – 0.3). *Ibid.* at 13. See generally Judit Kanthak & Nityanand Jayaraman, *Ships for Scrap III: Findings of a Greenpeace Study on Workplace and Environmental Contamination in Alang-Sosiya Ship Breaking Yards, Gujarat, India*, Steel and Toxic Wastes for Asia (The Netherlands: Greenpeace International, 2001), online: Greenpeace Nederland <http://www.greenpeaceweb.org/shipbreak/shipsforscrap3.pdf>; Eco Matser, Howard Liu & Marietta Harjono, *Ships for Scrap IV: Findings of a Greenpeace Visit to Four Shipbreaking Yards in China*, Steel and Toxic Wastes for Asia (The Netherlands: Stichting Greenpeace Nederland, 2001), online: Greenpeace Nederland <http://www.greenpeaceweb.org/shipbreak/shipsforscrap4.pdf>.

[13] EC, *Council Regulation 259/93 of 1 February 1993 on the supervision and control of shipments of waste within, into and out of the European Community*, [1993] O.J. L 30/1 [*WSR 259/93*]; EC, *Regulation (EC) 1013/2006 of the European Parliament and of the Council of 14 June 2006 on shipments of waste*, [2006] O.J. L 190/1 [*WSR 1013/2006*].

[14] Stopford, *supra* note 9 at 77–112.

[15] Katrine Vetaas Vedeler, *From Cradle to Grave – Value Chain Responsibility in the Ship Scrapping Industry* (Master Thesis, Norwegian School of Economics and Business Administration, 2006) at 21, online: BORA-NHH <http://bora.nhh.no/>.

[16] EC, Directorate-General Energy & Transport, *Oil Tanker Phase Out and the Ship Scrapping Industry: A Study on the Implications of the Accelerated Phase out Scheme of Single Hull Tankers Proposed by the EU for the World Ship Scrapping and Recycling Industry* (EC, 2004) at 42, online: European Commission <http://ec.europa.eu/transport/maritime/safety/doc/prestige/2004_06_scrapping_study_en.pdf>.

trade is seaborne,[17] during an upswing in the world economy, the demand for sea transport services also increases, pushing up the price for such services.[18] Increased freight costs leads to better prices for second hand ships, thereby revitalising this market segment and reducing the numbers that are sold as scrap, which in effect raises the demolition market price. Such a scenario leads to a corresponding spurt in shipbuilding activities[19] to bridge the gap between demand and supply,[20] eventually stabilising the market. Consequently, if the average age of ships scrapped in 2000 was 19 years,[21] by 2007, it had increased to 30 to 35 years.[22]

During a period of economic recession, as is happening presently, international trade stagnates. The result is overcapacity in the freight market, and more ships are sent to the scrap yards.[23] The mounting costs and dangers of maintaining

---

[17] It is estimated that world seaborne trade increased significantly in 2006, reaching 7.4 billion tonnes of loaded goods, the annual growth rate being 4.3 percent. The breakdown of world seaborne loaded goods by continent is as follows: Africa (10.7 percent), America (21.5 percent), Europe (19.6 percent), Oceania (9.1 percent), and Asia 39.1 percent (the largest share). For further details see UNCTAD Secretariat, *Review of Maritime Transport*, UNCTAD/RMT/2007 (2007) 4.

[18] *Ibid.*

[19] *Ibid.* at 42 (noting that the tonnage of containerships on order has tripled since December 2006 (an increase of 220 percent); orders for oil tankers increased by 193 percent; orders for dry bulk carriers by 154 percent; and orders for general cargo vessels by 102 percent).

[20] France, Interdepartmental Committee on the Dismantling of Civilian and Military End-of-Life Ships, *Le rapport de la Mission Interministérielle portant sur le Démantèlement des Navires civils et militaires en fin de vie*, at 13 (27 March 2007, Chair: Xavier de la Gorce), online: SGMer <http://www.sgmer.gouv.fr/IMG/pdf/2007-06-18_-_Rapport_MIDN_english_version.pdf> [*MIDN*] (pointing out that even though the global fleet rose from 650 million DWTs in 1995 to 920 million DWTs in 2006, a 40 percent increase in transport capabilities, it is still not sufficient to meet the increase in goods exchange entailed by globalisation).

[21] Sawyer, *supra* note 5 at 538.

[22] *MIDN*, *supra* note 20 at 12. This figure excludes the global fishing fleet which represents some 45,000 high seas units. Assuming that ships have an average operational life of about 30 years, nearly 1,500 ships are to be dismantled every year. However, the reality is different: between 1994 and 2004, the number of dismantled ships varied from 500 to 1,200 ships. During the same decade, the destruction of transport capabilities fluctuated between 15 million deadweight tonnes (DWTs) at the beginning of the period, with a peak of 30 million DWTs in 1999, followed by a slow decrease to seven to six million DWTs in 2005/2006 with about 300 ships per year. Rising freight prices and an inability to meet the rising demand for new ships has resulted in a higher average age for ships, which has risen from 26 years between 1994 and 1999 to current averages of 29 year old for oil-tankers, 30 year old for cargos and 34 year old for liners or ferries. *Ibid.*

[23] Vedeler, *supra* note 15 at 21. For instance, OSLO-listed Eitzen Group has cancelled 11 new building orders with a yard in Japan. See Craig Eason, "Eitzen Cancels a Third of its Newbuild Orders" *Lloyd's List* (8 April 2009), online: Lloyd's List <http://www.lloydslist.com/ll/news/viewArticle.htm?articleId=20017637394>. Craig Eason, "Srab Walks Away from Newbuilding Order" *Lloyd's List* (20 March 2009), online: Lloyd's List <http://www.lloydslist.com/ll/news/viewArticle.htm?articleId=20017630516> (reporting that the Swedish shipowner Srab walked away from a three vessel newbuilding order ignoring the three million Euros down payment that it made to the Turkish yard after values of the three 8,400 DWT vessels dropped by up to six million Euros each).

obsolete ships[24] together with the phasing out of single-hulls,[25] is leading to an anticipated glut of ships that will need recycling services in the coming years.[26]

There are few practical alternatives to shipbreaking. Lay-up only postpones the issue and opportunities for converting ships into museums, storage facilities or tourist attractions is generally limited.[27] If not scrapped, ships have to be scuttled,[28] *i.e.,* they are decontaminated and sunk in a carefully chosen location to form an artificial reef.[29] However, scuttling can result in expensive clean-up costs[30] and there is always the possibility of hidden dangers.[31] As well, there is no opportunity to utilise the steel and other materials. This leaves shipbreaking as the primary acceptable mode for ship disposal.[32] It is therefore prudent to ensure that shipbreaking practices are streamlined lest ships be dumped at sea or abandoned in internal waters or in ports where they may pose a risk to human health and the environment.[33]

---

24 Matt Cohen, "U.S. Shipbreaking Exports: Balancing Safe Disposal with Economic Realities" (2005) 28 Environs Envtl. L. & Pol'y J. 237 at 238 (noting that since ships must renew their seaworthiness certificate after 25 to 30 years of service, it becomes difficult to meet the requirements for a seaworthy certificate and, consequently, ships are decommissioned). See also Revised List of Certificates and Documents Required to be Carried on Board Ships, MSC/Circ.1151, FAL.2/Circ.87, MEPC/Circ.426, 17 December 2004 (KR-CON).

25 For the full timetable see IMO Meeting Adopts Accelerated Single-hull Tanker Phase-out, New Regulation on Carriage of Heavy Fuel Oil, MEPC 50th Sess. (1 & 4 December 2003), online: International Maritime Organization <http://www.imo.org/About/mainframe.asp?topic_id=758&doc_id=3341>.

26 Greenpeace, "Destination Unknown: European single hull oiltankers: no place to go" *Greenpeace International*, (December 2004) at 3, online: Greenpeace International <http://www.greenpeace.org/raw/content/india/press/reports/destination-unknown.pdf>. See also Craig Eason "DNV Predicts Fleet Overcapacity Could Reach 10,000 Vessels" *Lloyd's List* (10 March 2009), online: Lloyd's List <http://www.lloydslist.com/ll/news/viewArticle.htm?articleId=20017626328>.

27 See generally IMO Guidelines on Ship Recycling: Res. A.962(23) Adopted on 5 December 2003 (Agenda Item 19), A 23/Res.962, 4 March 2004, ¶ 1.2 (KR-CON) [IMOGSR].

28 *Ibid.*

29 *ARSBC Artificial Reefs in British Columbia*, online: Artificial Reef Society of British Columbia <http://www.artificialreef.bc.ca/OurReefs/index.htm> (noting that since 1991, the Artificial Reef Society of British Columbia has sunk seven ships off British Columbia to form artificial reefs).

30 Hess *et al.*, *supra* note 6 at 59-80.

31 Marianne White, "Warships sunk for artificial reefs may pose PCB hazard" *The Province [of Canada]* (8 January 2008), online: The Province <http://www.canada.com/theprovince/news/story.html?id=5ef3226a-d6ca-43c4-abeb-275a4b2c20f8> (reporting that environmental groups are worried about the former 'HMCS Saguenay', a warship sitting on the ocean floor outside Lunenburg, Nova Scotia. This ship is believed to contain PCB-contaminated wiring).

32 Sawyer, *supra* note 5. IMOGSR, *supra* note 27.

33 *Convention on the Prevention of Marine Pollution by Dumping of Wastes and Other Matter*, 29 December 1972, 1046 U.N.T.S. 120, 11 I.L.M. 1291 (entered into force 30 August 1975), online: International Maritime Organization <http://www.imo.org/includes/blastDataOnly.asp/data_id%3D16925/LC1972.pdf>, replaced by *1996 Protocol to the Convention on the Prevention of Marine Pollution by Dumping of Wastes and Other Matter, 1972 and Resolutions Adopted by the Special Meeting*, 7 November 1996, 36 I.L.M. 1 (entered into force 11 March 2008).

Almost every part of the ship – the hull, machinery, equipment, fittings, generators, batteries, hydrocarbons, and even furniture – can be re-used. Virtually nothing goes to waste. Thus, ship recycling is considered an integral element in a sustainable development strategy providing jobs, raw materials for construction, and economic benefits.[34] Initially, however, there was little knowledge regarding the true nature and character of the deadly cocktail of hazardous and carcinogenic substances and chemicals used in shipbuilding. Construction parameters were not designed to meet the burial requirements of a ship. Consequently, huge quantities of asbestos and other hazardous materials were employed in the construction of vessels built primarily between the 1960s and the early 1980s.[35] These materials pose the biggest obstacle to clean shipbreaking as will be discussed below.[36]

Technically the best and safest known insulation and fire resistant material, asbestos was used in various substances until its hazardous properties came to be known.[37] Consequently, it has been remarked that anyone who has entered a building structurally more complex than "a dung-smeared hut" has probably been exposed to asbestos.[38] Asbestos exposure can be lethal; it can lead to asbestosis, lung cancer and mesothelioma.[39] In ships, asbestos is generally used as a thermal and fire insulator, and can be found in gaskets, pipes, bulkheads and walls. It is also used as a noise and vibration dampener and is seen in high concentrations in the engine room. Due to vibrations of the ship or during repair of asbestos-containing construction materials, asbestos fibres break free into the air endangering the health of seafarers and others on board.[40]

---

[34] *Ship Recycling*, online: EMSA <http://www.emsa.eu.int/end185d012d002.html>; *Ship Recycling*, online: U.S. Department of State <http://www.state.gov/r/pa/ei/pix/b/sat/82342.htm>.

[35] EC, Commission, *Green Paper on Better Ship Dismantling* (Brussels: COM (2007) 269 Final), at 2, online: European Commission <http://ec.europa.eu/environment/waste/ships/pdf/com_2007_269_en.pd f> [Commission, *Green Paper*].

[36] The IMO Guidelines identified the following items on a ship that can contain potentially hazardous substances: electrical equipment (e.g., transformers, batteries, accumulators); coolers; scrubbers; separator; heat exchangers; storage facilities for production and other chemicals; tanks, diesel tanks including bulk storage tanks; stored solvents, and other chemical stocks; paints; electrical cabling installed before 1975 (plastic covering may contain PCBs); sacrificial anodes; fire extinguishing and firefighting equipment; piping, valves and fittings; pumps and compressors; engines and generators; oil sumps; hydraulic systems; and light fittings and fixtures. IMOGSR, *supra* note 27 s. 4.2.

[37] Sawyer, *supra* note 5 at 539. See also Report of the Correspondence Group, *supra* note 7 at 23 (pointing out that most ships that were built in the 1970s, are the ones that are now being made available for scrapping, and these contain asbestos in high concentrations. Generally, steam ships contain more asbestos. Typically a 200,000 DWT steam-powered very large crude carrier (VLCC) would contain 6,000–7,500 kgs of asbestos or asbestos containing material).

[38] Katie Nester, "*Norfolk & Western Railway Company v. Ayers*: Asbestosis-Inflicted Plaintiffs and Fear of Cancer Claims", Notes, (2004) 23 St. Louis U. Pub. L. Rev. 367 at 367.

[39] *Ibid.* (noting that the turn of the century saw close to 200,000 cases in state and federal courts in the United States).

[40] See generally Eero Pukkala & Heikki Saarni, "Cancer Incidence among Finnish Seafarers, 1967–92" (1996) 7:2 Cancer Causes & Control 231 (JSTOR).

Growing realisation that asbestos is a silent killer on ships has even led to a specialised class of litigation – asbestos litigation.[41] The International Maritime Organization (IMO) has established guidelines to ensure the maintenance and monitoring of on-board materials containing asbestos.[42] Recently, an amendment was affected to regulation II-1/3-5.2, of the *International Convention for the Safety of Life at Sea, 1974* (SOLAS) which prohibits all new installations of asbestos on board ships, without exceptions from 1 January 2011.[43] Thus, during their life time these vintage ships take a heavy toll on human health and more importantly, in their death, they take more lives.[44]

Due to their high heat resistance, superior insulating and fire-retardant properties, and their ability to take on a variety of physical forms that range from oily liquids to waxy solids, polychlorinated biphenyls (PCBs), once termed as an "industrial miracle", have had multifarious industrial and commercial applications.[45] Discovered in the late 19th century, the massive production and use of the PCBs started in 1929.[46] So highly valued were these chemicals that some city codes in the United States even banned the use of electrical equipment in public buildings that did not contain them.[47] Many insurance companies required equipment in certain locations to have PCBs.[48] By the 1960s, however, there was growing awareness that the same qualities that made them attractive could pose serious environmental and health concerns since PCBs had the ability to enter the food chain.[49] A simple exposure by itself increases the risk of developing various

[41] Brian P. Brancato, "Asbestosis Litigation and Marine Insurance" (1995) 20 Tul. Mar. L.J. 105 (WLeC) (pointing out to the technical difficulties in asbestosis litigation which marine insurers face). See also *Consumer Education and Research Centre v. Union of India* (1995), [1995] A.I.R. S.C. 922 (India S.C.) (although not directed to the shipbreaking scenario, the Supreme Court of India has laid down guidelines to protect the occupational health of workers who are engaged in asbestos industries) [*Consumer Education and Research Centre*].

[42] Guidelines for Maintenance and Monitoring of On-board Materials Containing Asbestos, MSC/Circ.1045, 28 May 2002, ¶ 2 (KR-CON).

[43] See Adoption of Amendments to the International Convention for the Safety of Life at Sea, 1974, as Amended: Res.MSC.282(86) Adopted on 5 June 2009 (KR-CON).

[44] Asbestos is responsible for the most deaths in shipbreaking yards in the Third World. See Annie Zaidi, "Killer mineral" *Frontline [of India]* 23:20 (7 October 2006), online: Frontline <http://www.hinduonnet.com/fline/fl2320/stories/20061020002610000.htm>. See also *Consumer Education and Research Centre, supra* note 41. It is estimated that nearly 1,000–3,000 tonnes of asbestos is generated from the ship dismantling yards every year. Commission, *Impact Assessment, supra* note 12 at 14.

[45] Thomas R. Head, III, "PCBs-The Rise and Fall of an Industrial Miracle" (2005) 19:4 Nat. Resources & Env't. 15.

[46] *Ibid.*

[47] *Ibid.*

[48] *Ibid.*

[49] The Japanese were the first to discover that PCBs have devastating effects on human health following the Kanemi Rice Oil Incident of 1968. This led to the *Chemical Substances Control Law of 1973*. Later, the United States responded to this issue with the *Toxic Substances Control Act of 1976*. Ronald Star, "American and Japanese Controls on Polychlorinated Biphenyls (PCBs)" (1976) 1 Harv. Envtl. L. Rev. 561.

cancers and can impair the immune, reproductive, nervous, and endocrine systems.[50] It has been observed that "the story of PCBs and how they have spread throughout the planet and into the body fat of almost every living creature is one of the most fascinating and instructive Chapters in the history of the era of synthetic chemicals".[51]

In ships, PCBs are generally found in wiring and other electric products. When a ship is torn down, to recover copper from electric wiring, the plastic coated wires are usually burnt. As this happens on the coast, the PCBs are released into the oceans where its chemical composition is altered, enabling this carcinogen to accumulate in high concentrations in the flesh of fish and other marine animals.[52] Humans who eat PCB-contaminated fish, which may not always be the local population, face a similar health threat.[53]

For decades, tributyltin[54] (TBT), an organotin compound, was used in the anti-fouling paints of ships as a biocide to inhibit the growth of drag-inducing barnacles, algae, and other organisms on ship hulls.[55] However, its extreme stability and resistance to natural degradation in water renders it the most toxic substance ever deliberately introduced into the marine environment.[56] TBTs interfere with hormonal behaviour in marine fauna which when consumed by humans can debilitate their health.

Lead and mercury are the primary heavy metals of concern in shipbreaking. Lead is toxic and is commonly found in batteries, paints, and components of motors, generators, piping and cables.[57] Perhaps the oldest known industrial disease, lead poisoning affects the nervous system, and impairs hearing, vision and muscle coordination.[58] In children, it causes learning difficulties, mental

---

[50] David Dodds, "Breaking Up is Hard to Do: Environmental Effects of Shipwrecking and Possible Solutions under India's Environmental Regime", Comment, (2007) 20 Pac. McGeorge Global Bus. & D. L.J. 207 at 218.

[51] Theo Colborn, Dianne Dumanoski & John Peter Meyers, *Our Stolen Future* (New York: Penguin, 1997) at 89.

[52] *Ibid*. See also Report of the Correspondence Group, *supra* note 7 at 24.

[53] Dodds, *supra* note 50.

[54] For a detailed analysis on anti-fouling systems and shipbreaking, see Part V of Chapter Four titled, "Contemporary International Law on Ship Recycling".

[55] *Anti-fouling Systems*, online: International Maritime Organization <http://www.imo.org/> (provides an overview of anti-fouling systems). It is estimated that nearly 170-540 tonnes of TBT and 6,000 to 20,000 tonnes of environmentally harmful paints are released due to the breaking operations. Commission, *Impact Assessment*, *supra* note 12 at 14.

[56] S.M. Evans, T. Leksono & P.D. McKinnell, "Tributyltin Pollution: A Diminishing Problem Following Legislation Limiting the Use of TBT-Based Anti-fouling Paints (1995) 30:1 Mar. Pollution Bull. 14 (ScienceDirect); J.R., "Paint additive hammers coral" *Science News* 167:13 (26 March 2005) 206 at 206 (JSTOR) (noting that an important reason for coral decline can be TBT).

[57] Aage Bjørn Andersen, "An Issues Paper: Worker Safety in the Ship-breaking Industries" (2001) at 30 (ILO, Sectoral Activities Programme), online: International Labour Organization <http://www.ilo.org/public/english/dialogue/sector/papers/shpbreak/wp-167.pdf>.

[58] Paul Reznikoff, "Lead Poisoning", (1942) 42:10 Am. J. Nursing 1123 (JSTOR).

retardation, and delayed neurological and physical development. On board ships, mercury is found in thermometers, electrical switches, level switches and light fittings.[59] It is also a persistent bio-accumulative pollutant that affects the nervous system.

Concern also arises in conjunction with bilge and ballast waters.[60] Bilge water is the accumulation of potentially polluting liquids in the lowest part of the ship's hull (the bilge). This noxious water often contains oil, cargo residues, inorganic salts, arsenic, copper, chromium, lead, and mercury. During the dismantling of a ship, the quantity of bilge water tends to increase due to the accumulation of rain-water, cooling water, and containment water that are used during the shipbreaking process. Usually, the bilge water is spilt directly into the ocean during the wrecking operations, thereby threatening aquatic ecosystems, species and organisms.[61]

Ships on their final journey carry ballast water that is usually discharged in coastal waters prior to the vessels' beaching. Ballast water may contain organisms that endanger local marine biodiversity. Due to a lack of proper reception facilities at the yards, the accumulated sediments in the ballast tanks, which are exposed when the ship is cut open, is also dumped into the ocean.[62]

Apart from the above, some of the other dangerous by-products of the shipbreaking process are oil sludge,[63] ozone depleting f-gases,[64] polyvinyl chloride (PVC),[65] polycyclic aromatic hydrocarbons (PAHs),[66] radioactive materials,[67]

---

[59] UNEP, Conference of the Parties to the Basel Convention on the Control of Transboundary Movements of Hazardous Wastes and their Disposal, 6th Mtg., (9–13 December 2002) Technical Guidelines for the Environmentally Sound Management of the Full and Partial Dismantling of Ships, UNEP/CHW.6/23, 8 August 2002 at 44, online: Basel Convention <http://www.basel.int/meetings/cop/cop6/cop6_23e.pdf> [Technical Guidelines for Full and Partial Dismantling of Ships].

[60] For further discussion on the nature and impact of ballast waters on marine environment, see Tony George Puthucherril, "Ballast Waters and Aquatic Invasive Species: A Model for India" (2008) 19 Colo. J. Int'l Envtl. L & Pol'y 381 (WLeC).

[61] Technical Guidelines for Full and Partial Dismantling of Ships, *supra* note 59 at 28.

[62] Andersen, *supra* note 57 at 32.

[63] The oil sludge from end-of-life ships alone total between 400,000 and 1.3 million tonnes per year. Commission, *Impact Assessment*, *supra* note 12 at 14.

[64] *Ibid.* (pointing out that about 150–400 tonnes of ozone depleting f-gases are released from cooling systems annually due to careless ship dismantling contributing to worldwide climate change problems). In fact, ship-borne CFCs contribute 10 percent of the global emissions. Andersen, *supra* note 57 at 31.

[65] PVC can be found in cables, floor coverings, plastic devices, etc. Consisting of more than 50 percent chlorine, when burnt, PVC produces large quantities of hydrogen chloride gas that, if inhaled, forms hydrochloric acid in the lungs. In addition, carbon monoxide, dioxins and chlorinated furans are also released. Dioxins and furans are particularly toxic. Andersen, *ibid.* at 28.

[66] PAHs are the largest single class of carcinogens known today. They are formed by incomplete decomposition of any organic material containing carbon and hydrogen, e.g., oil products and residues, or due to the combustion of oil. PAHs are persistent and have long-term effects on both environment and human health. *Ibid.* at 28–29.

[67] Radioactive materials on board a ship can be present in liquid level indicators, smoke detectors or emergency signs. Even though these generate low-level radioactive waste, their handling and

perfluorooctane sulfonates,[68] and brominated flame retardants.[69] It is estimated that between 2006 and 2015, nearly 5.5 million tonnes of materials of potential environmental concern will end up in the dismantling yards, highlighting the need for sustainable ship recycling practices.[70]

Shipbreaking entails a series of activities that begin with the design and construction of a ship, extends to its operational life, finally culminating in its deconstruction. These involve shipowners, builders, naval architects, repairers, classification societies, material suppliers, etc.[71] However, the actual process of dismantling starts only months before a ship sets sail to a recycling yard, when the shipowner decides that it is no longer economical to maintain the ship.[72] The major players involved in this phase, their stakes and the asymmetry in power relations that exists between them are examined in turn below.

Even though they are the most important element in the ship recycling business, the *shipbreakers* have the least voice. Often described as the victims of this industry, the unorganised shipbreakers risk their life daily by breaking ships for a pittance, raising serious human rights issues as this activity infringes on their rights to life, health, livelihood and human dignity.[73]

The sole object of the *ship recycler* is to ensure maximum return on his/her investment. They are highly organised and have their own associations to protect their interests.[74] Regarding introduction of new technology and improved work

---

disposal has to be under strict conditions, since ionising radiation can prove lethal to human health and the environment. *Ibid.* at 32.

[68] Perfluorooctane sulfonates (PFOS) are mainly found in fire-fighting aqueous film forming foams (AFF). Even though the use of PFOS has been phased out in new AFF, PFOS may still be used. It is highly toxic and can affect aquatic life. Recycling of Ships: Comments to the Report of ISRWG 2 – Proposal to Include Three Hazardous Materials in the Draft Convention, Submitted by Norway, IMO/MEPC 56/3/10, 16 May 2007, ¶ 2 (KR-CON).

[69] Brominated flame retardants are persistent, bioaccumulative and toxic. Found even in remote areas like the Arctic, they are mainly used for cryogenic insulation in liquefied gas tank carriers. *Ibid.*, ¶ 3.

[70] Commission, *Green Paper*, *supra* note 35 at 2.

[71] See generally *Recycling of Ships: Report of the Correspondence Group*, Submitted by the Co-ordinator of the Correspondence Group, IMO/MEPC 47/3/3, 30 November 2001, annex 1 (KR-CON).

[72] Recycling of Ships: Proposal on a "Phased Approach to Ship Dismantling", Submitted by the Netherlands, IMO/ MEPC 52/3/1, 6 August 2004, ¶ 1 (KR-CON).

[73] See Recycling of Ships: The Need to Develop a New Legally-binding Instrument that will Build and Improve upon Existing Environmental Justice Legislation, Submitted by Greenpeace International, IMO/ MEPC 54/3/5/Rev.1, 1 February 2006, ¶ 6 (KR-CON) (noting that the people who lose their lives due to shipbreaking activities are hardly mentioned in official reports; they are reduced to mere numbers and statistics).

[74] For further information, see *Ship Recycling Industries Association (India)*, online: Ship Recycling Industries Association (India) <http://www.sriaindia.com/>. There are nearly 32 ship recycling companies in Bangladesh that are organised under the banner of the Bangladesh Shipbreakers Association. This organisation is powerful and very rarely is there any intervention by the government in its activities. Commission, *Impact Assessment*, *supra* note 12 at 14.

practices, the recyclers shy away from responsibility, claiming that the primary duty to ensure safe and clean conditions in the demolition yards rests with the government within whose jurisdiction these units operate. The economic uncertainties that surround the operation of this industry also encourage them to evade responsibility.[75]

The decision of a *shipowner* to sell is purely economic and what happens thereafter is not their concern. Even though 'responsible ship scrapping' is slowly gaining currency among shipowners,[76] their decisions are motivated by the desire to maximise profits from the scrap. In most cases, shipowners do not have any contractual nexus with the recycling facility. They sell their ships to *cash buyers*, who are generally dedicated brokers and are integral to the ship recycling chain. The cash buyers choose the most appropriate recycling facility, takes decisions regarding recycling,[77] and often assume the role of an owner.[78]

The *shipping industry* represented by several international non-governmental organisations (NGOs) like the International Chamber of Shipping, the Baltic and International Maritime Council and the International Ship Recycling Association, also has a significant role in the ship recycling business. The role of each of these NGOs is detailed below.

The International Chamber of Shipping (ICS)[79] represents the "voice of national shipowners' associations". In 1999, the ICS established the Industry Working Party on Ship Recycling (IWPSR)[80] in response to growing concerns regarding

---

[75] UNESCO, *Impacts and Challenges of Large Coastal Industry: Alang-Sosiya Ship-Breaking Yard, Gujarat, India*, Coastal Region and Small Island Papers 17 (Paris: UNESCO, 2004) at 17, online: UNESCO <http://www.unesco.org/csi/pub/papers3/assby.pdf>. See also Letter from Ship Recycling Industries Association (India) to H.K. Dash Saheb, Vice Chairman of the Gujarat Maritime Board (18 October 2006), online: Ship Recycling Industries Association (India) <http://www.sriaindia.com/home.html> (a representation against the Notification dated 30 September 2006, on the "So called Revival Package" at Alang).

[76] EC, Directorate General Environment, *Ship Dismantling and Pre-cleaning of Ships: Final Report (EC, 2007)* at 67, online: European Commission <http://ec.europa.eu/environment/waste/ships/pdf/ship_dismantling_report.pdf>.

[77] Recycling of Ships: National Workshop on the Development of the International Convention for the Safe and Environmentally Sound Recycling of Ships, Submitted by India, IMO/MEPC 57/3/5, 25 January 2008, ¶ 14 (KR-CON).

[78] Recycling of Ships: Normal Recycling Procedures, Submitted by ICS, IMO/MEPC 55/INF.12, 18 August 2006, ¶ 6–7 (KR-CON).

[79] For further information, see generally *International Chamber of Shipping*, online: International Chamber of Shipping <http://www.marisec.org/ics/ICS%20brochure.pdf>.

[80] The Industry Working Party is comprised of representatives from the Baltic and International Maritime Council (BIMCO), the International Association of Dry Cargo Shipowners (INTERCARGO), International Association of Independent Tanker Owners (INTERTANKO), International Chamber of Shipping (ICS), International Tanker Owners Pollution Federation (ITOPF), International Transport Workers' Federation (ITF), and the Oil Companies' International Marine Forum (OCIMF). The European Community Shipowners' Associations (ECSA) and the International Association of Classification Societies (IACS) participated as observers. For further details, see *ibid.*, ¶ 1.2.

the working conditions and environmental standards in the recycling yards.[81] The IWPSR has developed a code of practice, which primarily deals with issues that shipowners themselves can reasonably address to ensure safer and cleaner scrapping.[82] Nonetheless, ICS considers working conditions, employee health, and safety as issues to be determined by legislators, employers and administrators.[83]

The world's largest private shipping non-government organisation, the Baltic and International Maritime Council (BIMCO),[84] controls nearly 65 percent of the world merchant fleet.[85] The 2007 BIMCO policy on ship recycling emphasised that ship recycling is the most environmentally friendly way of disposing ships and, if handled properly, is a green industry.[86] Prior to this, in 1987, the BIMCO introduced the first standard contract for the sale of ships for recycling, the SALESCRAP 87.[87] This has been revised and renamed as DEMOLISHCON.[88] Even though BIMCO supports the implementation of a legally-binding IMO instrument on ship recycling, it is of the view that the responsibility to certify that ships are ready for recycling should lie with the flag state.[89]

A recent development that might significantly impact the development of the ship recycling industry is the establishment of the International Ship Recycling Association (ISRA)[90] in October 2007. Based in the Netherlands,[91] ISRA, which

---

[81] *Industry Code of Practice on Ship Recycling*, (August 2001) ¶ 1.1, online: Basel Convention <http://www.basel.int/ships/docs/08e.pdf>.

[82] *Ibid.* See also Recycling of Ships, Submitted by the International Chamber of Shipping (ICS) on Behalf of the Industry Working Party on Ship Recycling, IMO/MEPC 47/3/2, 30 November 2001, ¶ 5 (KR-CON) (outlining some best practices like creation of an inventory of potentially hazardous materials on board, identification of hazardous areas on vessels, delivering vessels in gas free condition, etc.). Recycling of Ships: Comments on the Report of the Correspondence Group to MEPC 49, Submitted by India, IMO/MEPC 49/3/3, 13 May 2003, ¶ 1.7 (KR-CON) (noting that there has been little adherence by shipowners to the Industry Code of Practice and to DEMOLISHCON contract). See also *ibid.*, ¶ 1.5.

[83] For further details, see generally Recycling of Ships: Environmental Contamination in Alang-Sosiya Shipbreaking Yards, Submitted by Greenpeace International, IMO/MEPC 46/INF.21, 15 February 2001 (KR-CON).

[84] For further information, see BIMCO <http://www.bimco.org/>.

[85] Directorate General Environment, *supra* note 76 at 87.

[86] For further details, see *BIMCO Policy on Ship Recycling*, online: BIMCO <http://www.bimco.org/Members%20Area/News/General_News/2007/01/0126BIMCO%20Policy%20on%20Ship%20Recycling.aspx>.

[87] *DEMOLISHCON: SALESCRAP 87 Contract Revised and Re-named DEMOLISHCON*, online: BIMCO <http://www.bimco.org/Corporate%20Area/Documents/Document_samples/Sundry_Other_Forms/DEMOLISHCON.aspx>.

[88] For the text of DEMOLISHCON, see BIMCO Standard Contract for the Sale of Vessels for Demolition and Recycling Code Name: "DEMOLISHCON", online: BIMCO <http://www.bimco.org/~/media/9698E1F48AFE4174B3761F31CA80E361.ashx>.

[89] Directorate General Environment, *supra* note 76 at 87.

[90] Its goals, *inter alia*, include: promotion of sustainable ship recycling, changing the image of the dismantling industry, and exploring alternatives to beaching. *ISRA*, online: I.S.R.A. <http://www.isra-dis.nl/>.

[91] *Ibid.*

describes itself as the voice of green recycling companies, seeks to expand its membership to include ship recycling companies from around the world, including those from the Indian subcontinent. However, as a condition of membership, prospective candidates must upgrade their yards to the ISRA standards and commit themselves to necessary investments that such upgrading will incur.[92] ISRA is seeking consultative status with the IMO.[93]

Often mute spectators to the drama that unfolds within their territory, *Third World governments* are mainly interested in the revenue generated from the shipbreaking industry. Their policies often foster a favourable climate for its growth. Even though they have an interest to ensure that the industry respects human rights and operates in an environmentally friendly manner, economic benefit usually receives precedence.[94]

In most countries, the *judiciary* has a constitutional duty to uphold basic human, economic and environmental rights and the rule of law. Increasingly, the courts have been called on to intervene in ship recycling matters.[95] Activism or in-activism by the judiciary, as will be seen in the subsequent discussion,[96] can have serious repercussions on the growth and development of this industry.

Even though there is green recycling capacity in *OECD countries*,[97] the economics of shipbreaking does not facilitate the development of an indigenous and intensive ship scrapping programme in these industrialised countries.[98] Consequently, most ships that reach the Indian subcontinent for scrapping belong to shipowners in OECD member states. Despite growing awareness in the OECD regarding the human rights and environmental consequences of these exports,[99]

---

[92] *Ibid.*

[93] James Mc Farlane, Address (Speech on the 3rd International Intercessional Meeting of the IMO Working Group on Ship Recycling in Nantes, France, (21–25 January 2008), online: I.S.R.A <http://www.isradis.com/documents/Speech%20by%20board%20member%20James%20Mc%20Farlane.pdf>.

[94] For further analysis, see Part III, below.

[95] For instance see *Research Foundation for Science Technology and Natural Resource Policy v. Union of India* (2007), [2007] S.C. 890 (India S.C.) (indlaw.com).

[96] For further discussion, see Chapter Three entitled, "Limitations of a National Response to Regulate the Global Shipbreaking Industry: A Study of the Indian Experience".

[97] Denmark, Danish Environmental Protection Agency, *Shipbreaking in OECD*, Working Report No. 18, Danish Environmental Protection Agency, 2003 (by Frank Stuer-Lauridsen, Nikolai Kristensen & Jesper Skaarup Cowi A/S), online: Miljøministeriet <http://www2.mst.dk/Udgiv/publications/2003/87-7972-588-0/pdf/87-7972-589-9.pdf>.

[98] U.K., H.C., "Environment, Food and Rural Affairs Committee on Dismantling Defunct Ships in the UK", HC 834 (18th Report of Session 2003–2004) 8, online: United Kingdom Parliament <http://www.publications.parliament.uk/pa/cm200304/cmselect/cmenvfru/834/834.pdf> (pointing out that in Western Europe while a vessel can earn only USD10 per tonne, in India and in Bangladesh, it could be as high as USD390-410 per tonne) ["Dismantling Defunct Ships in the UK"].

[99] See generally L. Christopher Noland, "The Ghouls That Won't Go Away – The Dire Environmental Consequences Posed by the Ghost Fleet in the James River", Note, (2006) 30 Wm. & Mary Envtl. L. & Pol'y Rev. 513 (WLeC) (arguing that the United States government and MARAD take up responsibility for disposing their fleets domestically).

the general perception, particularly among the environmental non-governmental organisations (ENGOs), is that these countries are not doing enough to respect the principle of not transferring harm to developing nations.

Since the birth of modern shipping, to reduce operational costs, taxes and the rigours of governmental regulation, shipowners have often registered their ships under a flag of convenience. Even though *flag states* also apply international standards pursuant to IMO conventions, enforcement is generally lax as these countries generally have "insufficiently developed maritime administration".[100] Taking advantage of this situation, many shipowners register their ships under such open registers as ships approach the end-of-life stage. As such ships might also have changed hands, it becomes difficult to pinpoint responsibility for ensuring clean scrapping.

Among the major *international agencies*, three intergovernmental organisations, the International Maritime Organization, the International Labour Organization (ILO) and United Nations Environment Programme (UNEP), have an interest in ship recycling. In addition to individual measures, they have established a Joint Working Group (JWG) concerned with ship recycling.

Traditionally, the role of the IMO, the specialised agency of the United Nations to ensure "safe, secure and efficient shipping on clean oceans", has been limited in respect of land-based activities. Ship recycling was generally treated as a domestic affair. However, if the recent IMO initiatives on ship recycling receive worldwide support, land-based recycling facilities will be brought under its purview. Never before this have land-based activities, except for those in ports, been considered as part of an IMO convention.[101] The IMO does not deal with or regulate the safety and environmental protection aspects in relation to shore-based ship-repair and shipbuilding activities. This raises the question, should the IMO concern itself with a ship when its lifetime is over?[102] As mentioned earlier, ship recycling is the culmination of a series of stages, each governed by comprehensive legal frameworks developed by the IMO covering aspects relating to safety, environmental protection, technical co-operation, maritime security and efficiency.[103] It is therefore imprudent to view shipbreaking in isolation from the

---

[100] Edgar Gold, Aldo Chircop & Hugh Kindred, *Maritime Law* (Toronto: Irwin Law, 2003) at 186–87.

[101] Directorate General Environment, *supra* note 76 at 5.

[102] Recycling of Ships: Submitted by India, IMO/MEPC 46/7/1, 14 February 2001, ¶ 4–5 (KR-CON).

[103] *Introduction to IMO*, online: International Maritime Organization <http://www.imo.org/>. See also Recycling of Ships: Examination of the Practical Application of the Draft International Convention for the Safe and Environmentally Sound Recycling of Ships — A Timeline for Recycling a Ship According to the Draft Convention's Current Regulations, Submitted by the United Kingdom, IMO/MEPC 56/3/22, 18 May 2007 (KR-CON) (calculating 125–153 days as the total time required under the New Draft convention on Ship Recycling to complete transactions, beginning from the decision to recycle a ship until completion of the dismantling at the recycling yard and delivery of the salvaged material to the mill or processor).

preceding stages in the life of a ship thereby rendering the IMO as the most important player among the international agencies that have a stake in ship recycling.[104]

Since the 1980s, the ILO has been concerned about the labour standards in shipbreaking yards.[105] Even though it has not adopted any mandatory instrument to revamp work conditions in these yards, the various conventions and recommendations on occupational safety and health applicable generally to the worker class can be extended to improve the work conditions in the ship recycling yards.[106]

UNEP's interest in shipbreaking stems from its interest in transboundary movement of toxic wastes for their disposal. Even though there is disagreement on the role of the UNEP Basel Convention[107] in ship recycling, until recently it was the only legal instrument in force that could minimise the trade in toxic ships. Several of the principles in the Basel Convention, such as "environmentally sound management" and "prior written notification", are central to a sustainable shipbreaking strategy. However, the ship recycling industry considers the Basel Convention as being "ill-suited"[108] to regulating the ship scrapping business.

These three intergovernmental organisations have undertaken a number of joint initiatives with regard to ship recycling. Pursuant to the 51st session of the Marine Environment Protection Committee of IMO (51 MEPC 2004), the Seventh Meeting of the Conference of the Parties to the Basel Convention (7 COP 2004), and the 291st session of the ILO Governing Body (2004), a JWG was established as a platform for "consultation, co-ordination and co-operation" in relation to their work programmes and activities on ship recycling, with the primary objective to avoid duplication of work and overlapping of roles.[109] Furthermore, to

---

[104] IMOGSR, *supra* note 27.

[105] For further details, see *Safe Work Shipbreaking*, online: International Labour Organization <http://www.ilo.org/public/english/protection/safework/sectors/shipbrk/index.htm>.

[106] Nikos Mikelis, "Developments and Issues on Recycling of Ships" (Paper presented to The East Asian Seas Congress, Haikou City, China, 12 December 2006) at 3, online: International Maritime Organization <http://www.imo.org/includes/blastDataOnly.asp/data_id%3D17980/Developments.pdf>. See also Safety and Health in Shipbreaking: Guidelines for Asian Countries and Turkey, ILO, Interregional Tripartite Meeting of Experts on Safety and Health in Shipbreaking for Selected Asian Countries and Turkey, Bangkok, 7–14 October 2003, MESHS/2003/1, online: International Labour Organization <http://www.ilo.org/public/english/standards/relm/gb/docs/gb289/pdf/meshs-1.pdf>.

[107] *Basel Convention on the Control of Transboundary Movements of Hazardous Wastes and their Disposal*, 22 March 1989, 1673 U.N.T.S. 126, 28 I.L.M. 657 (entered into force 5 May 1992) [*Basel Convention*]. "Dismantling Defunct Ships in the UK", *supra* note 98 at 9 (quoting the Environment Agency that applying the *Basel Convention* to shipbreaking is extremely problematic).

[108] See generally France, Interdepartmental Committee on the Dismantling of Civilian and Military End-of-Life Ships, *Le rapport de la Mission Interministérielle portant sur le Démantèlement des Navires civils et militaires en fin de vie*, annex V (27 March 2007, Chair: Xavier de la Gorce), online: SGMer <http://www.sgmer.gouv.fr/IMG/pdf/Annex_5_Ill_suited_Basel_Convention.pdf>.

[109] See generally *Report of the Working Group*, ILO/IMO/BC WG 1/8, 18 February 2005 (KR-CON). The first session of the JWG was held from 15 to 17 February 2005 at the IMO Headquarters. The second session was held from 12 to 14 December 2005 at the United Nations Office in Geneva. The third session of the JWG was held from 29 to 31 October 2008. For further

promote "a coordinated approach" to address the issues faced by the ship recycling industry, the Basel Convention Secretariat, in collaboration with the ILO and IMO, initiated the Global Programme for Sustainable Ship Recycling.[110] This programme is based on four fundamental principles: a life-cycle approach,[111] inclusion,[112] collaboration,[113] and continuity.[114] It seeks to develop model facilities, introduce certification schemes, undertake policy development, establish linkages, and initiate training and workshops.[115] In January 2008, initial discussions on the programme concept were held in Dhaka, Bangladesh involving an array of stakeholders.[116]

ENGOs like Greenpeace[117] and the Basel Action Network,[118] organised under the "NGO Platform on Shipbreaking",[119] have highlighted the plight of the

---

details see *Sector Meetings*, online: International Labour Organization <http://www.ilo.org/public/english/dialogue/sector/techmeet/iloimobc 08/index.htm>. See also <http://www.basel.int/ships/jimbwg.html>

[110] The Programme has three main objectives: First, it seeks to promote sustainability of the ship recycling industry by enhancing the application of internationally recognised standards relating to occupational safety, health and environmental protection; secondly, it promotes effective implementation of the future IMO convention by building upon existing technical assistance activities promoting the guidelines of the IMO, ILO and Basel Convention, and elements of the new convention (as developed); and thirdly, promotion of an integrated approach to the ship recycling industry by addressing infrastructural and other `needs in the participant countries in and beyond the ship recycling yards. *Global Programme for Sustainable Ship Recycling*, online: Basel Convention <http://www.basel.int/ships/gpssr/index.html>. See also Donata Rugarabamu, "Inaugural Discussions on the Global Programme on Sustainable Ship Recycling" (PowerPoint presented to the Inaugural Discussions on the Global Programme on Sustainable Ship Recycling, Dhaka, Bangladesh, 12 January 2008), online: Basel Convention <http://www.basel.int/ships/gpssr/index.html>.

[111] "Preparatory Discussions on Promoting Sustainable Ship Recycling through the Global Programme: Discussion Document", Global Programme on Sustainable Ship Recycling, Dhaka, Bangladesh, 13 January 2008 at 1, online: Basel Convention <http://www.basel.int/ships/gpssr/index.html> (the "life-cycle approach" recognises that addressing procedures and activities only within the recycling yard is insufficient. For safe and environmentally-sound ship recycling, there has to be appropriate infrastructure even beyond the yards).

[112] *Ibid.*, (highlighting the need to include ship recycling in national development and poverty reduction strategies).

[113] *Ibid.*, (emphasising on close collaboration between the stakeholders to ensure sustainable global ship recycling).

[114] *Ibid.*, at 2 (this principle emphasises the need to build upon the work already done to achieve goals).

[115] Susan Wingfield, "Introduction to the Global Programme Concept" (PowerPoint presented to the Preparatory Discussions on Promoting Sustainable Ship Recycling through the Global Programme, Dhaka, Bangladesh, 13 January 2008), online: Basel Convention <http://www.basel.int/ships/gpssr/in dex.html>.

[116] See generally *Global Programme for Sustainable Ship Recycling*, *supra* note 110.

[117] For further information on the Greenpeace campaign against unsustainable shipbreaking, see *Greenpeace*, online: Greenpeace <http://www.greenpeaceweb.org/shipbreak/>.

[118] *About the Basel Action Network – BAN*, online: Basel Action Network <http://www.ban.org/main/about_BAN.html>.

[119] The NGO Platform on Shipbreaking includes: Greenpeace, the Bellona Foundation, the European Federation for Transport and Environment (T&E), Basel Action Network (BAN), the International Federation for Human Rights (FIDH), the North Sea Foundation, the International Ban Asbestos Secretariat, Ban Asbestos France, Ban Asbestos Network India, Young Power in Social

shipbreakers in the ship demolition yards and the environmental vandalism in these areas. Due to their vigilance and activism, they have mobilised public opinion and uncovered several surreptitious transfers. Wherever possible, they have initiated legal action. These organisations have also exerted influence on the international legal regime on shipbreaking.[120]

It is clear from this brief review that shipbreaking is an integral component of the shipping industry and perhaps the only viable alternative to ensure that redundant ships are not abandoned to rust away and contaminate the environment. However, the process of breaking is laden with dangers as it involves release of lethal toxins. The imbalances of power between the different players compound the situation since many of the crucial actors have a very limited voice in influencing the growth and development of the industry. Thus, for shipbreaking to become a truly sustainable activity, it will be necessary to have a level playing field where all stakeholders can participate in a dialogue to address their legitimate concerns.

III. Shipbreaking Yards on the Indian Subcontinent or Toxic Hotspots?

The major players in the shipbreaking business on the Indian subcontinent are India, Pakistan and Bangladesh. The socio-economic conditions in these countries are more or less identical – humungous population, depressing levels of poverty, untrained human resources, high levels of illiteracy and unemployment, widespread environmental pollution, natural resources degradation, and lack/non-enforcement of laws.[121] In the race to integrate with a competitive global economy, these countries have often given precedence to short-term economic

---

Action Bangladesh (YPSA), Bangladesh Environmental Lawyers Association and Corporate Accountability Desk – The Other Media India. For further details, see *Breaking News,* NGO platform on shipbreaking <http://www.shipbreakingplatform.com/>. See also Recycling of Ships: Urgent Considerations for Correcting Deficiencies in the Draft Convention on the Safe and Environmentally Sound Recycling of Ships, Submitted by Greenpeace, IMO/MEPC 55/3/7, 10 August 2006 (KR-CON). For a copy of the text of the Joint Declaration on Implementing Urgent Global Solutions to the Shipbreaking Crisis, see Recycling of Ships: The Need to Develop a New Legally-binding Instrument That Will Build and Improve upon Existing Environmental Justice Legislation, Submitted by Greenpeace International, IMO/MEPC 54/3/5/Rev.1, 1 February 2006, annex (KR-CON) (this declaration was signed by 25 human health, environmental and human rights NGOs and trade unions).

[120] For instance see generally Recycling of Ships: Analysis of Inconsistencies between the IMO Guidelines on Ship Recycling and the Basel Convention, Submitted by Greenpeace International, IMO/MEPC 51/INF.11, 6 February 2004 (KR-CON).

[121] For an overview of the socio-economic conditions of these countries, see *The Oxford Encyclopedia of Economic History*, e-reference ed., Joel Mokyr ed. (USA: Oxford University Press, 2005) "India"; *The Oxford Encyclopedia of Economic History*, e-reference ed., Joel Mokyr ed. (USA: Oxford University Press, 2005) "Pakistan"; *The Oxford Encyclopedia of Economic History*, e-reference ed., Joel Mokyr ed. (USA: Oxford University Press, 2005) "Bangladesh".

gains at the expense of the environment and human resources.[122] Since the shipbreaking industry provides ready employment to unskilled hands, high quality steel which helps conserve energy while reducing the need for more mining of the ore, and revenue to governments, it occupies a central place in the national development strategies of these countries.[123] However, as will be seen below, these benefits come at a price.

A. *Bangladesh: The Market Leader*

Bangladesh and India long shared a duopoly over the ship scrapping business. However, Bangladesh has emerged as the undisputed leader, performing over 70 percent of the dismantling business.[124] Interestingly, the birth of the ship demolition industry in Bangladesh is reminiscent of a fairy tale. In the 1960s, a cyclone, which killed thousands, brought a Greek ship, 'M.D. Alpine', to the shores of Chittagong.[125] At first, objects and other materials from the ship were

---

[122] The annual economic growth rate of Bangladesh is four to five percent, however, the economic loss due to environmental damage is 2.7 percent of its GDP. "Bangladesh Country Environmental Analysis" (2006) Bangladesh Development Series Paper No: 12 (The World Bank Office, Dhaka) at 1, 93, online: World Bank <http://siteresources.worldbank.org/BANGLADESHEXTN/Resources/295759-1173922647418/complete.pdf>. See generally South Asia Environment and Social Development Unit, *Pakistan Strategic Country Environmental Assessment*, World Bank Report No. 36946-PK, Vol. 1: Main Report (21 August 2006), online: World Bank <http://siteresources.worldbank.org/SOUTHASIAEXT/Resources/Publications/448813-1188777211460/pakceavolume1.pdf> (the World Bank calculates that environmental degradation annually costs Pakistan at least six percent of its GDP).

[123] *Role in Steel Economy*, online: Ship Recycling Industries Association (India) <http://www.sriaindia.com/steel_economy.html> (noting that the capital investment required for producing two million tonnes of steel through ship recycling is INR300 crores (approx. USD62 million, the exchange rate USD1= INR48.24 as on 13 August 2009) when compared to over INR6,000 crores (approx. USD1 billion) required *via* the alternative mode). To produce two metric tonne (mt) of steel, a steel plant would require about 3,508,000 tonnes of iron ore, 3,094,000 tonnes of fuel, 80,610 tonnes of chemicals, 50–120 Mm^3 of water, involving a cost of INR1,000 crores (approx. USD208 million) and generating about 878,967 tonnes of solid waste. To produce the same quantity of steel *via* the shipbreaking process there is no requirement for iron ore, fuel and process chemicals. It only requires 1825 m^3 of water and generates 12,500 tonnes of solid wastes involving a cost of INR100 crores (approx. USD20 million). Atul Sharma, "Implementation of Recent Directives of the Hon'ble Supreme Court of India Regarding Sustainable and Environment Friendly Ship Recycling in India" (PowerPoint presented to the National Workshop on International Convention for Safe and Environmentally Sound Recycling of Ships, Maritime Training Institute, Mumbai, India, 9 January 2008) [unpublished], online: Gujarat Maritime Board <http://www.gmbports.org/international_workshop.htm>.

[124] In 2005, the market share of Bangladesh was 63.7 percent, which when translated in terms of tonnage is 3,607,749 DWT, the number of vessels being 70. In 2006, the market share went up to 68.14 percent. In terms of tonnage this is 4,997,688 DWT, the number of vessels being 167. Technical Guidelines for Full and Partial Dismantling of Ships, *supra* note 59 at 38. See also France, Interdepartmental Committee on the Dismantling of Civilian and Military End-of-Life Ships, *Le rapport de la Mission Interministérielle portant sur le Démantèlement des Navires civils et militaires en fin de vie*, annex II at 3 (27 March 2007, Chair: Xavier de la Gorce), online: SGMer <http://www.sgmer.gouv.fr/IMG/pdf/Annex_2_Dismantling_countries.pdf> [*MIDN*, Annex II].

[125] Young Power in Social Action, *Workers in Shipbreaking Industries: A Base Line Survey of Chittagong (Bangladesh)* (Bangladesh, 2005) at 22, online: Young Power in Social Action <http://www.ypsa.org/publications/shipbreaking_baseline_survey.pdf>.

carted away by the local inhabitants, leaving behind a shell. Realisation soon dawned that the true value lay in the scrap steel obtainable by breaking the hull.[126] Another milestone was the scrapping of a Pakistani ship that was badly damaged in the Liberation War of 1971. Since then, the industry has become an important lifeline for the Bangladeshi economy, with nearly 200,000 Bangladeshis benefiting either directly or indirectly from it. Bangladesh does not have any local iron ore resources, and therefore, it relies mainly on the import of obsolete ships to feed its steel demands. So crucial are wrecked ships to this country, that they provide nearly 80 percent of the steel requirements.[127] This has forced Bangladesh to increase the price of scrap steel to such a level as to capture the market.[128]

Even though shipbreaking operations are performed at several sites along the Bangladesh coast, the major locus of this activity is in the Sitakund area of Chittagong[129] where more than 30 shipbreaking yards operate.[130] The infrastructure in these yards is primitive and the industry here, as in Pakistan and India, has the dubious distinction of being one of the most unsafe occupations.[131] Greenpeace estimates that on average, one shipbreaker dies in the yards every week and one gets injured every day.[132] The environmental consequences of this industry are equally alarming.[133]

---

[126] *Ibid.*

[127] Commission, *Green Paper*, *supra* note 35 at 2.

[128] Technical Guidelines for Full and Partial Dismantling of Ships, *supra* note 59 at 38. See also *MIDN*, Annex II, *supra* note 124 at 8 (noting that Bangladesh has literally asphyxiated other competitors by continuously increasing the price of obsolete ships). For example in March 2006, Bangladesh bought an oil tanker for USD380/t, and by December 2006, it was willing to pay USD487/t, a price increase of USD100 in a matter of nine months). Many factors enable Bangladesh to offer such high prices. For instance, in Bangladesh the steel is mainly treated "cold" in the rerolling mills which enables the recycler to save on energy-intensive and expensive re-melting in electrical furnaces. Commission, *Impact Assessment*, *supra* note 12 at 10.

[129] *MIDN*, Annex II, *ibid.* at 7 (pointing out that there are 32 dismantling sites set up over eight km^2 in Sitakund close to Chittagong).

[130] "Bangladesh Country Environmental Analysis", *supra* note 122 at 22.

[131] See generally Greenpeace-FIDH in cooperation with YPSA, End of Life Ships: The Human Cost of Breaking Ships, Jacqueline Schuiling (Netherlands: Greenpeace International, 2005), online: Greenpeace International <http://www.greenpeace.org/raw/content/international/press/reports/end-of-life-the-human-cost-of.pdf>.

[132] See generally *ibid.*, at 15. However, there is discrepancy between the official figures and those provided by the state. See also Sawyer, *supra* note 5 (noting that the major specialty of the demolition yards in Bangladesh is the scrapping of oil tankers, which are the most dangerous to scrap due to the presence of residual oil and trapped gases that could ignite from the use of acetylene torches. Many have died due to suffocation and explosions); Bishwajit Roy, "Health Problems Among the Workers in the Ship Breaking Industry", online: BILS <http://66.102.1.104/scholar?hl=en&lr=&q=cache:LBrRF QaDcUQJ:www.bils-bd.org/Journal_Archive_pages/2004/April_June/English/Health_Problems.pdf+Health+Problems+among+the+Workers+in+the+Ship+Breaking+industry>. As per admission by the shipbreakers during 2005–2007, there have been 32 deaths. Md. Ali Shahin, "Shipping Hazardous Waste" (PowerPoint presented to the Preparatory Discussions on Promoting Sustainable Ship Recycling through the Global Programme, Dhaka, Bangladesh, 13 January 2008), online: Basel Convention <http://www.basel.int/ships/gpssr/index.html>.

[133] See generally Md. M. Maruf Hossain & Mohammad Mahmudul Islam, *Ship Breaking Activities and Its Impact on the Coastal Zone of Chittagong, Bangladesh: Towards Sustainable Development*,

Despite almost 40 years of operation, there is to date no legal framework to specifically regulate the shipbreaking business which remains market driven.[134] Many existing laws relating to industry, environment, labour, as well as import and export regulations, are indirectly applied to the ship scrapping business.[135] The lack of a specific law, however, leads to confusion on regulatory responsibilities. This is compounded by the several government departments that manage the industry rendering enforcement and monitoring difficult.[136] Adverse publicity and the growing realisation that re-casting present practices will serve the larger interests of the nation has led to government initiatives for a policy on safe ship scrapping.[137] Courts in Bangladesh have also begun monitoring the yards, though not to the same extent as their Indian counterparts.[138]

B. *Shipbreaking in Pakistan*

Not much is known about the shipbreaking industry in Pakistan, except that it was the first country in Asia to set up a dismantling industry independent of shipbuilding.[139] The activity began on the Gadani beach near Karachi before Pakistan

---

(Chittagong: Advocacy & Publication Unit, YPSA, 2006), online: Young Power in Social Action <http://www.ypsa.org/publications/Impact.pdf>.

[134] Shahin, *supra* note 132.

[135] Sawyer, *supra* note 5 at 548 (observing that "the race to reach the bottom of environmental and safety standards has been fully realised in Bangladesh where regulations are non-existent"). Among the applicable laws, the most important that relate to the environment is the *Environment Conservation Act 1995* and *Environment Conservation Rules 1997*. As far as labour is concerned, *Labour Law 2006*, the *Employment of Labour (Standing Orders) Act 1965*, the *Workmen's Compensation Act 1923* are relevant. Md. Arifur Rahman & Muhammed Ali Shahin, "Shipbreaking Industry in Bangladesh Needs Modernisation and Policy Planning" *The Bangladesh Observer* (9 November 2007) at 4. The *Merchant Shipping Act 1884* (as amended) lays down "instructions" in relation to the "breaking of vessels". Permits are required for beaching, and thereafter surveys are conducted to verify the safety measures. The *Factories Act, inter alia*, prescribes the legal weight limits. Notice M 12/9 of 24 October 1985 of the Mercantile Marine Department lays down certain requirements like the kinds of minimum equipment needed, appointment of a minimum of three supervisors, etc. See also Report of the Correspondence Group, *supra* note 7 at 27; Md. Shahjahan "Ship Recycling in Bangladesh", (PowerPoint presented to the Inaugural Discussions on the Global Programme on Sustainable Ship Recycling, Dhaka, Bangladesh, 12 January 2008), online: Basel Convention <http://www.basel.int/ships/gpssr/index.html> (noting that under the *Environment Conservation Rules 1997*, shipbreaking falls under Orange B category and therefore environmental clearance is mandatory under the *Environment Conservation Act 1995* and *Environment Conservation Rules 1997*).

[136] Rahman & Shahin, *ibid.*

[137] *Ibid.* Bangladesh, Water Resources Planning Organisation, *Institutional Aspects of Ship Breaking Industry in Bangladesh*, WP044 (Dhaka: Ministry of Water Resources, 2005) (Prepared by Khurshed Alam), online: Integrated Coastal Zone Management, Bangladesh <http://www .iczmpbangladesh.org/rep/wp044.pdf>.

[138] Bangladesh Environmental Lawyers Association v. Bangladesh, [2003] Writ Petition No. 2911, online: Bangladesh Environmental Lawyers Association <http://www.belabangla.org/html/pil.htm>.

[139] See *MIDN*, Annex II, *supra* note 124 at 5.

became independent.[140] Thereafter, sporadic acts of breaking matured into a full-scale industry. In the early 1980s, the government announced a series of measures enabling Gadani to enter onto the list of leading shipbreakers in the world.[141] During this period, when activity was at its peak, it is believed that the sector employed over 30,000 workers directly, with another 500,000 earning their livelihood from related activities. Conditions in Gadani are no different from the other regions on the subcontinent; shipbreaking continues to be performed at the most basic level. Even though the labour force at Gadani is the cheapest in Asia,[142] giving it an edge over other players, the industry is on the decline, primarily due to rising scrap prices and high import duties.[143]

C. *Shipbreaking in India*

Historically, shipbreaking in India was confined to the dismantling of small barges and coastal wrecks. Realising its economic potential, in 1978, the Union Government adopted a policy for importing ships for scrapping by the Metal Scrap Trade Corporation and also set up a shipbreaking development fund.[144] The industry was soon recognised as a manufacturing process, which facilitated its growth.[145]

Initially, most shipbreaking activities were concentrated in the docks at Mumbai city, in Maharashtra State.[146] Due to space constraints and the need to diversify operations, studies to find alternative locations were commissioned. Alang in Gujarat State, washed by the Gulf of Cambay, was found appropriate for the

---

140 Alauddin Masood, "Ship-breaking attracting entrepreneurs" *Dawn the Internet ed.* (24 December 2001), online: Dawn: the Internet Edition <http://www.dawn.com/2001/12/24/ebr20.htm>.

141 Stopford, *supra* note 9. In fact, the scrapping sites comprised nearly 100 plots measuring about 2,500 square yards. See also *Gaddani Ship-breaking*, online: Lasbela Government <http://www.lasbeladistrictgovt.com/gaddani/ship%20breaking.htm>.

142 *MIDN*, Annex II, *supra* note 124 at 6.

143 In 1999, Gadani had a 15 percent share of the dismantling market. With the government decision to hike the import duty to 45 percent, the industry declined. In 2005, Gadani demolished only six ships, *i.e.*, 1.3 percent of the market. Faced with virtual obliteration from the scene, industrialists and others lobbied the government to reduce taxes. The impact was immediate. In 2006, Pakistan was able to purchase 16 ships (273,230 DWT) enabling it to overtake China and move into the third position with 3.73 percent of the market. *Ibid.* "Tax evasion only hope for industry's survival" *International The News [of Pakistan]* (29 March 2008), online: International the News <http://thenews.jang.com.pk/daily_detail.asp?id=103695> (quoting the chairman of the Pakistan Shipbreaking Association that the only hope left for a revival depends on whether the government lifts the 15 percent sales tax).

144 UNESCO, *supra* note 75 at 15.

145 For the text of the Environmental Guidelines for Shipbreaking Industries, see Recycling of Ships: Information on the Environmental Guidelines for the Ship Recycling Industry in India, Submitted by India, IMO/MEPC 47/INF.3, 6 November 2001, annex (KR-CON).

146 For an overview of shipbreaking activities in Maharashtra, see generally Odd Harald Røst, "Shipbreaking: Lives at stake" *Metal World* 4:2 (June 2003) 12, online: International Metalworkers' Federation <http://www.imfmetal.org/main/files/2-2003.pdf>.

beaching of heavy ships. In the initial days,[147] the developmental activities at Alang were largely unplanned and unregulated. This is attributable to the poor industrial climate in India at that time and the fact that government policies for new industries were at a nascent stage.[148] The Gujarat Industrial Development Corporation monitored activities. These were later transferred to the Gujarat Maritime Board (GMB).[149] Shipbreaking activities in India are carried out not only at Alang, but also in the States of West Bengal (Kolkata),[150] Kerala (Beypore),[151] and Maharashtra (Mumbai).[152] Other coastal states seeking to cash-in on the economic opportunities that this activity offers are throwing open their beaches to accommodate dying ships.[153] However, the heart of this industry in India, and perhaps the world, remains in Alang.

The Alang-Sosiya shipbreaking yard started functioning from 13 February 1983, when the first vessel, 'MV Kota Tenjong', beached at the yard.[154] The yard has witnessed spectacular growth, spreading to over 10 kilometres along the Alang coast and accommodating hundreds of dying ships in its 173 plots.[155] On average, 200 ships are broken annually, the highest number being 361 recorded for the year 1998.[156] The industry produces about 3.5 million tonnes of steel equivalent a year, which averages the production of a major Indian steel plant.[157]

---

147 UNESCO, *supra* note 75 at 15.

148 *Ibid*. at 9.

149 See *Gujarat Maritime Board Act 1981* (No. 30 of 1981, Gujarat State, India), Preamble (provides for the constitution of a Maritime Board vesting in it the administration, control and management of minor ports in the State of Gujarat).

150 Santanu Sanyal, "Kolkata Dock to re-build ship-breaking potential" *The Hindu Business Line [of India]* (11 July 2005), online: The Hindu Business Line <http://www.thehindubusinessline.com/2005/07/11/stories/2005071100330600.htm>.

151 For an overview of the shipbreaking unit, see Ship Breaking Unit (Kozhikode) Port Based, online: Steel Industrials Kerala Limited <http://www.steelinkerala.com/html/s0220div.htm#div>.

152 In Maharashtra, shipbreaking activities take place in the Mumbai Port. For further details, see Røst, *supra* note 146. See also India, Maharashtra Pollution Control Board, *Second ATR-Part V*, (16 March 2004) at 6, online: Maharashtra Pollution Control Board <http://mpcb.mah.nic.in/images/2ndanex5.pdf>.

153 K. Venkateshwarlu, "Moves afoot to set up ship-breaking unit near Kakinada" *The Hindu [of India]* (10 October 2004), online: The Hindu <http://www.thehindu.com/>. For the details of the shipbreaking unit in Kakinada, Andhra Pradesh State, see Ship Breaking Unit at Kakinada Port, online: Department of Ports <http://www.andhraports.com/kkdshipbreakingunit.htm>.

154 UNESCO, *supra* note 75 at 15.

155 For details on the infrastructure at Alang, see *Alang Ship Recycling*, online: Gujarat Maritime Board <http://www.gmbports.org/alang_alang.htm>. It has 163 plots having an area ranging between 1,350 sq. metres to 3,600 sq. metres. There are also ten plots that can accommodate VLCCs each having an area of 6,000 sq. metres. *Ibid.*

156 India, *Report of the Committee of Technical Experts on Ship Breaking Activities* (Chairman: Dr. Prodipto Ghosh, New Delhi: Ministry of Environment & Forests, 30 August 2006) [unpublished] at 10 [on file with the author] [*Technical Committee Report*].

157 India, Government of India Ministry of Labour and Employment, *The Second National Commission on Labour Report* (New Delhi: Ministry of Labour and Employment, 2002), at 630, online: Government of India Ministry of Labour <http://www.labour.nic.in/lcomm2/nlc_report.html>

Alang is estimated to have provided approximately 23 million tonnes of steel between 1997 and 2007.[158]

1. *The Final Journey to Alang*

Generally, the shipbreaking process starts once the ships are purchased by the recycler either directly or through brokerage services. The sale of a ship involves considerable commercial risk. If the market falls at any time between the conclusion of the contract and the delivery of the vessel, the end buyer may seek to re-negotiate the sale terms.[159] Accordingly, shipowners, for whom selling decommissioned ships is a rare event, find the safest practise is to sell their ships to an intermediate cash buyer specialised in the business. The cash buyer in turn sells the ship to the recycling yard.[160] In fact, so integral is the cash buyer in this system that more than 95 percent of sales are concluded *via* this entity.[161]

Ships are sold for recycling by owners either on the basis of "as is, where is, in any geographical location" or delivered at the outer anchorage of a recycling facility. In the first situation, the cash buyer becomes the legal owner of the vessel during its last voyage to the recycling facility. In the second, the vessel is delivered to the cash buyer and then, within hours, days or weeks, it is re-delivered to the ship recycling facility. During this period, the ship is not re-flagged by the cash buyer and beaching is usually in a flagless status.[162]

Usually, the price is paid in United States dollars per light displacement tonnage (LDT).[163] Factors such as the amount and quality of the steel, the place where the ship was built, the presence of equipment and other items on board that could be reused, etc. are relevant to determining the price of a ship.[164] Ironically, the presence of toxic substances does not in any way influence the price.[165]

(pointing out that the steel produced at Alang is almost equivalent to that produced by the Tata Iron and Steel and Company).

158 *Research Foundation for Science Technology and Natural Resource Policy, supra* note 95.

159 *Ibid.*

160 *Ibid.*

161 See generally Recycling of Ships: Further Development of the Draft Convention, Submitted by India, IMO/MEPC 57/3/16, 8 February 2008 (KR-CON).

162 *Research Foundation for Science Technology and Natural Resource Policy, supra* note 95.

163 The price of scrap steel has long hovered around USD150/LDT (the low being USD100/LDT and the high around USD200/LDT). However, in recent times, the strong demand for scrap steel, coupled with the low supply of ships, has pushed the price to record levels of USD500/LDT for average vessels and more. A very large crude oil tanker with a cargo carrying capacity of 300,000 tonnes and around 45,000 LDT can fetch USD5 to USD10 million or even more. Commission, *Green Paper, supra* note 35 at 5.

164 For example, as there is no market for Russian equipment, breaking a Russian ship entails 'high-weight loss' and consequently, these ships are not preferred. At the same time, ships constructed in the United States give the highest return in terms of the quality and value of the steel recovered. However, this renders them expensive. Japanese ships are considered to be the cheapest and quickest to break. UNESCO, *supra* note 75.

165 Commission, *Green Paper, supra* note 35 at 5.

Once a deal is struck, "the race against rust" begins. If the recycler is to recover the investment, arrangements should be made for an early and efficient dismantling. Even on its final journey, the ship may carry cargo, unless towed, and once this is offloaded, it sails straight to the scrap yard. On reaching the Arabian Sea, the ship is anchored in international waters off Alang pending clearances.[166] Once approved, the wait begins for the high tide, to propel the ship at full speed onto the shore of the appropriate plot.[167]

2. *Placing Profits before People: The Human Costs of Shipbreaking*

After beaching, the autopsy begins and from this stage, the shipbreakers are exposed to a series of hazards as the dead ship is rarely pre-cleaned. One of the more traditional modes of cleaning is to drill holes into the beached ship through which sea water enters, washing the oil-contaminated tanks at high tides.[168] The primary breaking takes place in the inter-tidal zone, where the bow and the stern are cut open to access objects of value.[169] Thereafter, the hull plating, large segments of the ship's structure are opened and sequentially extracted and are either winched or towed ashore.[170] Oils, gaseous wastes, asbestos, etc., are removed.[171] The ship is manually torn down by the shipbreaker. It is estimated that nearly four to five months are required to scrap a ship of an average size.[172] Once on the beach, the recovered scrap is cut to size by using gas torches. The materials are sorted into hazardous and non-hazardous categories. While the hazardous materials are disposed of,[173] the scrap steel and other objects of value are loaded on to trucks by the workers who usually carry it on their heads or shoulders, to be transported to the re-rolling mills.[174] As the demolition advances, the ship is

---

166 Meanwhile, administrative procedures are set in motion with several teams simultaneously inspecting the vessel. The first inspection is by the GMB in consultation with the Atomic Energy Regulatory Board for any radioactive/nuclear wastes, followed by the Gujarat Pollution Control Board to certify that the ship does not contain any hazardous wastes. The permission for beaching and cutting is granted thereafter. The ship is also inspected by a government-approved technician for hazardous materials and residues and by a non-ferrous metal dismantler to assess the best way to break it. UNESCO, *supra* note 75.

167 Twice a month, at full moon and new moon, when the tides are at their highest points, the ship is driven at full speed to the highest land point on the beach. Dodds, *supra* note 50 at 213. Prior to breaking, the tanks are discharged to make the ship light to enable it to climb as high as possible on to the beach. Consumables and other equipment like electronic goods are also removed so that they are not damaged during beaching. Technical Guidelines for Full and Partial Dismantling of Ships, *supra* note 59 at 33.

168 Commission, *Impact Assessment*, *supra* note 12 at 14.

169 *Technical Committee Report*, *supra* note 156 at 12.

170 Technical Guidelines for Full and Partial Dismantling of Ships, *supra* note 59 at 33.

171 *Technical Committee Report*, *supra* note 156 at 12–13.

172 UNESCO, *supra* note 75.

173 *Technical Committee Report*, *supra* note 156 at 14.

174 Dodds, *supra* note 50 at 214.

dragged further up the shore to give way to the beaching of subsequent candidates.[175] In all these stages, as will be seen in the following paragraphs, unpredictable danger lurks that can jeopardise the life and limb of the shipbreaker.[176]

The occupation of a shipbreaker on the Indian subcontinent is one of the most dangerous in the world. The shipbreakers generally do not have any protective gear, or when provided, the quality is substandard. Furthermore, due to the intense humidity and heat, the workers experience considerable difficulty when wearing protective clothing. Most work is done with bare hands, sledgehammers, crowbars, flashlights, and gas torches.[177] In this scheme of things, the shipbreaker emerges more or less as a replaceable commodity, less human, and the basic guarantees of life and dignity are not for him/her. This is so because, even if one worker is lost, there are thousands willing to take his/her place. Moreover, the workers at these scrap yards are unorganised and generally are not visible to law.[178] With no trade unions, they have limited bargaining power.

The work force at Alang is comprised mainly of migrant male labourers hailing from the neighbouring impoverished States of Orissa, Jharkhand and Bihar, as the local workers find the job too risky.[179] With the decline in shipbreaking activities in the past few years, there is no continuous employment for workers. It is estimated that the shipbreakers only work for 180 days a year.[180] Since alternate

---

[175] UNESCO, *supra* note 75 at 16–17.

[176] In the inter-tidal zone, some of the major dangers encountered are inhalation of toxic fumes and gases from the flame due to the cutting operations, particularly if the activity is carried out in confined spaces; risk of fire and explosion; falling objects, trips and slips; inhalation of air containing free asbestos fibres during removal of asbestos; exposure to hazardous liquids and gases like oil, cleaning compounds, and redundant gases like ammonia, PCBs; drowning of workers in case of floating or partially floating structures; accidents due to darkness, etc. In secondary breaking, the primary hazards include risk of fire and explosion, inhalation of toxic fumes, handling of hazardous liquids, exposure to asbestos fibres during removal of asbestos, etc. *Technical Committee Report*, *supra* note 156 at 12, 14. See also Report of the Correspondence Group, *supra* note 7 at 16 (for the years 1997 to 1999, there were 102 deaths at Alang; the number of injured stood at 100).

[177] *MIDN*, *supra* note 20. See also *Technical Committee Report*, *ibid.* at 16 (the average annual incidence of fatal accidents in the shipbreaking industry at Alang during the period 1995–2005 is two per 1,000 workers). The all India incidence of fatal accidents for the same period in the mining industry, which is considered to be one of the most accident prone industries, is 0.34 per 1,000 workers. The analysis of fatal accidents reveals that the most frequent cause of injuries were falls from height (21 percent), fire (20 percent), being struck by falling objects (18 percent), gassing (11 percent), striking against objects (11 percent), and explosions (6 percent). *Ibid.*

[178] *Halsbury's Laws of India*, vol. 17 (India: Butterworths, 2003) "Employment-I (Unorganised Labour)" at 1 ¶ 130.001.

[179] Greenpeace-FIDH in cooperation with YPSA, *supra* note 131 at 31. Most of the migrant workers at Alang/Sosiya come from different states in India. The state distribution is as follows: Uttar Pradesh (41 percent); Bihar (22 percent); Orissa (31 percent); other states – Jharkhand, Maharashtra, Gujarat, Rajasthan, Punjab, West Bengal (six percent). *Technical Committee Report*, *supra* note 156 at 46. Young Power in Social Action, *supra* note 125 at 16 (in Bangladesh, the majority of the workers also hail from the poverty stricken parts of northern Bangladesh).

[180] *Technical Committee Report*, *ibid.* See also International Metalworkers' Federation, *Status of Shipbreaking Workers in India: A Survey*, (2006) at 20, online: International Metalworkers' Federation <http://www.imfmetal.org/main/files/06042810465779/Shipbreaking_survey.pdf>.

employment is almost impossible find, the shipbreakers generally head back to their native states.[181] Although workers are provided with identity cards containing personal details, including blood group,[182] the migratory nature of the employment gives rise to several discrepancies and workers often lose benefits such as provident funds.[183] Illiterate or having very low levels of education, the shipbreakers at Alang come from families that live below the poverty line.[184] Not aware of their basic rights, they can be easily exploited. However, in the absence of employment in their villages, people continue to migrate to Alang.[185] Generally, the shipbreakers are not adequately trained for the job, although training programmes have been introduced on the use of safety equipment in recent years. Whether these programmes are adequate to ensure a safe working climate is debatable.[186]

The shipbreakers at Alang fall into the age group of 19 to 45 years.[187] They work for 10 hours a day, including two hours of compulsory overtime. There are 11 categories of recognised shipbreakers: *jodiwala* (workers who carry heavy iron plates), *battiwala* (these workers cut the ship with oxygen torches), helpers (assist the *battiwalas*), common labourers, *mukadam* (contractors), supervisors, winch operators, crane drivers, fitters, carpenters and foremen.[188] To this list, one must add the *godadiwala* (asbestos worker).[189] Women rarely find employment in these yards due to the strenuous nature of the work,[190] and employment of children is prohibited.[191] Regarding daily wage rates, 81 percent of the shipbreakers earn a wage that ranges slightly more than USD1–2[192] (INR50–100); 11 percent earn

---

181 *Technical Committee Report, ibid.*

182 See also Report of the Correspondence Group, *supra* note 7 at 29.

183 UNESCO, *supra* note 75 at 40.

184 Illiterates constitute 23 percent of the work force. Among the literates, 32 percent of the workforce has attended primary school, another 32 percent middle school, 10 percent senior secondary level, and three percent higher secondary level and graduation. *Technical Committee Report, supra* note 156 at 47.

185 Prafulla Das, "Broken lives" *Frontline [of India]* 23:2 (28 February 2006), online: Frontline <http://www.flonnet.com/fl2302/stories/20060210007301600.htm>.

186 Greenpeace-FIDH in cooperation with YPSA, *supra* note 131 at 31. Personal protective equipment in general and appropriate personal protective equipment in particular, is not provided to the labourers. Minimum protective gear such as helmets and gumboots appear in sufficient quantities and of proper quality when there is an inspection or a safety audit. There is no work plan, and there are no fixed working hours. Sometimes labourers work in the yard for 24 hours without getting any wages for overtime. Shipbreakers get no compensation for the loss of fingers, toes or even hands.

187 International Metalworkers' Federation, *supra* note 180 at 4.

188 UNESCO, *supra* note 75 at 20.

189 *Technical Committee Report, supra* note 156 at 75.

190 International Metalworkers' Federation, *supra* note 180 at 4.

191 *Child Labour (Prohibition and Regulation) Act 1986* (No. 61 of 1986, India), online: India Code <http://indiacode.nic.in/>. Young Power in Social Action, *supra* note 125 (noting that in Bangladesh, child labour in the yards is prevalent).

192 International Metalworkers' Federation, *supra* note 180 at 19. This conversion uses the exchange rate USD1 = INR44.0454, as on 30 January 2006, the year in which this report was published. See also *Technical Committee Report, supra* note 156 at 46.

between USD2–3[193] (INR101–150); seven percent earn between USD3–5[194] (INR151–200); and only one percent earn between USD5–6 (INR201–250).[195] The yards are highly labour intensive, and there is very little mechanisation.[196]

Not only are the working conditions deplorable, the general amenities available to the work force are equally substandard. They live in makeshift housing either on or near the site, continually exposing themselves to harmful chemicals. The sanitary conditions are also unhygienic. A combination of all these factors, coupled with unsafe drinking water, rampant prostitution, and poor health care,[197] renders the situation appalling. Those who manage to survive the yards fall prey to lethal diseases ranging from asbestosis to acquired immune deficiency syndrome (AIDS).[198] The average life expectancy of the shipbreakers is reported to be 40 to 50 years.[199] Generally, injured shipbreakers are not paid any compensation. In cases of death, depending on the strength, if any, of representation by the family, compensation could be as low as USD341 (INR15,000) to as high as USD2,273 (INR100,000).[200] The common saying amongst the shipbreakers, *Alang se Palang* (which when translated means, from Alang to death bed) aptly captures the situation.[201]

3. *Environmental Consequences*

Breaking ships is one of the least environmental friendly activities affecting both aquatic and terrestrial environments. In all places that have had thriving

---

193 *Ibid.*

194 *Ibid.*

195 *Ibid.*

196 The non-mechanised process based on manpower used on the Indian subcontinent has a productivity of a few dozen tonnes/man/year compared to the highly mechanised process used in Western countries with a capability of around 1,000 tonnes/man/year. An intermediate process with limited equipment but significant manpower, such as that used in Turkey, China and even in some North American yards, has productivity of a few hundred tonnes/man/year. *MIDN, supra* note 22 at 13.

197 See generally Lyla Bavadam, "Profits over safety" *Frontline [of India]* 23:2 (28 January 2006), online: Frontline <http://www.flonnet.com/fl2302/stories/20060210007701100.htm> (Bavadam quotes a doctor from nearby Bhavnagar who used to treat cases from Alang to prove that the facilities for onsite emergency care are negligible. Even an intravenous drip for bleeding fractures was not available for the workers. The result was that a majority of patients died on the way to the hospital). However, the GMB claims that there are facilities for primary treatment at the Red Cross Hospital together with a well-equipped burn ward. A mobile medical van with doctor service has also been provided. A full-fledged hospital is being built by a public trust. Periodical health camps are held in association with shipbreakers. *Alang Ship Recycling: Compliance*, online: Gujarat Maritime Board <http://www.gmbports.org/alang_shecomp.htm>.

198 Health awareness projects through Bhavnagar Blood Bank in association with Gujarat State AIDS Control Society are being implemented. *Goal and Objective*, online: Project AIDS Alang <http://www.projectaidsalang.org/>. There have been 38 confirmed cases of AIDS, and every week some 50 to 55 cases of sexually-transmitted diseases being reported. *Technical Committee Report, supra* note 156 at 47.

199 International Metalworkers' Federation, *supra* note 180 at 12.

200 For the exchange rate used, see *supra* note 192; International Metalworkers' Federation, *ibid.*

201 Bavadam, *supra* note 197.

shipbreaking businesses, particularly those areas where the ship is broken after it is grounded on the beach, the environmental damage in coastal areas has been severe.[202] This is because there is limited scope for containing spillage of chemicals and hazardous substances in the inter-tidal zones.[203] At Alang, there is a high degree of contamination by heavy metals[204] and plastic debris.[205] Asbestos is omnipresent.[206] Since shipbreaking commenced, the groundwater table has plummeted from 15 metres to nearly 130 metres.[207] The quality of the water from the coastal aquifers is also suspect due to leachate, worsening a persistent problem of salinity.[208] Increased human activity has also resulted in the contamination of surface water sources.[209] There is severe air pollution in the yards.[210] Until recently, the common practice was to burn wastes in the open, triggering the release of harmful chemicals into the air. The noise levels during the breaking operations are also very high and have caused several accidents.[211] The steel re-rolling mills also contribute significantly to the pollution load. Once the steel plate is removed from the ship, it is immediately put into the furnace without any cleaning. This releases toxins, exacerbating the levels of pollution.[212] For a long time, there were

---

202 India, *Report of the Committee to Review the Coastal Regulation Zone Notification 1991* (Chairman: Prof. M.S. Swaminathan, New Delhi: Ministry of Environment & Forests, February 2005) at 73 [on file with the author] (identifying shipbreaking as one of the major reasons for coastal pollution); Kurunthachalam Kannan *et. al*, "Butyltins in Muscle and Liver of Fish Collected from Certain Asian and Oceanian Countries" (1995) 90:3 Envtl. Pollution 279 at 284 (ScienceDirect) (pointing out that elevated concentrations of butyltins in fish from Bangladesh can be associated with ship scrapping activities). In India, fish collected from Mumbai and Kolkata also showed higher concentrations of butyltin than those from inland areas. *Ibid.* Ship scrapping, anti-foulant sources and sewage disposal have been identified as potential sources of butyltin exposure. *Ibid.* at 287.

203 See also Report of the Correspondence Group, *supra* note 7 at 18.

204 See generally M. Srinivasa Reddy *et al.*, "Distribution, Enrichment and Accumulation of Heavy Metals in Coastal Sediments of Alang–Sosiya Ship Scrapping Yard, India" (2004) 48 Mar. Pollution Bull. 1055 (ScienceDirect) (observing that the enrichment of heavy metals at Alang-Sosiya is relatively high).

205 See generally M. Srinivasa Reddy *et al.*, "Description of the Small Plastics Fragments in Marine Sediments Along the Alang-Sosiya Ship-Breaking Yard, India" (2006) 68 Estuarine, Coast. & Shelf Sci. 656 (ScienceDirect) (reporting that on an average there is 81 mg of small plastic fragments per kg of sediment at Alang-Sosiya).

206 See generally Kanthak & Jayaraman, *supra* note 12 (confirming asbestos in five out of the six samples collected from Alang. It was even found in the living quarters of the workers).

207 UNESCO, *supra* note 75 at 19.

208 *Ibid.* at 24.

209 *Ibid.*

210 The ambient air quality monitoring carried out by the Gujarat Pollution Control Board at the Alang shipbreaking yard for 2005 reveals that the average annual suspended particulate matter was 375.68 $\mu g/m^3$ while the annual national ambient air quality standard was 360 $\mu g/m^3$. Similarly, the respirable particulate matter was 203.50 $\mu g/m^3$, the national average being 120 $\mu g/m^3$. Units for both sulphur dioxide and nitrogen dioxide stood at 13.20 $\mu g/m^3$ while the national average was 80 $\mu g/m^3$. *Technical Committee Report*, *supra* note 156 at 36.

211 UNESCO, *supra* note 75 at 34 (noting that shipbreaking being labour intensive requires a high level of teamwork and coordination which could be difficult due to the noise conditions).

212 See generally Bavadam, *supra* note 197.

no mechanisms to ensure the proper disposal of hazardous wastes. However, due to the intervention by the Supreme Court of India, systems for hazardous waste management have been set up.[213] Nevertheless, the fact that the recycling operations continue to be performed in the inter-tidal zone undermines efforts aimed at pollution control.

D. *Shipbreaking in China*

China was once the world leader in shipbreaking, and by 1993 nearly half of all the ocean-going ships were scrapped there. Shipbreaking takes place mainly in the 90 breaking yards situated on the deltas and the lower reaches of the Pearl and Yangtze rivers. The major shipbreaking yards are in Zhang Jiagang in Jiangsu province. There are additional yards in Fujian and Guangdong provinces.[214] Due to growing concerns about the environmental impacts of this industry, certain regulatory measures were adopted. India quickly exploited this situation, dislodging China from the top slot.[215] The shipbreaking practices in China are rated as being superior to its Asian counterparts as there is significant green recycling capacity[216] and the breaking is carried out in accordance within the terms of a regulatory framework.[217] It is believed that more and more responsible

---

[213] For further details, see Part III of Chapter Three titled, "Limitations of a National Response to Regulate the Global Shipbreaking Industry: A Study on the Indian Experience"; Directorate General Environment, *supra* note 76 at 66 (noting that nearly 22 companies in Alang have received ISO 14001 and/or OSHA 18001 certification).

[214] Samantha L. Jones, "A Toxic Trade: Ship Breaking in China" (A China Environmental Health Project Fact Sheet, 1 February 2007), online: Woodrow Wilson International Center for Scholars <http://www.wilsoncenter.org/topics/docs/shipbreaking_feb1.pdf>.

[215] See Report of the Correspondence Group, *supra* note 7 at 28.

[216] The processing of steel plates takes place on concrete surfaces rather than in the vicinity of intertidal zones. The workforce is also better trained, afforded greater protection and there is a higher degree of mechanisation in the processes. Kanthak & Jayaraman, *supra* note 12 at 6.

[217] In 1982, China promulgated its Marine Environment Protection Law. Based on this law, a series of other regulations and laws were enacted, including the *Regulations on Prevention of Environmental Pollution by Ship-Breaking 1988*. This regulation applies to all units and individuals engaged in shipbreaking activities on the coast and/or on the water within China's jurisdiction. The establishment of shipbreaking programmes has been integrated into the comprehensive plan of the local areas. The activity is prohibited in the protection zone for drinking water sources, pumping salt water for desalting, salt-fields, important fishing areas, seaside resorts, scenic or historic sites, and other areas which need particular protection. An environmental impact statement approved by the relevant authority is to be formulated before the establishment of a shipbreaking enterprise. The unit is to establish installations for preventing pollution, cleaning up inflammable, explosive and toxic substances in the ship before its dismantling. Zou Keyuan, "Implementing Marine Environmental Protection Law in China: Progress, Problems and Prospects" (1999) 23:3 Mar. Pol'y 207 at 214 (ScienceDirect). See also Report of the Correspondence Group, *supra* note 7 at 28 (pointing out that the Chinese National Environment Protection Bureau has also produced a "Technical Manual on Preventing Pollution from Ship Demolition").

shipowners prefer the Chinese yards due to their superior facilities[218] and their huge capacity.[219]

E. *Emerging Scenario*

It is clear that the shipbreaking industry transcends national boundaries. From the time a decision is made to scrap a ship until it reaches its final resting place, and even thereafter, there are several transactions involving various actors, sometimes spanning several continents. Ship recycling is primarily centered in Asia, particularly on the Indian subcontinent. The trio comprised of Bangladesh, India and Pakistan commands almost 90 percent of the business today.[220] To recapitulate the preceding discussion, firstly, very little capital investment is required to start this business on the Indian subcontinent. All that is needed are good beaching facilities. But since ships are broken on the inter-tidal zone, the terrestrial and aquatic environments are severely polluted. The dismantling sites have other advantages as well. The abundant supply of cheap labour willing to carry out the hazardous work, even without minimum physical protections, lowers operational costs considerably. Accordingly, the working conditions in these yards are sub-standard, posing serious human rights issues.

The next important factor is the thriving market for scrap steel and for the second-hand materials that are scooped out of the ship. Perhaps the most crucial element that has facilitated the entrenchment of this industry on the subcontinent is the lack of strict environmental and labour regulations and lax enforcement of existing laws on the subject which enable ship recyclers and owners to avoid their responsibilities to their workers and to protect the environment. Shipbreaking, as seen on the Indian subcontinent, represents a classical case of negative externalities. This is so because, even though this activity results in pollution to the air, to the soil and to the oceans, making them less useful to the local communities, to the workers and others, including future generations, those responsible for this vandalism are legally not required to remedy the damage.[221] Consequently, the

---

[218] See *P&O Nedlloyd and China Partnership in Ship Recycling*, online: Hong Kong Shipowners Association <http://www.hksoa.org/member/Technical%20Papers/Tech/Tom%20Peter%20Blankestijn.pdf> (pointing out that although the Chinese yards provide low salaries, there is no child labour, good housing and food, structured staff training, protection equipment, good environmental controls, a limited number of accidents, programmes for cleaning the yards, and plans for improvement).

[219] Greenpeace, "Special Report, Shipbreaking: Toxic Waste in Disguise the China Connection" at 2, online: Greenpeace International <http://www.greenpeace.org/raw/content/international/press/reports/shipbreaking-toxic-waste-in-d.pdf> (noting that China has an unused capacity of at least 10 mdwt (million dry weight tonnes), almost the same capacity that was active in India when Alang was at its peak).

[220] *MIDN*, Annex II, *supra* note 124 at 2.

[221] Jeffery L. Harrison, *Law and Economics*, 2d ed., West Nutshell Series (St. Paul, MN: West Group, 2000) at 43.

ship recycler is saved from investments in technology needed for safer and cleaner shipbreaking. The amount of toxins present in a ship does not influence its value, and this allows the recycler to offer huge prices for the obsolete tonnage. For these reasons shipbreaking, even after decades of operation on the subcontinent, continues at a primitive level.

In sharp contrast, responsible shipping companies prefer to have their ships recycled in the green yards in China.[222] This development will significantly influence the future course of this industry. A central feature of the IMO-sponsored international regime for the safe and environmentally sound recycling of ships is that parties are to ensure the authorisation of ship recycling facilities under their jurisdiction.[223] Unless changes are made, it can be expected that several of the yards on the Indian subcontinent will not meet the IMO standards required for authorised status. This might push these players out of business, and ships that require scrapping services may head to the better equipped Chinese yards. Therefore, it could be in their best interests for the players on the Indian subcontinent to initiate programmes to augment the infrastructure in their yards so as to attain a "green" status.

IV. Shipbreaking in the OECD Member States

Comprised of 30 industrialised countries committed to the principles of democracy and a market economy, the OECD member states include some of the former leading shipbreakers from Europe and Asia.[224] Even though the industry migrated to the Indian subcontinent, OECD member states such as Belgium,[225] the

[222] *P&O Nedlloyd and China Partnership in Ship Recycling*, *supra* note 218.

[223] Article 6 of the *Hong Kong International Convention for the Safe and environmentally Sound Recycling of Ships, 2009*. For the text of the convention, see International Conference on the Safe and Environmentally Sound Recycling of Ships: Agenda Item 8, Adoption of the Final Act and Any Instruments, Recommendations and Resolutions Resulting from the Work of the Conference: Hong Kong International Convention for the Safe and Environmentally Sound Recycling of Ships, 2009, Text Adopted by the Conference, IMO/SR/CONF/45, 19 May 2009, online: SJÖFARTSVERKET <http://www.sjofartsverket.se/pages/19514/sr-conf-45.pdf> (opened for signature 1 September 2009).

[224] See *supra* notes 8–11 and accompanying text.

[225] The *Van Heyghen* recycling yard is located at the Gand port. Even though Gand is capable of handling large vessels, it primarily dismantles fishing vessels and ships up to 5,000 tonnes. It holds an ISO 9001 certification. France, Interdepartmental Committee on the Dismantling of Civilian and Military End-of-Life Ships, *Le rapport de la Mission Interministérielle portant sur le Démantèlement des Navires civils et militaires en fin de vie*, annex III at 2 (27 March 2007, Chair: Xavier de la Gorce), online: SGMer <http://www.sgmer.gouv.fr/IMG/pdf/Annex_3_Dismantling_site_in_Europe_and_O ECD.pdf> [*MIDN*, Annex III]; Danish Environmental Protection Agency, *supra* note 97 at 2.

Netherlands,[226] the United Kingdom,[227] Italy,[228] Denmark,[229] Spain,[230] Canada,[231] and the United States[232] still have recycling capacity that is green, albeit on a limited scale. In fact, the OECD countries constitute the primary exporters of toxic ships to China and to the Indian subcontinent.

There are three facets of the shipbreaking industry in OECD countries and its status as the leading exporter of toxic ships that warrant further exploration. First, the ship recycling industry in Turkey is taken as a representative example to illustrate the principal dilemmas that confront this industry in general in the OECD. Second, an exposition on the European Union WSR is provided. Despite being one of the toughest laws employed to curb the illicit trade in toxic ships, its practical success has been limited. What are the reasons for its ineffectiveness? Finally, the example of the controversial export of United States' ghost ships to the United Kingdom is used to highlight the difficulties encountered by national regulators in the OECD countries in balancing their conflicting interests, *i.e.,* their commitment to environmental justice and protection of human rights in developing countries, as well as their inability to dispose their rusting ships domestically.

A. *Turkey: The Major Player in the OECD*

Though ship dismantling in Turkey can be traced back to the Ottoman Empire,[233] it was only after 1974 that the industry assumed importance with the establishment

---

[226] *MIDN*, Annex III, *ibid* at 1, 4. The major dismantling site in the Netherlands is *Scheepssloperij*. There are plans for a shipbreaking facility called Ecodock. For further details, see *About Ecodock*, online: nv ecodock <http://www.ecodock.info/uk/overecodock.php>.

[227] There are three major ship dismantling sites in the United Kingdom: Swan Hunter, in Newcastle is mainly involved in the dismantling of North Sea platforms; Harland and Wolff, in Belfast; and Able UK at Hartlepool Teeside which was to dismantle the ghost ships. *MIDN*, Annex III, *ibid* at 4.

[228] The Simont site in the port of Naples also holds an ISO 9001 certification. It mainly dismantles naval warships. See *Ibid*. at 3; Danish Environmental Protection Agency, *supra* note 97 at 2.

[229] The principle Danish shipbreaker is Fornaes Shipbreaking with a yard in Grenaa. The annual capacity is up to 10,000 GT (gross tonne) (17,000 DWT) with a maximum length of 100 m, roughly corresponding to 4–5 vessels of 2,000 GT. Most vessels broken until now have been smaller fishing vessels. Denmark has a dormant capacity, e.g., in the dry dock in Nakskov and in Frederikshavn. *MIDN*, Annex III, *supra* note 225 at 24–25.

[230] Directorate General Environment, *supra* note 76 at 87 (*Desguaces de la Arena* (*Soto del Barco*) in Spain can dismantle vessels up to 220 m length); Danish Environmental Protection Agency, *supra* note 97 at 24.

[231] See generally OECD, Environment Policy Committee, *Barriers to the Implementation of Environmentally Sound Management Practices for Small and Medium Size Metal Recycling Enterprises, Case Study: Canadian Ship Dismantling Enterprise*, Working Group on Waste Prevention and Recycling, Doc. No. ENV/EPOC/WGWPR/RD(2002)7/FINAL (2003), online: OECD <http://www.oecd.org/dataoecd/61/35/2502052.pdf>.

[232] Mark J. Kaiser, "A Review of Ship Breaking and Rig Scrapping in the Gulf of Mexico" (2008) 39 Ocean Devel. & Int'l Rev. 178 at 185.

[233] *Ship Recyclers' Association of Turkey*, online: Ship Recyclers' Association of Turkey <http://www.gemisander.org/index.php?lang=en> (providing an overview of the Turkish ship recycling industry).

of recycling yards at Aliağa, Izmir.[234] Within a decade, the government legalised the industry, and by 1986, Turkey became the third leading shipbreaking country in the world.[235] With players on the subcontinent proving more competitive, Turkey now ranks fourth among the shipbreaking countries in the world.[236]

Despite its membership in the OECD, it is widely believed that Turkey was being used by other more developed nations as a dumping ground for their toxic ships, just like its counterparts in Asia. A Greenpeace study in 2002 slammed Turkey for the high contaminants in the marine environment and unsafe work conditions in its yards.[237] Since that time, the situation has improved considerably, facilitated by the decision to turn back the asbestos contaminated 'Sea Beirut' to France.[238] Several of the recycling companies in Aliağa have improved conditions in their yards, and some even have international certification for environmental and occupational health.[239]

Turkey has several advantages that make it a preferred destination for shipowners who require environmentally-friendly scrapping. It is situated at a vantage point[240] and can be easily accessed from both the European Union and the Middle East. Shipbreaking at Aliağa uses a modified slipway recycling,[241] which

---

[234] *Ibid.*

[235] *Ibid.*

[236] In 2006, Turkey demolished only seven vessels for a market share of 0.55 percent. *MIDN*, Annex II, *supra* note 124 at 3.

[237] See generally Erdem Vardar & Marietta Harjono, *Ships for Scrap V: Greenpeace Report on Environmental, Health and Safety Conditions in Alia?ga Shipbreaking Yards, Izmir, Turkey*, Steel and Toxic Wastes for Asia (Greenpeace Mediterranean, 2002), online: Greenpeace Nederland <http://www.greenpeaceweb.org/shipbreak/shipsforscrap5.pdf>. See also *Report of the Special Rapporteur on the Illicit Movement and Dumping of Toxic and Dangerous Products and Wastes on the Enjoyment of Human Rights, Fatma Zohra Ouhachi-Vesely, On Her Mission to Turkey (10 to 19 March 2004)*, UNESC, 61st Sess., E/CN.4/2005/44 (15 February 2005), annex, at 4, online: NGO platform on shipbreaking <http://www.shipbreakingplatform.com/dmdocuments/other_documents/Special%20Rap porteur.Report.Turkey.pdf>.

[238] *Cemsan Ship Dismantling Metal and Steel Industry Trade Ltd. Co. v. Ministry of Environment, Ankara* (2003/1184), Case No. 2002/496 (Izmir 2nd Admin. Ct.) (unofficial translation), online: Basel Convention <http://www.basel.int/ships/Turkey300903e.pdf>. The main issue centered on the decision to stop the import of the French ship 'The Sea Beirut', which was brought to Aliağa for dismantling on the ground that it contained high levels of asbestos. Following protests by Greenpeace, samples were analysed which revealed high amounts of asbestos leading to the stopping of the import. In the absence of any document to prove prior removal of hazardous waste, the court upheld the decision to stop the import. *Ibid.*

[239] Directorate General Environment, *supra* note 76 at 63.

[240] See generally *Ship Recyclers' Association of Turkey*, *supra* note 233.

[241] The considerable depth near the coast enables even large vessels to be run or towed ashore. The vessel is cut from the bow with large sections extracted and pulled onto the cutting area on land by powerful winches while smaller sections and equipment are removed by cranes and cut in special areas, some of which have impermeable concrete slabs. Gökdeniz Neşer, "The Shipbreaking Industry in Turkey: Environmental, Safety and Health Issues" (2008) 16 J. Cleaner Production 350 at 352 (ScienceDirect). A slipway is basically a sloping concrete slab leading from above high water mark to a depth where the ship is floated off the cradle and launched. For dismantling, the ship is hauled up the slipway and is dismantled internally and the hull cut down. U.K., Department for

reduces chances of environmental pollution. A regulatory framework,[242] improved worker protection,[243] greater mechanisation in the processes,[244] and being part of the OECD[245] give it a competitive edge from the standpoint of responsible scrapping. Despite these benefits, Turkey has not been able to attract obsolete tonnage commensurate to its capacity. The primary reason for this situation is the price that it offers for the scrap.[246] While the price for scrap steel in Turkey hovers around USD200 per light weight tonne, the scrap yards on the Indian subcontinent are able to offer double that amount. This dilemma is true of the ship recycling industry in general in the OECD, rendering it an unattractive proposition to have ships recycled there.

B. *The European Union Waste Shipment Regulations*

The European Union has been in the forefront of preventing trade in hazardous wastes and has one of the most stringent laws on this subject. The WSR 259/93 on the "supervision and control of shipments of waste within, into, and out of the European Community",[247] which came into force on 6 May 1994, is the "most comprehensive and restrictive plan adopted in the European Union covering all waste shipments, including hazardous wastes".[248] In July 2007, this Regulation was replaced by WSR 1013/2006.[249] Even though WSR 259/93 has been replaced, it was operational for more than a decade and was invoked to curtail the movement of several toxic ships from the European Union to the Indian subcontinent. An examination of this law as it applies to shipbreaking is useful in identifying the gaps and the implementation hurdles that regulators have encountered that facilitate the trade in toxic ships. The news features of WSR 1013/2006 will also be reviewed briefly.

Environment, Food and Rural Affairs, *Overview of Ship Recycling in the UK*: *Guidance* (London: Crown Copyright, Department for Environment, Food and Rural Affairs, 2007) at 13, online: Defra <http://www.defra.gov.uk/environment/waste/strategy/pdf/shiprecycle-strategy-overview.pdf>.

242 Neşer, *ibid.* (pointing out to a series of laws - *Ship breaking Regulation, the Labour Health and Work Safety Regulation, Hazardous Chemicals Regulation, Control of Hazardous Wastes Regulation* – to regulate this industry).

243 The ship recycling sector in Aliağa has created jobs for nearly 1,800 workers and has benefited another 8,000 people indirectly through ancillary industries. The workers are provided training. Periodical health checks are conducted, and no serious occupational disease have been encountered. See generally *Ship Recyclers' Association of Turkey*, *supra* note 233.

244 Neşer, *supra* note 241 (pointing out that an asbestos removal centre has been set up in Aliağa).

245 *MIDN*, Annex II, *supra* note 124 at 7.

246 *Ibid.* at 6.

247 See *WSR 259/93, supra* note 13. See Mark E. Allen, "Slowing Europe's Hazardous Waste Trade: Implementing the Basel Convention into European Union Law", Comments, (1995) 6 Colo. J. Int'l Envtl. L. & Pol'y 163 at 170 (initially, the Council of the European Community adopted several directives to create a framework for regulating hazardous waste disposal and their transboundary movement. Since these were not being enacted into the laws of the member states, the Commission decided to promulgate a regulation covering all waste shipments).

248 Allen, *ibid.*

249 See *WSR 1013/2006, supra* note 13.

The WSR 259/93 covers five categories of waste shipments. Of these, the rules relating to exports from European Union to non-member countries are most relevant for ship recycling. Waste shipments are divided into two categories: wastes that are intended for recovery[250] and those for disposal.[251] If the export of a ship for dismantling is an export for disposal, then such transfers are prohibited, unless it is to a European Free Trade Association (EFTA) country that is also a party to the Basel Convention.[252] Export of ships for recovery are also prohibited, except to countries to which the decision of the OECD Council of 30 March 1992 on the control of trans-frontier movements of wastes destined for recovery operations applies and to countries that are parties to the Basel Convention or other bilateral or multilateral agreements that are compatible with the European Union legislation.[253] In the context of shipbreaking, this classification is crucial.[254]

The WSR 259/93 includes several of the requirements mandated by the OECD decision (in relation to recovery) and the Basel Convention (in relation to disposal).[255] Accordingly, as in the OECD decision, it categorises wastes into three lists: green, amber, and red.[256] Shipment of wastes listed in annex II, or the Green List, that are intended for recovery are generally excluded from the regulatory requirements.[257] Interestingly, ships have been included on the Green List if "vessels and other floating structures for breaking up, [are] properly emptied of any cargo and other materials arising from the operation of the vessel which may have been classified as a dangerous substance or waste".[258] As the term "properly emptied" has not been defined, there is confusion as to what exactly it means.[259] Moreover, the opening terms of annex II reads:

> Regardless of whether or not wastes are included on this list, they may not be moved as green wastes if they are contaminated by other materials to an extent which (a) increases

---

[250] See *WSR 259/93*, *supra* note 13 art. 2(k) (wherein recovery is defined as under article 1(f) of Council Directive 75/442/EEC). For the text of the Directive, see EC, *Council Directive 75/442/EEC of 15 July 1975 on waste*, [1975] O.J. L 194/39.

[251] See *WSR 259/93*, *ibid.*, art. 2(i) (wherein disposal is defined as under article 1(e) of Council Directive *75/442/EEC*).

[252] *Ibid.*, art. 14.

[253] *Ibid.*, art. 16.

[254] *Stichting Greenpeace Nederland, a Foundation that has its Seat in Amsterdam v. the State Secretary for Housing, Spatial Planning and the Environment* (21 February 2007), Case No. 200606331/1 (Council of State, Administrative Jurisdiction Division), online: Basel Convention: Dismantling of Ships – Relevant Caselaw <http://www.basel.int/ships/relevcaselaw.html> (holding that such exports are exports for disposal). See also Directorate General Environment, *supra* note 76 at 30.

[255] *WSR 259/93*, *supra* note 13 art. 2(r). See also Hilary Stone, "Flouting the Law – European Shipment of Hazardous Waste" (2007) 21:1 Nat. Resources & Env't 49 (HeinOnline).

[256] *WSR 259/93*, *ibid.*, annex II, III, IV.

[257] *Ibid.* art. 1(3)(a).

[258] *Ibid.*, annex II, GC 030.

[259] Directorate General Environment, *supra* note 76 at 28.

> the risks associated with the waste sufficiently to render it appropriate for inclusion in the amber or red lists, or (b) prevents the recovery of the waste in an environmentally sound manner.[260]

This is important for shipbreaking because even though a ship does not carry hazardous wastes as cargo, it can still be brought under the terms of WSR 259/93 if its in-built wastes are of such a quantity as to transform its character.[261]

Pursuant to an amendment to WSR 259/93[262] to implement the Basel Ban,[263] annex V lists the wastes that fall under this export ban. Even though end-of-life ships are not explicitly listed, unless ships for scrapping have been built recently, they are likely to contain hazardous substances that are listed under the annex and would therefore fall under its purview.[264]

In spite of being one of the most comprehensive laws controlling the shipment of hazardous and non-hazardous wastes,[265] shipping interests in the European Union have developed techniques to by-pass the rigour of this law. They resort to reflagging end-of-life ships and to corporate mechanisms that enable real owners to cloak their identity, thereby posing serious enforcement challenges.[266]

---

260 *WSR 259*/93, *supra* note 13 annex II.

261 *Upperton Ltd., with offices in Mauritius v. The Minister of Housing, Spatial Planning and the Environment*, (19 June 2002), LJN-No.: AE4310 Case No.: 200105168/2 (Council of State, The Hague), reproduced in Recycling of Ships: Decision Concerning the Export of the "Sandrien" for Ship Recycling, Submitted by Greenpeace International, IMO/MEPC 48/INF.11, 10 July 2002, annex (KR-CON).

262 EC, *Council Regulation (EC) 120/97 of 20 January 1997amending Regulation (EC) 259/93 on the supervision and control of shipments of waste within, into and out of the European Community*, [1997] O.J. L 22/14.

263 See generally EC, *Council Decision 97/640/EC of 22 September 1997 on the approval, on behalf of the Community, of the amendment to the Convention on the control of transboundary movements of hazardous wastes and their disposal (Basel Convention), as laid down in Decision III/ of the Conference of the* Parties, [1997] O.J. L 272/45.

264 Asbestos is mentioned under entries A 2050, 17 06 01 and 17 06 05 while PCBs are under entries 16 01 09, 16 02 09 and 16 02 10. See Directorate General Environment, *supra* note 76 at 27.

265 See Allen, *supra* note 247 at 174.

266 Directorate General Environment, *supra* note 76 at 6. Between 2001 and 2003, 14 percent of the ships sent for scrapping flew the flags of European Union member states and 18 percent the flags of states that acceded to the European Union in 2004. At least 36 percent of world shipping tonnage in 2006 was owned by companies domiciled in the European Union. Approximately 100 warships and other government vessels flying European Union flags, most of them French and British, are expected to be decommissioned in the next 10 years. Commission, *Green Paper*, *supra* note 35 at 2. The Green Paper also calls for more controls by waste shipment and port authorities in European ports, targeting ships above a certain age (25 years) or where there are other indications that they are intended for dismantling. *Ibid*. at 9. In 1999, on being informed that the owner of the bulk carrier 'Forthbank' wanted it to be scrapped in Alang, inspections were carried out by the Flemish Waste Department. This led to the detainment of the 'Forthbank' on the grounds that it was not properly emptied of hazardous materials thereby violating WSR 259/93 and the Basel Convention. Thereafter, on the basis of its insurance papers, the 'Forthbank' departed Antwerp, sailing for another three years, finally to be scrapped in Chittagong. See generally Recycling of Ships: Comments on the Report of the Correspondence Group: Selected Cases of Decommissioning of Vessels Indicating the

Although the scrap volume from the merchant fleet under European Union member state flags ranges from 1.3 to 2.1 million LDT/year, the actual available scrap volume is difficult to estimate because of this reflagging.[267] At the same time, the scrap volume from government vessels (mainly warships, government-owned vessels, and oil rigs) is limited to 40,000 LDT/year and the supply is inconsistent.[268] The sporadic supply of ships makes it difficult to establish a viable dismantling capacity in the European Union. The expense involved in safely breaking a ship and the non-availability of a steady market for second-hand fixtures extracted from these ships also impedes the establishment of a ship dismantling industry in the European Union. As a result, the European Union remains as one of the major exporters of toxic ships.

Unlike its predecessor, WSR 1013/2006 emphasises the need "to ensure the safe and environmentally sound management of ship dismantling … to protect human health and the environment".[269] It recognises the work of the IMO, the ILO and the Basel Secretariat to establish mandatory guidelines at the international level to provide efficient and effective solutions to the problems posed by ship dismantling.[270] The WSR 1013/2006, which has as its main objective protection of the environment,[271] also prohibits the export of wastes for disposal except to EFTA countries that are also parties to the Basel Convention.[272] Exports for recovery to non-OECD countries are permissible except for seven categories of waste.[273] In comparison to WSR 259/93, this regulation incorporates many "rigorous controls". However, WSR 1013/2006 also fails to adequately address concerns relating to reflagging end-of-life ships.

C. *Shipbreaking in the United States: Exorcising the Ghost Fleet*

Due to stringent environmental and safety requirements mandated by the laws of the United States, US-flagged ships constitute an insignificant fraction of total sea traffic.[274] The majority of ships that fly the United States' flag are those that are owned and operated by the United States government rendering it, in the process, the owner of the largest fleet of obsolete ships that need disposal.[275]

Need for Mandatory Requirements, Submitted by Greenpeace International, IMO/MEPC 49/3/2, 9 May 2003, annex (KR-CON).

[267] Directorate General Environment, *ibid.* at 6.

[268] *Ibid.*

[269] See *WSR 1013/2006*, *supra* note 13 Preamble ¶ 35.

[270] *Ibid.*

[271] *Ibid.*, art. 49.

[272] *Ibid.*, arts. 34,40.

[273] *Ibid.*, art. 36.

[274] Cohen, *supra* note 24 at 244.

[275] *Ibid.* at 245.

Until the 1980s, most United States' ships were disposed of domestically. During the Cold War period, there was a massive build up of the navy, the army and the air force. Since the hardware had to be battle ready, there was a sharp decline in the number of federal ships that were sent for scrapping.[276] This down-swing in supply, coupled with rising environmental awareness of the harmful consequences of ship scrapping, led to stringent regulations that precipitated the collapse of the domestic shipbreaking industry.

With the Cold War coming to an end, the United States government decided to streamline the naval fleet.[277] However, domestic dismantling capabilities had almost vanished. Consequently, a large number of leaking and environmentally hazardous ships had to be mothballed and placed in storage with the National Defence Reserve Fleet (NDRF) created under the *Merchant Ship Sales Act of 1946*.[278]

Termed the "ghost fleet",[279] or the "ticking time bombs",[280] due to their deteriorating condition (it is said that even a hammer could penetrate the hulls of some of these ships),[281] this flotilla was seen as a ready reserve to meet national exigencies. The NDRF has two divisions – the Ready Reserve Force (RRF)[282] and the non-Ready Reserve Force (non-RRF).[283] Ships within the NDRF that cannot meet

---

[276] The Reagan administration invested almost USD2 trillion in a massive military build-up, which also included an expansion of the naval fleet to about 600 vessels. Most of the available vessels were converted to military use. With the Cold War coming to an end in 1989, the Reagan administration was faced with the difficult issue of finding ways to downsize the inventory of ships that had become useless overnight. See Takako Morita, "N.I.M.B.Y. Syndrome and the Ticking Time Bomb: Disputes over the Dismantling of Naval Obsolete Vessels", Note, (2005) 17 Geo. Int'l Envtl. L. Rev. 723 at 726 (WLeC).

[277] *Ibid.*

[278] Pub. L. No. 321, § 11, 60 Stat. 41 (1946) (codified as amended at 50 U.S.C. App. 1735–46 (2000)).

[279] David Killion, "Trans-Atlantic Ghost Busting: The Failed Attempt to Dispose of the Chesapeake Ghost Fleet in the United Kingdom", Comment, (2008) 42 U. Rich. L. Rev. 731 at 733 (WLeC).

[280] Morita, *supra* note 276.

[281] Basel Action Network, "Needless Risk: The Bush Administration's Scheme to Export Toxic Waste Ships to Europe" (20 October 2003) at 21, online: Basel Action Network <http://www.ban.org/Library/Needless%20Risk%20Final.pdf>.

[282] The RRF component established in 1976 is a reserve of vessels maintained in a condition that would allow these ships to be activated in 20 days or less in case of a national emergency. See U.S., Maritime Administration, *Transfer of National Defense Reserve Fleet Vessels from the James River Reserve Fleet for Disposal at Able UK Facilities, Teesside, UK Environmental Assessment: Final* (Department of Transportation, 2004) at 1, online: U.S. Department of Transportation <http://www.marad.dot.gov/Programs/Scrapping/web/02-MARAD%20UK%20FinalEA%20Text_cameraready_062904.pdf> [*Environmental Assessment: Final*]; U.S. Department of Transportation Maritime Administration, *The National Defense Reserve Fleet, Ready Reserve Force*, online: U.S. Department of Transportation <http://www.marad.dot.gov/Ship%20Disposal/PRESS_NDRF_RRF_4qtr05.pdf> (noting that RRF ships (termed as "America's Sea Power Reliant Partner") are frequently used. For instance, in 2003, 40 RRF ships were used in support of Operations Enduring Freedom and Iraqi Freedom).

[283] The non-RRF vessels receive minimal maintenance and would require at least 30 days to be activated (MARAD, 1997). *Environmental Assessment: Final*, *ibid.*

the Ready Reserve Fleet requirements are designated for disposal and are placed in the non-retention fleet located at the James River Reserve Fleet in Virginia, the Beaumont Reserve Fleet in Texas, and the Suisun Bay Reserve Fleet in California.[284]

Under the *Federal Property and Administrative Services Act of 1949*,[285] the duty to dispose of government-owned merchant vessels of 1,500 GT or more rests with the Department of Transportation, Maritime Administration (MARAD). Prior to 1994, MARAD sold these obsolete ships to the highest bidder (domestic or foreign).[286] Since the disposal rates were not commensurate with the volume that was available for scrapping, Congress passed the *National Maritime Heritage Act of 1994*,[287] directing MARAD to dispose of all of its surplus ships by 30 September 1999.[288]

In response to growing concerns about the horrific working conditions in the shipbreaking yards in Asia, the United States Environmental Protection Agency (EPA) issued a letter advising MARAD not to export ships containing toxic wastes.[289] However, faced with fast approaching deadlines within which the scrapping was to be completed, MARAD negotiated an agreement with the EPA in 1997 to resume the service of foreign shipbreakers. However, the exports were suspended due to concerns over worker safety[290] and, in 1998, a one-year moratorium on the overseas sale of obsolete naval ships for scrapping was imposed.[291]

Realising that it would be impossible to complete the scrapping within the prescribed timeframe,[292] the deadline was extended to 30 September 2006.[293] More importantly, the moratorium was reversed.[294] Finally, in December 2002, the *Bob Stump National Defense Authorization Act of Fiscal Year 2003*[295] authorised MARAD to export the decaying ships. MARAD and the EPA initiated a pilot programme under which a USD17.3 million contract was awarded to the

---

[284] Cohen, *supra* note 24 at 246.

[285] 40 U.S.C.A. § 101.

[286] It is estimated that between 1987 and 1994, MARAD exported 128 vessels to China, India, Mexico, and Taiwan. Morita, *supra* note 276 at 731.

[287] Pub. L. 103–451, 108 Stat. 4769 (1994).

[288] *Ibid.*

[289] Morita, *supra* note 276 at 724.

[290] *Ibid.* at 725.

[291] *Ibid.*

[292] Noland, *supra* note 99 at 514 (pointing out that nine spills occurred between 2000 and 2003 in respect of the James River Reserve Fleet. The largest spill, from the U.S.S. Donner in August of 2000, poured 1,000 gallons of oil into the James River). See also U.S., Maritime Administration, *National Defense Reserve Fleet (NDRF), Suisun Bay, CA, Vessel Environmental Review*, (Oakland: 2007) at ES–6, online: BayAreaNewsGroup <http://bayareanewsgroup.com/multimedia/cct/multimedia/pdf/suisunpollute.pdf> (recommending corrective actions to arrest exfoliation of paint into the environment).

[293] Noland, *ibid.* at 523–24.

[294] Killion, *supra* note 279 at 741.

[295] Pub. L. No. 107–314, 116 Stat. 2458 (2002).

Post-Service Remediation Partners to tow 13 NDRF ships from the James River Fleet to the Able UK facility in Teesside, United Kingdom, where Able UK was to recycle them.

Initially, four ships were to be exported and once Able UK satisfactorily completed the dismantling of this set, an additional nine were to follow. An action was filed by the Sierra Club and the Basel Action Network alleging that MARAD and EPA had violated the *Toxic Substances Control Act* (TSCA)[296] and other environmental statutes.[297] Though a temporary restraining order was granted "in part and denied in part",[298] the primary fallout of the order passed subsequently,[299] virtually stalled any hopes to develop an ambitious export programme.[300] At the

---

[296] 15 U.S.C.A. § 2601 (West 2004).

[297] *Basel Action Network v. Maritime Administration*, 285 F.Supp.2d 58 (D.D.C., 2003). Plaintiffs filed a motion for a temporary restraining order (TRO) to enjoin the defendants from exporting 13 defunct naval ships to the United Kingdom for recycling on the grounds that the export violates three statutes: the TSCA (this statute prohibits export of PCBs without an exemption from the EPA, granted pursuant to a formal rulemaking. In this case, the EPA did not engage in rulemaking; rather their decision was based on an "enforcement discretion" letter stating that the EPA would not enforce the PCB export ban against MARAD as long as certain conditions relating to disposal of ships were met), the NMHA (this Act according to the plaintiffs requires that MARAD, as the best value alternative, use a different disposal option for these ships since towing for 45 days across the North Atlantic would pose significant environmental risks), and finally, the *National Environmental Policy Act of 1969* (NEPA) (this Act directs MARAD and EPA to do an environmental assessment (EA) before the export. MARAD prepared reports in 1994 and 1997 on its ship disposal programme and, in addition, submitted reports to Congress which it asserted were the "functional equivalent" of an EA. However, to the plaintiffs, the EPA had defaulted on the EA requirement). These arguments were considered against the standards required for a TRO. As far as the issue of rulemaking was concerned, the court held that the EPA was well within its discretionary power. Regarding NMHA, the court held that the defendants had purged the ships from all PCBs prior to transit. Moreover, they also produced the International Load Line Exemption Certificates from the United States Coast Guard showing that the vessels were seaworthy. As for the NEPA claim, the court determined that the reports submitted to the Congress constituted the "functional equivalent" of an EA with respect to the four ship pilot programme. However, with respect to the remaining nine ships, since these were not part of the pilot programme, supplemental EAs were required and accordingly, the court passed a temporary order blocking their export. *Ibid.*

[298] *Ibid.*

[299] *Basel Action Network v. Maritime Administration*, 370 F.Supp.2d 57 (D.D.C., 2005). After issuance of the TRO with respect to nine of the 13 ships, which was modified to a preliminary injunction, the parties filed cross-motions for summary judgment. This was dismissed by the court on the grounds that the NEPA did not apply beyond United States' territorial waters. MARAD had satisfied its obligation under NEPA to take a "hard look" at the impact of its decision in the United States. Further, the court found that the environmental organisations had failed to wait the required 60-day period to file suit after giving notice of intent to sue under TSCA.

[300] The *Report to Congress on the Progress of the Vessel Disposal Programme* comments on the fallout of the decision thus:

> A favourable March 2, 2005 ruling by the U.S. District Court for the District of Columbia did not provide MARAD with relief from the requirements of TSCA. While the court concluded the EA prepared by MARAD fully met its obligations under the National Environmental Protection Act … and dismissed the plaintiff's complaint, the court's ruling does not remedy the underlying environmental issues under TSCA that triggered the legal action initially and

same time, another legal challenge unfolded in the United Kingdom, compelling Able UK to reapply for its licenses.[301] With the Hartlepool Borough Council refusing the licenses, the contract stood cancelled.[302] This left the fate of the four ships that had crossed the Atlantic for recycling in a limbo.

Clearly, MARAD's attempt to exorcise the ghost fleet by exporting the ships failed. The episode however served to strengthen the domestic dismantling capacity in the United States, with nearly eight dismantling facilities (from the initial three) being set up, and 72 ships removed as of 2006.[303] As MARAD continues to explore all feasible disposal options, the number of decaying ships needing disposal continues to increase.[304]

Unlike the general pattern, the ghost ship controversy involved the transfer of ships for dismantling between developed countries *inter se*.[305] All the ships were basically empty structures not carrying any cargo or waste, and they were subjected to major decontamination before being sent for scrapping. However, as with other ships of their time, they had asbestos and PCBs as part of their structure. The United Kingdom had the technology to dispose of these hazards. However, the controversy erupted mainly because of the confusion over "which

---

does not preclude plaintiffs or other citizens from immediately filing another civil action against MARAD to deny the export of obsolete ships.

U.S., Maritime Administration, *Report to Congress on the Progress of the Vessel Disposal Programme* (Department of Transportation, 2007) at 16-17, online: U.S. Department of Transportation <http://www.marad.dot.gov/Ship%20Disposal/ReportToCongressFinal03_19_07.pdf> [*Report to Congress,* 2007].

[301] The litigation in the United Kingdom consisted of two cases. The first related to Able UK's waste management license, while the second was concerned with Able UK's permission from the City of Hartlepoole. See *Queen on the Application of Friends of the Earth Ltd. v. Environment Agency* (2003), [2003] EWHC 3193 (Admin), [2004] Env. L.R. 31 (Q.B.D. Admin) (WL) (quashing the modifications to the waste management license of Able UK increasing the quantity of waste it could process under the license on the grounds that these modifications to the license made it materially different from the original license). In *Queen on the Application of Gregan v. Hartlepool BC* (2003), [2003] EWHC 3278 (Admin), [2004] J.P.L. 1088 (Q.B.D. Admin) (WL) (ruling that the planning permit obtained by Able UK in 1997 to build a dry dock applied only to dismantling of marine structures such as oil rigs and not to ships).

[302] Basel Action Network, Press Release, "Toxic Ghost Fleet Ship Export to UK Stopped" (31 May 2007), online: Basel Action Network <http://www.ban.org/ban_news/2007/070531_toxic_ghost_fleet.html> (reporting that since Able UK was unable to obtain the permits required to conduct its business in Teesside even after three and a half years, MARAD decided to annul the contract).

[303] *Report to Congress,* 2007, *supra* note 300 at 9 (there are still 118 vessels awaiting disposal and, on an average, an additional 13 are being added to the disposal queue every year).

[304] *Ibid.* at 15–17.

[305] Earthjustice, Press Release, "Legal Challenge Keeping "Ghost Fleet" Ship Scrapping Jobs in the US" (15 October 2004), online: Earthjustice <http://www.earthjustice.org/news/press/004/legal_challenge_keeping_ghost_fleet_ship_scrapping_jobs_in_the_us.html> (quoting Jim Puckett of BAN, this was a stealth attempt to set a legal precedent to open the floodgates for exports to China, India, or Bangladesh). See also Cohen, *supra* note 24 at 263.

steps were completed under some laws, but not yet satisfied by other laws".[306] This emphasises the need for clear domestic legal frameworks grounded on an overarching international law on the subject to facilitate greater co-operation between the players and to ensure more transparency in this area of international industry regulation.

D. *Perpetuating Environmental Racism*

It is obvious that although shipbreaking is performed in several OECD countries, it is not on the same scale and size as that on the Indian subcontinent. There are several reasons for this paradox, the most important being cost.[307] Accordingly, merchant ships that have reached their end-of-life are reflagged. In such a situation, the regulatory responsibility over the vessel is transferred to the new flag, which often does not possess the necessary wherewithal to ensure the safe disposal of these ships. For the industrialised countries, it seems that this outsourcing is not a cause for concern, as these ships no longer remain in their backyards.[308] By transferring the burden of managing wastes indirectly to the Indian subcontinent *via* a different set of players, even though legal *per se*, in effect, the industrialised nations are not respecting the principle of "not transferring harm to developing countries".[309] In this sense, they may be criticised for perpetuating environmental racism.[310]

V. Conclusion

Concentrated in some of the poorest regions of the world, shipbreaking activities play an important role in supporting national economies in the Third World.

[306] Viola Blayre Campbell, "Ghost Ships and Recycling Pollution: Sending America's Trash to Europe" (2004) 12 Tulsa J. Comp. & Int'l. L. 189 at 226.

[307] The difference in labour costs, environmental and health requirements, and revenue from recycling and second-hand materials enables operators in South Asia to offer far better prices than what competitors can offer in other countries. For instance, Bangladesh, due to lax labour and environmental mandates, can pay USD450–USD500/t for a ship, while Chinese facilities with slightly better standards offer half this price and United States operators' a tenth. Commission, *Green Paper*, *supra* note 35 at 6.

[308] Kimberly K. Gregory, "The Basel Convention and the International Trade of Hazardous Waste: The Road to the Destruction of Public Health and the Environment is Paved with Good Intentions" (2001) 10:2 Currents 80 (WLeC). See also Morita, *supra* note 276.

[309] UNCED, *Rio Declaration on Environment and Development*, 1992, 31 I.L.M. 874, Principle 2. See also *Trail Smelter Arbitration (U.S. v. Can.),* U.N. Rep. Int'l Arb. Awards 1905 (1949), 3 reprinted in (1941) 35 Am. J. Int'l L. 684 (finding Canada liable for the emission of sulphur-dioxide fumes which caused significant damage to agriculture and forest lands in the United States).

[310] Hugh J. Marbury, "Hazardous Waste Exportation: The Global Manifestation of Environmental Racism", Note, (1995) 28 Vand. J. Transnat'l L. 251 at 291 (WLeC) (noting that hazardous waste exporting is environmental racism on a global scale).

Apart from providing revenue to cash strapped governments and high quality steel at cheap prices, the shipbreaking industry provides millions of unskilled labourers with a chance to earn their livelihood in situations of limited economic opportunities. As well, by ensuring the timely removal of outdated tonnage, this industry makes a vital contribution to sustainable development globally.

Despite these benefits, there are several shortcomings that dent shipbreaking's claim to be a sustainable activity. The recycling facilities for ships on the Indian subcontinent are not only big cemeteries for dying ships; they are also a graveyard for the shipbreakers who toil hard to dismember them. Safety and health conditions in the scrap yards are woefully inadequate. Due to accidents and explosions, many die or are maimed for life. Thousands contract irreversible diseases many years after having worked in these yards. There is also severe ecological destruction that continues to threaten eco-stability and long-term economic growth. However, governments in these countries are reluctant to change, as they want as few impediments as possible in providing an appropriate climate for the growth and development of this industry. They ignore the fact that while assisting the developed world to get rid of their liability for short-term economic gain, they are in the long run irreparably damaging their own environment. Shipowners and governments in the First World, who derive the maximum benefit during the lifetime of a ship, generally ignore their contributions to the appalling conditions that exist in the ship demolition yards. They prefer to pass on responsibility to the new flag states, the ship recyclers, governments in the Third World and, finally, to the poor and the disadvantaged in these countries.

There is a huge quantity of decommissioned tonnage that is expected to hit the market in the coming years. It is doubtful whether the recycling yards on the Indian subcontinent will be able to absorb this volume by utilising present practices. If steps are not initiated to refurbish and modernise their recycling yards, not only will the Indian subcontinent be courting a human and environmental disaster of colossal proportions, but private shipowners will be forced to maintain ghost ships that can pose serious risk to the marine environment and to safety.

Given the global character of this industry, there is a unique opportunity for an international legal regime to be forged that prescribes uniform standards. Its functioning must be based on co-operation between the industrialised and the developing shipbreaking countries, taking into consideration the peculiar problems these developing countries face while initiating programmes to secure sustainable ship recycling. Changes to the conditions of their scrap yards and the standards of operation they follow will help render this industry a sustainable marine economic use around the world.

Chapter 3

Limitations of a National Response to Regulate the Global Shipbreaking Industry: A Study of the Indian Experience

I. Introduction

The shipbreaking industry functions with minimal or almost 'no' legal regulation on the Indian subcontinent. This *laissez-faire* environment has entrenched this industry in this part of the world. However, the consequences of this development have been severe. While shipowners and recyclers make windfall profits, hundreds of human lives have been lost, thousands of individuals have been maimed, and there has been severe degradation of the marine environment.[1] Among the three leading players – Pakistan, Bangladesh and India – not much is known about the state of relevant legal regulation in Pakistan. In Bangladesh, it is believed that legal regulation continues at a nascent stage.[2] In India, recent initiatives to develop legal standards for the industry have had a profound impact on shipbreaking operations in that country. India's largest shipbreaking yard, Alang, was a name once synonymous with shipbreaking itself. In its heyday, it provided employment to more than 35,000 workers, breaking almost one ship a day.[3] Since 2003, Alang has been in decline. In 2005–2006, only 101 ships were broken providing employment to only 5,000 workers.[4] One of the major reasons for this downturn can be traced to the attempt by the Supreme Court of India to introduce a regulatory framework to ensure cleaner and safer scrapping.[5] Since then, the industry has migrated to neighbouring Bangladesh,

---

[1] See generally Greenpeace-FIDH in cooperation with YPSA, *End of Life Ships: The Human Cost of Breaking Ships*, Jacqueline Schuiling (Netherlands: Greenpeace International, 2005), online: Greenpeace International <http://www.greenpeace.org/raw/content/international/press/reports/end-of-life-the-human-cost-of.pdf>.

[2] *Ibid.* at 48.

[3] India, *Report of the Committee of Technical Experts on Ship Breaking Activities* (Chairman: Dr. Prodipto Ghosh, New Delhi: Ministry of Environment & Forests, 30 August 2006) [unpublished] at 10 [on file with the author] [*TEC Report*] (noting that between 1998–99 Alang broke nearly 361 ships employing 35,000 workers).

[4] *Ibid.*

[5] See also *Recycling of Ships: Report on the Current Status of Ship Recycling at Alang in India*, Submitted by India, IMO/MEPC 53/INF.12, 13 May 2005 (KR-CON) (pointing out that India's unilateral strict measures is diverting the recycling industry to non-regulating countries). India has lost its number one position and approximately 30,000 direct and 70,000 indirect workers have lost their

even as attempts are underway in India to dilute the rigour of the regulatory regime to re-attract the obsolete tonnage.

Unilateral national approaches to legal regulation of shipbreaking that are not founded on an international legal framework are incapable of regulating an industry that is essentially global in character. Analysis of the Indian regulatory experience concerning the ship scrapping industry, which includes the legislative response focussing on the rules and regulations applicable in Alang and the jurisprudence developed by the Indian Supreme Court on shipbreaking, reveals the limited success national governments have had in adequately regulating this industry. The arrival of the toxic ships, the 'Riky', the 'Clemenceau', and the 'Blue Lady' in Indian waters and the foot-dragging response by both the executive and the judiciary in spite of the comprehensive regulatory framework, demonstrates the sway of market forces and the inability of national laws to regulate this global industry. An examination of these facets of the Indian regulation of the shipbreaking industry shows that there is the need for a sound international legal regime to facilitate movement from shipbreaking to sustainable ship recycling.

II. The Legislative Framework for Ship Recycling: Rules and Laws Applicable in Alang

In addition to the specific laws that have direct application to the shipbreaking industry in India, a plethora of general laws indirectly apply. These laws primarily address worker's rights and ensuring safe working conditions and secondly, environmental protection.

A. *Protecting Labour and the Environment*

The shipbreakers of Alang belong to the category of unorganised workers who are generally not visible to the law.[6] Accordingly, it is difficult to identify an employer and hence an employer-employee relationship. This makes it problematic for law to influence the allocation of rights and responsibilities and to build a system of

jobs. Further, economic consequences include the closure of almost 65 percent of scrap-based rolling mills and 75 percent of oxygen plants. Transporters to and from Alang are not getting even a single trip in a week, as against four in the past. *Ibid.*, ¶ 6. See also Recycling of Ships: Mandatory Requirements for the IMO Guidelines, Submitted by India, IMO/MEPC 53/3/6, 13 May 2005, ¶ 1 (KR-CON) (noting that India has been in the forefront to enforce the "gas-free-for-hot-work", resulting in many ships being sent to alternative destinations for recycling).

[6] International Metalworkers' Federation, *Status of Shipbreaking Workers in India: A Survey* at 1, online: International Metalworkers' Federation <http://www.imfmetal.org/main/files/06042810465779/Shipbreaking_survey.pdf> (noting that about 98 percent of the workers engaged in the ship recycling industry are unorganised workers).

social security for employees.[7] The overwhelming majority of Indian labour laws do not apply to the workers in the unorganised/informal sector. In fact, none of the laws which form the basis of the social security system cover the unorganised sector in toto.[8] The *Factories Act*,[9] the *Minimum Wages Act*,[10] the *Payment of Wages Act*,[11] the *Workmen's Compensation Act*,[12] the *Contract Labour (Regulation and*

---

[7] India, Government of India Ministry of Labour and Employment, *The Second National Commission on Labour Report* (New Delhi: Ministry of Labour and Employment, 2002) at 594–605, online: Government of India Ministry of Labour <http://www.labour.nic.in/lcomm2/nlc_report .html> [*Second National Commission on Labour Report*]. See also Kamala Sankaran, "Protecting the Worker in the Informal Economy: The Role of Labour Law" in Guy Davidov & Brian Langille, eds., *Boundaries and Frontiers of Labour Law* (Portland: Hart Publishing, 2006) 205.

[8] *Second National Commission on Labour Report, ibid.* at 703.

[9] *The Factories Act 1948* (No. 63 of 1948, India), online: India Code <http://indiacode.nic.in/>. Under this Act, a factory is a premise where a manufacturing process is carried on by employing 10 or more workmen with the aid of power or 20 or more workmen if the process is carried on without the aid of power. Most of the shipbreaking yards at Alang are registered under the *Factories Act*. Government permission is necessary to locate the factory and a license is required to run the factory. Sections 21–41 deal with safety precautions and security of workers. Sections 11–20 deals with cleanliness in the factory, waste disposal, temperature control, air pollution, lighting, drinking water and toilet facilities for workers, and health and environment related matters. Adult worker can be required to work only up to 48 hours in a week. No one can be made to work for more than five hours continuously. The overtime payable is at double the normal rate of wages. There are strict regulations regarding employment of young workers. A child up to 14 years of age cannot work in a factory. The Act also has provisions for extra care to be taken in case of hazardous processes (Chapter IV A). The implementation of the Act is left to the jurisdiction of state governments. Unfortunately, the implementation is unsatisfactory. Also, contract and ad hoc workers are not given the benefits that are given to permanent workers.

[10] *The Minimum Wages Act 1948* (No. 11 of 1948, India), online: India Code <http://indiacode .nic.in/>. An important legislation that can benefit unorganised labour, this Act seeks to ensure that market forces do not determine wages in industries where the workers are poor, vulnerable, unorganised, and without bargaining power. The Act requires that the government fix minimum wages for certain employments mentioned in Parts I and II of the Schedule appended to it. Government can add any employment to this list.

[11] *The Payment of Wages Act 1936* (No. 4 of 1936, India), online: India Code <http://indiacode .nic.in/>. This Act was adopted to regulate unauthorised or arbitrary deductions from wages and it seeks to ensure correct and timely payment of wages. In any establishment employing less than 1,000 persons, the wages are to be paid within seven days of the expiry of the "wage period". No deductions can be made from the wages of a worker excepting the ones permitted by the Act. In any case, the total amount of deductions during a wage period is not to exceed half the wages payable. However, because of a stipulated wage limit for the purpose of applicability, 95 percent of the unorganised workers are excluded from its coverage.

[12] *The Workmen's Compensation Act 1923* (No. 8 of 1923, India), online: India Code <http:// indiacode.nic.in/> [*Workmen's Compensation Act*]. This Act provides for payment of compensation to workmen for injuries sustained in accidents. The employer is liable to provide monetary compensation to the worker or dependant in case of death or disablement provided it occurs "out of and in the course of employment". The amount of compensation is determined on the basis of the nature of the injury, the average monthly wages received and age of the workman. Interestingly, the Act also lays down rules regarding payment of compensation in cases where an employee contracts any disease specified in Schedule III to the Act (entitled "List of Occupational Diseases") as an occupational disease peculiar to that employment, provided that the worker was employed for a continuous period of six months. In such cases, the contracting of the disease shall be deemed to be an injury by accident and unless the contrary is proved, the accident shall be deemed to have arisen out of and in the course of the employment.

Abolition) Act,[13] and the *Inter-State Migrant Workmen (Regulation of Employment and Conditions of Service) Act*[14] apply selectively to certain categories of unorganised labour. In addition to these laws, the *Public Liability Insurance Act*,[15] though not a labour law *per se*, is also relevant. This law seeks to provide for public liability insurance to persons affected by accidents while handling hazardous substances.[16] In spite of this impressive legislative edifice, it is doubtful whether the workers at Alang could benefit from these laws.[17]

Many statutes on the protection of the environment are applicable to the shipbreaking industry as well. The three most relevant are the *Air (Prevention and Control) of Pollution Act*,[18] the *Water (Prevention and Control of Pollution) Act*,[19]

---

[13] *The Contract Labour (Regulation and Abolition) Act 1970* (No. 37 of 1970, India), online: India Code <http://indiacode.nic.in/>. Its scope for unorganised labour is limited. This Act regulates the employment of contract labour in certain establishments and provides for its abolition under certain circumstances. It is applicable if the principal employer engages 20 or more contract workers in an establishment. However, contractors evade the Act by employing less than 20 workers.

[14] *The Inter-State Migrant Workmen (Regulation of Employment and Conditions of Service) Act 1979* (No. 30 of 1979, India), online: Office of Labour Commissioner <http://labour.delhigovt.nic.in/act/details_acts/interstatemigrnatwirkmen/index.html>. The vast majority of the unorganised sector in India constitutes migrant workers who are recruited from various parts of the country through contractors to work outside the state in which they are resident. Being illiterate and poor, they are subjected to abuse. Even though the Act provides for various benefits like non-discrimination in wages, holidays, journey allowance, regular payment of wages, provisions for suitable conditions of work, suitable residential accommodation, adequate medical facilities, and adequate protective clothing and equipment, etc., the major drawback of the Act is its applicability. Under this Act, 'inter-state migrant workman' is defined to mean, 'any person who is recruited by or through a contractor in one state under an agreement or other arrangement for employment in an establishment in another state'. Since such agreements or arrangements are not formally entered into, the migrant worker cannot claim benefit under the Act.

[15] *The Public Liability Insurance Act 1991* (No. 6 of 1991, India), online: India Code <http://indiacode.nic.in/>.

[16] *Ibid.*, preamble.

[17] UNESCO, *Impacts and Challenges of Large Coastal Industry: Alang-Sosiya Ship-Breaking Yard, Gujarat, India*, Coastal Region and Small Island Papers 17 (Paris: UNESCO, 2004) at 40, online: UNESCO <http://www.unesco.org/csi/pub/papers3/assby.pdf> (noting that the most difficult obstacle facing the implementation of labour laws is the lack of understanding among the workers as to their rights).

[18] *The Air (Prevention and Control) of Pollution Act 1981* (No. 14 of 1981, India), online: Ministry of Environment & Forests <http://envfor.nic.in/legis/air/air1.html>. This Act provides for the prevention, control and abatement of air pollution, and for the establishment of boards to carry out these purposes. Air pollution is defined in section 2(b) to mean the presence in the atmosphere of any air pollutant. An air pollutant is any solid, liquid or gaseous substance (including noise) present in the atmosphere in such concentrations as may be or tend to be injurious to human beings, living creatures, plants, property or environment. The Act contains stringent devices to prevent and control air pollution. Even an on-going industrial unit can be closed under the provisions of this Act, if it does not take steps to confirm to the requirements under the Act. For the Environmental Guidelines for Shipbreaking Industries issued by the CPCB, see Recycling of Ships: Information on the Environmental Guidelines for the Ship Recycling Industry in India, Submitted by India, IMO/MEPC 47/INF.3, 6 November 2001, snnex, ¶ 4.2 (KR-CON) [Environmental Guidelines for Ship Recycling in India] (calling upon the executing body responsible for running an incinerator for a cluster of units to obtain the relevant consent under the *Air Act*).

[19] *The Water (Prevention and Control of Pollution) Act 1974* (No. 6 of 1974, India), online: Ministry of Environment and Forests <http://envfor.nic.in/legis/water/wat1.html> [*Water Act*].

and the *Environment (Protection) Act*.[20] The *Disaster Management Act*,[21] which seeks to provide a legal framework for the effective management of disasters both natural and manmade,[22] is also relevant. Subordinate legislations such as the *Batteries (Management and Handling) Rules*,[23] the *Environment Protection Rules*,[24] the *Hazardous Wastes (Management and Handling) Rules*,[25] the *Coastal Regulation*

This Act provides for the prevention and control of water pollution and for the maintenance or restoration of the wholesomeness of water through the establishment of Water Boards for its effective implementation. The Act applies to streams which are defined under section 2 (j) to include sea or tidal waters. It prohibits any person from knowingly causing or permitting poisonous, noxious, or polluting matter to enter into a stream, well, sewer or land. However, knowledge about harm is an important element that is difficult to prove, which has considerably reduced the efficacy of this law. See Tony George Puthucherril, *Water Resources Management Law: A Case Study with Reference to the State of Kerala* (M.Phil. (Law) Thesis, W.B. National University of Juridical Sciences, Kolkata, India, 2003) [unpublished] [Puthucherril, *Water Resources*]. See also Environmental Guidelines for Ship Recycling in India, *ibid.* ¶ 4.1 (calling upon individual recycling units to obtain the consent to operate under the *Water Act*).

[20] *The Environment (Protection) Act 1986* (No. 29 of 1986, India), online: Ministry of Environment and Forests <http://envfor.nic.in/legis/env/env1.html>. Enacted pursuant to the Stockholm Conference, 'environment' is defined under the Act to include water, air and land and the interrelationship which exists among and between water, air and land, and human beings, other living creatures, plants, micro-organisms and property. Hazardous substance is any substance or preparation which by reason of its chemical or physico-chemical properties or handling is liable to cause harm to human beings, other living creatures, plants, micro-organisms, property or the environment. Section 3 empowers the central government 'to take all such measures as it deems necessary or expedient for the purpose of protecting and improving the quality of the environment and preventing, controlling and abating environmental pollution.' This includes a range of measures such as laying down procedures and safeguards for the handling of hazardous substances; giving directions necessary for the prevention, control and abatement of environmental pollution, etc. Section 8 compels persons handling hazardous substances to comply with certain procedural safeguards.

[21] *The Disaster Management Act 2005* (No. 53 of 2005, India), online: National Institute of Disaster Management <http://www.nidm.gov.in/DM_act2005.pdf>.

[22] Section 2(d) defines disaster to mean "a catastrophe, mishap, calamity or grave occurrence in any area, arising from natural or manmade causes, or by accident or negligence which results in substantial loss of life or human suffering or damage to, and destruction of, property, or damage to, or degradation of, environment, and is of such a nature or magnitude as to be beyond the coping capacity of the community of the affected area". *Ibid.*

[23] *Batteries (Management and Handling) Rules 2001* (S.O. 432(E)/2001 India), online: Ministry of Environment and Forests <http://envfor.nic.in/legis/hsm/leadbat.html>. Amongst other duties, an importer is to ensure that used batteries are safely transported only to registered recyclers, and no damage is caused to the environment during its transportation. *Ibid.*, r. 4.

[24] *Environment Protection Rules 1986* (S.O. 844(E)/1986 India), r. 13, online: Ministry of Environment and Forests <http://envfor.nic.in/legis/env/env4.html> (empowers the central government to prohibit and restrict the handling of hazardous substances including their imports in different areas).

[25] *Hazardous Wastes (Management and Handling) Rules 1989* (S.O. 594(E)/1989 India). These Rules specifically prohibit the import of hazardous wastes from any country to India for dumping or disposal. *Ibid.*, r. 11(1). However, import of hazardous wastes as raw material for recycling or reuse is permitted. *Ibid.* For the procedures relating to import of hazardous wastes see, *ibid.*, r. 13. Movement of hazardous wastes from or to the country is illegal without the prior permission of the central government or if the permission has been obtained through falsification, misrepresentation or fraud or if it does not conform to the shipping details provided in the document. In such cases, the hazardous wastes are to be shipped back within thirty days either to the exporter or to the exporting country. If there is no such possibility, it is to be disposed of within thirty days, with the costs to be borne by the exporting country. *Ibid.*, r. 15.

Zone Notification,[26] the *Manufacture, Storage and Import of Hazardous Chemical Rules*,[27] the *Noise Pollution (Regulation and Control) Rules*,[28] the *Ozone Depleting Substances (Regulation and Control) Rules*,[29] the *Plastics Manufacture, Sale and Usage Rules*,[30] the *Chemical Accidents Emergency Planning, Preparedness and Response Rules*,[31] the *Environmental Impact Assessment Notification*,[32] also have indirect application. Generally, the State Pollution Control Boards (SPCB) are

---

[26] *The Coastal Regulation Zone Notification 1991* (S.O. 114 (E)/1990 India), online: Ministry of Environment and Forests <http://envfor.nic.in/legis/crz/crznew.pdf>. The Coastal Regulation Zone Notification (CRZN) classifies the coastal area around the country into four zones and lists the activities permitted in each of these zones. The CRZN regulates the establishment of new industries and expansion of existing industries, except those directly related to the water front or directly needing foreshore facilities. *Ibid*., n. 2(i). Among the 13 activities that are prohibited within the CRZN relevant to shipbreaking is the manufacture, handling, storage, or disposal of hazardous substances. *Ibid*., n. 2(ii). See also Environmental Guidelines for Ship Recycling in India, *supra* note 18 ¶ 3 (noting that the siting of shipbreaking industries should be in strict conformity with the CRZN Rules). See also *Draft Notification 2008 (Coastal Zone Management)* (S.O. 1761(E)/2008 India), online: Ministry of Environment and Forests <http://envfor.nic.in/legis/crz/so-1761(e).pdf>. The object of this law is to protect and ensure the "sustainable development of the coastal stretches and marine environment through sustainable coastal zone management practices based on scientific principles". *Ibid*., cl. 2. This Draft Notification divides the coastal zone into four zones, namely, CZM-I, CZM-II, CZM-III, and CZM-IV. In the CZM-III, shipbreaking requires approval of the Union Ministry of Environment and Forests subject to an environmental impact assessment and environmental management plan.

[27] *Manufacture, Storage and Import of Hazardous Chemical Rules 1989* (S.O. 966(E)/1989 India), r. 18, online: Ministry of Environment and Forests <http://envfor.nic.in/legis/hsm/hsm2.html> (prescribes detailed rules in relation to the import of hazardous chemicals).

[28] *Noise Pollution (Regulation and Control) Rules* 2000 (S.O. 123(E)/2000 India), Sch. (specifies the ambient air quality standards in respect of noise which is 75 dB (A) Leq for day time (6.00 a.m. to 10.00 p.m.) and 70 dB (A) Leq for the night (10.00 p.m. to 6.00 a.m.)). See also Environmental Guidelines for Ship Recycling in India, *supra* note 18 ¶ 6.2.4 (limiting the noise level to 90 dB (A) at 1 metre distance. It also requires that the workers be provided with ear plugs).

[29] *Ozone Depleting Substances (Regulation and Control) Rules 2000* (S.O. 670(E)/2000 India), r. 8, online: Ministry of Environment and Forests <http://envfor.nic.in/legis/ods/odsrcr.html> (seeks to regulate the use of ozone depleting substances). *Ibid*., r. 5(1) (prohibits the import or export of any ozone depleting substance to any country specified in Schedule VI, unless a licence has been issued by the authority).

[30] *Plastics Manufacture, Sale and Usage Rules 1999* (S.O. 705(F)/1999), r. 6 (points out that recycling of plastic shall be strictly in accordance with the Bureau of Indian Standards Specification: IS 14534: 1998 titled "the guidelines for recycling of plastics").

[31] *The Chemical Accidents Emergency Planning, Preparedness and Response Rules 1996* (G.S.R. 347(E)/1996), r.3–r.10, online: Ministry of Environment and Forests <http://envfor.nic.in/legis/hsm/gsr347.htm> (provides for the constitution and lays down the functions of crisis groups at the central, state, district and local level to formulate responses in relation to chemical accidents).

[32] *Environmental Impact Assessment Notification 2006* (S.O. 1533/2006 India), online: Ministry of Environment and Forests <http://envfor.nic.in/legis/eia/so1533.pdf>. This new notification which supersedes the earlier notification titled 'Environmental Impact Assessment of Developmental Projects 1994' classifies projects for environmental clearance based on the spatial extent of potential impacts on human health and natural and manmade resources, into two categories. Those falling under category A are to be referred to the Union Ministry of Environment and Forests, and those under category B to the State Environment Impact Assessment Authority. Accordingly, all shipbreaking yards, including shipbreaking units, fall under Category A requiring clearance from the Union Ministry of Environment and Forests. *Ibid.*

responsible for implementing these environmental laws. However, given the poor administrative and logistical support available to these Boards, it becomes very difficult to monitor compliance, rendering many of these laws unenforceable.[33]

B. *The Ship Recycling Legal Regime*

In addition to this overarching legal framework concerning labour and environmental protection, the Central Pollution Control Board (CPCB) of India has issued "Environmental Guidelines for Shipbreaking Industries". These Guidelines provide guidance to shipbreaking industries to minimise the impact of their operations on the environment. Most strikingly, the Guidelines call for proper siting of ship recycling units and for preparing and implementing environmental management and disaster management plans.[34] Nationally, however, India is yet to legislate on shipbreaking.

Perhaps the most important law with a direct bearing on the issue of breaking ships is the provincial *Gujarat Maritime Board Ship Recycling Regulations 2003*[35] issued by the Gujarat Maritime Board (GMB). The Ship Recycling Regulation is applicable only to the ship recycling yards in the State of Gujarat, which includes the shipbreaking yard at Alang,[36] and it fortifies the existing legislative framework on worker safety, welfare and environmental protection[37] in the recycling yards. It also sets out duties for both the corporate ship recycler and the authorities overseeing the working of this industry.

Apart from the duty to obtain requisite permissions,[38] the ship recycler is to "adhere to and follow" relevant regulations, norms and guidelines set out under various International Maritime Organization (IMO) and International Labour

---

[33] Recently the Gujarat SPCB set up an office at Bhavnagar. Earlier, Alang fell under the Rajkot office, nearly 178 km away. The Bhavnagar office is understaffed and needs more monitoring tools. This severely affects the quality of inspection since a Gujarat SPCB official need to be present on every ship prior to it being allowed to be beached. An average inspection takes about six hours. Currently, with few ships arriving, an inspection can be carried out with some degree of thoroughness, but the integrity of this is likely to be compromised if the number of ships increases. See generally Lyla Bavadam, "Shipload of trouble" *Frontline [of India]* 24:22 (3 November 2007), online: Frontline <http://www.hinduonnet.com/fline/fl2422/stories/20071116505108300.htm>.

[34] For further details see Environmental Guidelines for Ship Recycling in India, *supra* note 18 ¶ 1.

[35] *Gujarat Maritime Board Ship Recycling Regulations 2003* (No. GMB/ALANG/73/110/2003/42 Gujarat, India) (published in Vol. XLIV The Gujarat Government Gaz. Extra Ordinary, 7 July 2003, 281–1) [*Ship Recycling Regulation*].

[36] *Ibid.*, reg. 1(ii). In this context, *The Gujarat Maritime Board (Conditions and Procedures for Granting Permission for Utilising Ship Recycling Plots) Regulation 2006* is also relevant. *The Gujarat Maritime Board (Conditions and Procedures for Granting Permission for Utilising Ship Recycling Plots) Regulation 2006* (No. GMB/Alang/134/2004 Gujarat, India) (published in Vol. XLVII The Gujarat Government Gaz. Extra Ordinary, 4 October 2006, 767–1) [*Ship Recycling Plot Regulation*]. Even though it primarily deals with granting permission for utilising ship recycling plots, the Regulation lays down several prescriptions for sustainable ship recycling.

[37] *Ship Recycling Regulation, supra* note 35 reg. 31.

[38] *Ibid.*, reg. 6(b).

Organization (ILO) conventions that have been ratified by the Indian government.[39] Before commencing work, the recycler is to ensure that the workers are provided with "proper training"[40] and with necessary instructions on fire fighting and the use of safety equipment.[41] Boards displaying precautions that need to be followed are also to be installed in the yard.[42] The recycler is to maintain an attendance register,[43] issue identity cards to the workers,[44] provide safety devices or tools,[45] arrange for compulsory insurance,[46] maintain a daily logbook on the progress of work,[47] introduce mechanisation,[48] and ensure that all equipments conform to the *Factories Act*.[49]

Consent for shipbreaking activities is also conditioned on the recycler obtaining necessary permissions under the different rules and following them to ensure safety.[50] In the event of an explosion, fire or accident, the recycler has to immediately

---

39 *Ibid.,* reg. 6(a). See also *Ship Recycling Plot Regulation, supra* note 36 reg. 13 (requiring a permission holder to comply with the requirements under the Ship Recycling Regulation).

40 The workers are to be provided with training for general and specific shipbreaking operations. The training has to be in the language that can be understood by the workers and it should be of such duration as to enable them to get acquainted with the knowledge and the risk factors associated with the operation. On successful completion of the training programme, there is to be certification. *Ship Recycling Regulation, ibid.,* reg. 7(i). It is interesting to note the heightened requirement in Schedule V(1) of the *Ship Recycling Plot Regulation*. It states that the permission holder is not to employ any worker unless the worker has undergone basic training by the recognised training institute/GMB in the field of work and is certified to that effect.

41 *Ibid.*, reg. 7(ix).

42 *Ibid.*, reg. 7(ii). The display has to be put up in a prominent place on the plot and should be in Hindi, Gujarati or any other language that can be understood by the workers. *Ibid.*

43 *Ibid.*, reg. 7(iii).

44 *Ibid.*, reg. 7(iv), (v).

45 *Ibid.*, reg. 7(viii). These include helmet, safety shoes, welding goggles, safety belts with safety line, handgloves, and self-contained breathing apparatus as provided under the relevant laws conforming to the relevant Indian standards specifications or equivalent standards as approved by the Chief Inspector of Factories.

46 *Ibid.*, reg. 7(xi).

47 *Ibid.*, reg. 7(vi).

48 *Ibid.*, reg. 7(x) (ship recycler is to ensure that necessary mechanical devices such as cranes, magnetic cranes, forklift, portable winches, etc. are employed during ship recycling activities). Schedule V(6) of the Ship Recycling Plot Regulation requires that the permission holder use modern technology for recycling activity such as use of a long nozzle cutting torch, mechanised hydraulic lifting equipment, gas detectors, mechanised blowers, and classified cutting equipment as prescribed by the GMB and according to the norms prescribed by the Basel Convention.

49 *Ibid.*, reg. 7(xii).

50 The recycler has to obtain necessary storage licenses for LPG/oxygen cylinders under the *Gas Cylinders Rules 1981*. Secondly, the storage should be in accordance with the relevant laws. Thirdly, no petroleum products or combustible material should be stored within the safety distance specified for LPG storage. Fourthly, fire-fighting devices, including water storage has to be ensured. Fifthly, portable fire extinguishers are to be kept on the plot at places that could facilitate easily access. Sixthly, minimum firefighting equipment like a fire extinguisher, fire fighting suits, foam sprayer, liquid foam, oxygen mask, sand box/buckets, and fire hose with complete attachment should be installed on the plot. Seventhly, there should always be sufficient space on the plot kept open for safe passage of fire fighting units and equipment. Eighthly, all equipment on the plot such as the crane, winch, chain rope and shackles, generator set, etc. should be maintained as per relevant laws. *Ibid.*, reg. 5.

report the matter in writing to the relevant authorities and must pay an *ex-gratia* compensation of INR20,000 (approx. USD434 at USD1 = INR46, as of 7 July 2003, the year in which the Ship Recycling Regulation was promulgated) to the next of kin of the deceased.[51] For an injured worker, the *ex-gratia* amount is fixed at INR5,000 (approx. USD109 at USD1 = INR46, as of 7 July 2003).[52] However, the GMB has the discretion to vary this amount.

Regarding measures to protect the environment and the duty of the ship recycler to ensure the same, the Ship Recycling Regulation mandates that the recycler obtain necessary permission from the SPCBs and ensure that the *Water Act 1974*, the *Air Act 1981*, *Hazardous Waste Management and Handling Rules 1989*, and the *Coastal Regulation Zone Notification 1991* are observed.[53] The recycler is to scrupulously adhere to SPCB instructions in matters relating to the collection and disposal of toxic materials and organic elements.[54] Steps are also to be taken to establish a solid waste management system on the plot.[55] The ship recycler is not to allow waste materials like oil cakes, thermocol, scrap iron and metallic pieces, glass wool, asbestos, etc., to be thrown directly into the sea or on to the shore.[56] Open burning of waste within the yards is prohibited,[57] and the residues obtained after cleaning the cargo holds of oil tankers should be disposed of in accordance with the SPCB directions.[58] The recycler is to keep the work place clean[59] and ensure that sea water is sprinkled regularly to minimise generation of dust during material handling.[60]

The Ship Recycling Regulation lays down detailed rules which the ship recycler has to follow while beaching ships. Applications for beaching are to be submitted to the port authority, along with necessary documentation, at least a day prior to the intended date of beaching.[61] The application can be presented on any

---

[51] *Ibid.*, reg. 25(b). In case of fatal accidents, the Ship Recycling Plot Regulation provides that the permission holder (the person who has the permission to utilise the shipbreaking plot for beaching and recycling of ships) has to pay INR100,000 (approx. USD2,222, as per the exchange rate USD1 = INR45 as of 2 January 2006, the year in which the Ship Recycling Plot Regulation was promulgated), to the GMB by way of penalty which would be deposited in the Worker's Welfare Fund maintained by the GMB. Also, the permission holder has to pay INR200,000 (approx. USD4,444, as per the exchange rate USD1 = INR45 as of 2 January 2006) to the heirs of the deceased person. *Ship Recycling Plot Regulation*, *supra* note 36 Sch. V(5).

[52] *Ship Recycling Regulation*, *supra* note 35 reg. 25(b), proviso. This is in addition to the compensation that may be received by the next of kin or the injured victim from insurance or any other enactment, including the *Workmen's Compensation Act*.

[53] *Ibid.*, reg. 21(a).

[54] *Ibid.*, reg. 21(c).

[55] *Ibid.*, reg. 23.

[56] *Ibid.*, reg. 21(b) (these are to be collected in a secured storage facility within the plot).

[57] *Ibid.*

[58] *Ibid.*, reg. 21(d).

[59] *Ibid.*, reg. 22.

[60] *Ibid.*, reg. 21(e).

[61] *Ibid.*, reg. 8(b)(i)–(iii). The application should be submitted along with relevant documents such as the permission/gas free certificate for entry of a vessel into dry/wet/floating dock and for man

day depending on the high tide.[62] Generally, only one ship is allowed to be beached at a time.[63]

The obligation on the recycler *vis-à-vis* cutting requires that the recycler install fire fighting pumps with water on the ship. One of these pumps should be situated near to the engine room and is to remain there till the ship is dismantled completely.[64] The ship recycler is also to appoint a "safety officer/safety supervisor" who is to supervise the cutting operation.[65] Only on completion of these steps can an application for cutting be submitted to the port authority,[66] along with necessary documentation.[67]

As mentioned earlier, the Ship Recycling Regulation also outlines the responsibilities on the part of the authorities.[68] The main regulatory responsibility of the port authority relates to the duty to grant permissions. Apart from this, the Ship Recycling Regulation empowers different sets of authorities to initiate steps to ensure safety. For instance, the port authority can suspend cutting operations in case of unsafe practices[69] and has overall supervisory powers over the ship recycling yards.[70] In addition, the GMB can appoint a maximum of 12 qualified safety officers for supervising the recycling activities and to attend to rescue operations.[71] The port authority, members of the safety rating team, the factory inspector, or safety officer can periodically visit the ship recycling plots and can bring to the notice of the safety officers/safety supervisors, unsafe practices that require corrective action.[72] The safety officers appointed by the GMB or by the recycling yard have specific functions. In the event of an accident, fire or explosion, either on the plot or on the ship, operations are to be suspended immediately for five days. During this period, the safety officers are to inspect the plot and determine the reasons that led to the accident. They can direct the plot holder to rectify these deficiencies to restore a "safe working environment".[73] If there is loss of life or serious bodily injury, an inquiry has to be commissioned, and an inquest report

---

entry issued under rule 43h of the *Petroleum Rules 2002*, original physical delivery certificate, etc. The Ship Recycling Plot Regulation under schedule V(3) provides that the permission holder has to provide all necessary documents of the ship that has arrived for beaching and also provide details regarding the identity of the sellers including that of the cash buyer.

62 *Ibid.*, reg. 9(a).
63 *Ibid.*, reg. 10.
64 *Ibid.*, reg. 13.
65 *Ibid.*, regs. 15–17.
66 *Ibid.*, reg. 11.
67 *Ibid.*, reg. 12.
68 *Ibid.*, reg. 4.
69 *Ibid.*, reg. 20.
70 *Ibid.*, reg. 18.
71 *Ibid.*
72 *Ibid.*, reg. 19.
73 *Ibid.*, reg. 25(a).

should be filed.[74] If it emerges that there has been negligence on the part of the plot holder/shipbreaker, certain penalties ensue.[75]

Another important responsibility relates to the duty to provide infrastructure in the yards. The GMB or the private operator has to provide basic infrastructure[76] for the ship recycling yards under their control.[77] At a minimum, this includes the establishment of a hydrant system for fire fighting, provision of potable water, drainage, street lighting, and a domestic waste management system.[78]

C. *Shortcomings to the Ship Recycling Regulation in India*

The Ship Recycling Regulation provides a broad framework that puts in place certain precautionary measures to minimise fire, explosions and accidents in the shipbreaking yards. Coming nearly three decades after the beginning of shipbreaking at Alang, it lays down a system to regulate shipbreaking activities.

There are, however, some shortcomings to the Ship Recycling Regulation. While it refers to the IMO and ILO conventions, it ignores the more important Basel Convention, which has direct relevance to ship recycling. Moreover, India is a party to the Basel Convention. Frequent reliance on general phraseology as opposed to specific language provides room for conflicting interpretations of this instrument. The subsequent Ship Recycling Plot Regulation, though dealing with an altogether different issue of allotment of plots for ship recycling, covers some of the omissions of the Ship Recycling Regulation. For instance, though it envisages training workers, the obligation to provide the same falls on the ship recycler. Given that most of the workers are illiterate and speak different languages as they hail from different states, training may not prove beneficial. As well, the quality of the training provided by the recycler can also be uneven. This concern is addressed by the Ship Recycling Plot Regulation, which mandates basic training relevant to the field of work by recognised training institutes or by the GMB and certification to that effect. Likewise, it is difficult to envision how the obligation imposed on the recycler to ensure mechanisation will be translated into practice. For the recycler,

---

[74] *Ibid.*, reg. 25(b). See also *ibid.*, reg. 25(c) (noting that penal actions under this Regulation shall run concurrently along with the penal action under other relevant statutes, regulations, etc.).

[75] *Ibid.*, reg. 25(b), proviso. In the case of a first incident, a penalty of INR500,000 (approx. USD10,417, as per the exchange rate USD1 = INR48, as of 2 January 2003) is attracted, and for every subsequent incident of fire, accident or explosion within a year leading to death or serious bodily injury, the penalty would be double. The amounts so collected are to be placed in a welfare account to be maintained by the port authority and must be used for the welfare of the workers. If more such incidents occur, then the permission for recycling activities stands cancelled for six months and would be granted thereafter only if the plot owner has enhanced the safety arrangements.

[76] *Ibid.*, reg. 29.

[77] *Ibid.*

[78] *Ibid.*

the fewer the number of machines and higher the number of human hands employed saves on capital expenditures and operational costs. The Ship Recycling Plot Regulation resolves this issue by leaving the extent of mechanisation to be determined by the GMB.

The Ship Recycling Regulation confers discretion on the Chief Inspector of Factories in relation to the quality of the safety equipment provided to the workers. There is the possibility of abuse of this discretion, which can contribute to accidents and fatalities. Even though the Ship Recycling Regulation lays down certain stipulations on environmental protection, the need to phase-out the present practice of beaching is absent. This is surprising since the maximum release of toxins into the marine environment arises from the cutting of the ship in the intertidal zone. Another important issue in the shipbreaking yards is the need for adequate emergency health care. Though the Ship Recycling Regulation refers to "basic infrastructure", it is silent on the need to improve health care facilities.

The Ship Recycling Regulation provides for *ex-gratia* and other payments in case of death or accidents. However, it has been observed that workers suffer from severe health complications long after their stint at Alang. The Ship Recycling Regulation does not envisage any protection for such workers. The Regulation is also silent on working hours, overtime, and improved living conditions in the yards, even though there is mention of penalties if workers are made to work beyond prescribed time and on holidays.[79] Supervisory powers have been conferred on the owner[80] at par with the port authority. This raises questions as to who will supervise the owner.

Perhaps the greatest flaw of the Ship Recycling Regulation is that it confers absolute powers on the recycler to appoint a safety officer/safety supervisor and then confers on this entity the power to inspect the plot in the event of an accident, major fire, or explosion and to identify its causes. In the long run, this provision could hamper the development of a safety culture in the yards. Further, the Ship Recycling Regulation provides several routes to the recycler or the shipowner to evade liability in case of an accident, explosion, or fire that has resulted in serious bodily injury or loss of human life. Rather than fixing liability on the operator of the hazardous process (recycler), and requiring this entity to prove absence of negligence, the Ship Recycling Regulation requires proof of negligence where the onus of proving such negligence lies on the "authorised person/s". This runs counter to the absolute liability principle, one of the central tenets in Indian environmental jurisprudence.[81]

---

[79] *Ibid.*, reg. 27(a).

[80] *Ibid.*, reg. 18. The owner is any person or body having a statutory right of ownership over the private ship recycling plot/yard. *Ibid.*, reg. 2(vi).

[81] *M.C. Mehta v. Union of India* (1986), [1987] A.I.R. S.C. 1086 (India S.C.) [*M.C. Mehta* (1986)].

Overall, the scope of the Ship Recycling Regulation is limited as it provides only a minimal legal framework for ensuring safe working conditions and environmentally friendly ship recycling operations. This regulation addresses only some of the contentious issues raised by this industry's operations, and these rather perfunctorily.

III. Shipbreaking and Sustainability through the Judicial Lens

The shipbreaking industry raises serious human rights as well as environmental protection concerns. Given the executive and legislative apathy to the problem, and the judiciary being one of the wings of the state constitutionally bound to safeguard the rights of citizens, courts are increasingly being called upon to redress human rights infringements in the ship demolition yards.[82] In this regard, the Supreme Court of India has played a decisive role, laying down a legal framework for the scrapping of ships, an achievement which perhaps no other constitutional court in the world can claim.

The judicial activism by the Supreme Court of India in the arena of ship recycling has had both positive and negative outcomes. Does the Indian judiciary sponsored model on shipbreaking have any lessons to offer to other countries? To understand this activism in context and to appreciate the tools and techniques adopted by the Supreme Court to pursue its constitutional goals in shipbreaking, analysis of the role of the judiciary in developing Indian environmental jurisprudence is necessary.

A. *Development of Indian Environmental Jurisprudence*

Environmentalism permeated the Indian body politic following the Stockholm Conference, which led to the 42nd constitutional amendment,[83] the establishment of the Ministry of Environment and Forests at the central government level,[84] and a spate of legislative activity on the environment front.[85] State governments also

---

[82] For instance see *Bangladesh Environmental Lawyers Association v. Bangladesh*, [2003] Writ Petition No. 2911, online: Bangladesh Environmental Lawyers Association <http://www.belabangla.org/html/pil.htm> (shipbreaking to be regulated by law).

[83] This amendment to the Constitution of India added Entries 17A & 17B (dealing with forests and protection of wild animals and birds) to List III of the Seventh Schedule. It also added article 48A which speaks about protection and improvement of the environment and safeguarding forests and wildlife.

[84] Government of India, *Ministry of Environment & Forests*, online: Ministry of Environment & Forests <http://envfor.nic.in/>.

[85] C.M. Abraham & Armin Rosencranz, "An Evaluation of Pollution Control Legislation in India" (1986) 11 Colum. J. Envtl. L. 101 at 104 (HeinOnline); Armin Rosencranz & Kathleen D. Yurchak, "Progress on the Environmental Front: The Regulation of Industry and Development in India" (1996) 19 Hastings Int'l & Comp. L. Rev. 489 at 492–99(WLeC).

contributed new laws, rules, regulations and policies.[86] Pollution control boards created at both the central and state government levels were entrusted with the primary responsibility of administering this statutory code.[87] In spite of all these initiatives, there was no commensurate change in practices, attitudes and administration, with the result that environmental degradation continued unabated, forcing the courts to intervene.

It was after the *Habeas Corpus*[88] fiasco that the Supreme Court of India started slowly but steadily to break itself free from procedural shackles. Two important developments which hastened this transformation were the demise of the strict rule of standing and the consequent expansion of the concept of *locus standi*.[89] Along with the abolition of these restrictive dogmas, the judiciary began to read new meaning and content into the constitutional guarantees, particularly article 21 (dealing with the right to life).[90] Life no longer meant mere animal existence; rather, it meant the right to live with human dignity and all that it encompasses.[91] Consequently, all problems which affected 'life' under article 21 came to be problems of constitutional law to be determined by the superior courts.

Judicial activism on the environmental front in India stems from the worst industrial accident in history, the "Bhoposhima",[92] which left thousands dead instantaneously.[93] This disaster laid bare the inadequacy of India's environmental

---

[86] For instance see *The Gujarat State Disaster Management Act 2003* (No. 20 of 2003, India), online: GSDMA <http://www.gsdma.org/dmact.pdf>; *The Karnataka Preservation of Trees Act 1976* (No. 76 of 1976, India), online: Karnataka Forest Department <http://www.karnatakaforest.gov.in/English/Acts_Rules/acts/KTP_Act_1976.pdf>; *The Andhra Pradesh Water, Land and Trees Act 2002* (No. 10 of 2002, India), online: ielrc.org <http://www.ielrc.org/content/e0202.pdf>; *The Kerala Ground Water (Control and Regulation) Act 2002* (No. 19 of 2002, India), online: The Official Web Portal of Govt. of Kerala <http://www.kerala.gov.in/dept_ground/annexure6.pdf>.

[87] The central and state pollution control boards have been constituted under the *Water Act*. See generally Puthucherril, *Water Resources*, *supra* note 19.

[88] *A.D.M. Jabalpur v. Shivakant Shukla* (1976), [1976] 2 S.C.C. 521(India S.C.) (justifying the imposition of internal emergency in 1975).

[89] Mijin J. Cha, "A Critical Examination of the Environmental Jurisprudence of the Courts of India" (2005) 10 Alb. L. Envtl. Outlook 197 at 207–12 (HeinOnline). See e.g., *M.C. Mehta (II) v. Union of India* (1988), [1988] 1 S.C.C. 471 (India S.C.) (conferring standing on a non-riparian to bring an action restraining the municipal corporation from discharging insufficiently treated sewage into the river). For an overview of public interest litigation in India, Pakistan and Bangladesh see Parvez Hassan & Azim Azfar Azim, "Securing Environmental Rights through Public Interest Litigation in South Asia", Comment, (2004) 22 Va. Envtl. L.J. 215 (WLeC).

[90] *Constitution of India 1950*, art. 21 ("[n]o person shall be deprived of his life or personal liberty except according to procedure established by law") [*Indian Constitution*].

[91] *Maneka Gandhi v. Union of India* (1978), [1978] A.I.R. S.C. 597 (India S.C.) (incorporating due process into the *Indian Constitution*); *Francis Coraile Mullin v. Administrator, Union Territory of Delhi* (1981), [1981] 1 S.C.C. 608 (India S.C.).

[92] See generally V.R. Krishna Iyer, "Bhoposhima: Crime without Punishment: Case for Crisis Management Jurisprudence" *Economic and Political Weekly* 26:47 (23 November 1991) 2705.

[93] Micheal R. Anderson, "Litigation and Activism: The Bhopal Case" (1993) Third World Legal Stud. 177 at 179 (HeinOnline). See generally, Jamie Cassels, "The Uncertain Promise of Law: Lessons from Bhopal" (1991) 29 Osgoode Hall L.J. 1 (HeinOnline).

laws. "To do a great right after all, it is permissible sometimes to do a little wrong"[94] was the rationale adopted by the Supreme Court to push through an inequitable settlement on the survivors of the Bhopal tragedy. Thus, the Bhopal litigation and the judicial response to it proved to be a black-spot on the Indian judiciary. Despite these failures, the post-Bhopal Supreme Court embarked upon an unparalleled strategy of judicial activism for developing environmental rights.[95]

As a first step towards securing environmental justice, the writ courts read into article 21 the right to a clean environment, thereby infusing it with a human rights dimension.[96] In its crusade to effectuate the right to a clean environment, the writ courts have also recognised concomitant rights like the right to clean drinking water,[97] the right to health,[98] tribal land rights,[99] rights of disaster victims,[100] and sustainable development,[101] raising all these to the status of fundamental human rights. By doing so, the court enhanced enforcement of environmental norms and standards.

The substantive and procedural implications of this new juridical basis meant that if an act impinged on the right to a clean and healthy environment, it afforded the common person, a swift and efficacious remedy under articles 32 and 226[102] of the Constitution. State and even non-state actors[103] were hauled up before the

---

[94] *Charanlal Sahu v. Union of India* (1989), [1990] 1 S.C.C. 613 ¶ 123 (India S.C.).

[95] M.C. Mehta, "Growth of Environmental Jurisprudence in India" [1999] Acta Jur. 71 at 72–73 (HeinOnline).

[96] *Subhash Kumar v. State of Bihar* (1991), [1991] A.I.R. S.C. 420 (India S.C.) (holding that the right to water and air is an enforceable fundamental right under article 21); *Shantistar Builders v. Narayan Khimalal Totame* (1990), [1990] 1 S.C.C. 520 (India S.C.) (stating that the right to a decent environment is covered under article 21).

[97] *F.K. Hussain v. Union of India* (1990), [1990] A.I.R. Ker. 321 (Kerala H.C.) (opining that, "the right to sweet water, and the right to free air, are attributes of the right to life, for these are the basic elements which sustain life").

[98] *L.K. Koolwal v. State of Rajasthan* (1986), [1988] A.I.R. Raj. 2 (Rajasthan H.C.) (holding that maintenance of health, preservation of sanitation and environment falls within the purview of article 21); *Murali S. Deora v. Union of India* (2001), [2001] 8 S.C.C. 765 (India S.C.) (the court prohibited smoking in public places to protect the right to life of passive smokers).

[99] *Manchegowda v. State of Karnataka* (1984), [1984] 3 S.C.C. 301 (India S.C.) (upholding laws that prohibit transfer of tribal lands to non-tribals); *Samatha v. State of A.P.* (1997), [1997] 8 S.C.C. 191 (India S.C.) (proscribing the grant of mining leases by the state government in the schedule areas to non-tribals).

[100] *Bipinchandra J. Diwan v. State of Gujarat* (2001), [2002] A.I.R. Guj. 99 (Gujarat H.C.) [*Diwan*] (holding that the right to assistance in calamity is an enforceable right).

[101] *N.D. Jayal v. Union of India* (2003), [2004] 9 S.C.C. 362 (India S.C.) (holding that 'sustainable development' is to be treated as an integral part of 'life' under article 21).

[102] Article 226 of the Constitution gives citizens the right to approach the High Court to enforce their fundamental rights, while article 32, which itself is a fundamental right, could be invoked by citizens for enforcement of rights conferred by Part III of the Constitution (namely, the fundamental rights). By recognising the right to a clean environment and related interests as part of article 21, citizens can now approach High Courts or the Supreme Court on any problem that affects the environment. *Indian Constitution, supra* note 90.

[103] State, under article 12 of the *Indian Constitution*, does not include private entities, and therefore fundamental rights cannot lie against to them. However, the Supreme Court through a process

courts to redress environmental wrongs. A related development was that rather than looking to environmental statutes for solutions, those aggrieved by environmental wrongs relied on the Constitution for securing the wholesomeness of the environment.[104] Activism on the environmental front resulted in judicial law-making to fill key normative gaps,[105] striking down of *ultra vires* laws,[106] and incorporation of concepts like sustainable development[107] and concomitant principles of polluter pays,[108] precaution,[109] inter-generational equity,[110] and public trust doctrine,[111] into Indian environmental jurisprudence. All this strengthened and developed effective injunctive, remedial, compensatory, exemplary, punitive and preventive remedies for the victims of environmental degradation.[112] Forwarding guidelines for effective implementation,[113] constituting green benches,[114] invoking

---

of constitutionally informed interpretation of environmental statutes has indirectly brought private entities within the discipline of fundamental rights. See generally *M.C. Mehta* (1986), *supra* note 81; *M.C. Mehta (Calcutta Tanneries Matter) v. Union of India* (1996), [1997] 2 S.C.C. 411 (India S.C.) (directing re-location of the Calcutta tanneries and imposing a pollution fine on them).

[104] Cha, *supra* note 89.

[105] *T.N. Godavarman Thirumulkpad v. Union of India* (1996), [1997] 2 S.C.C. 267 (India S.C.) (providing a definition to the term 'forest' under the *Forest Conservation Act, 1980*).

[106] *Indian Council for Enviro-legal Action v. Union of India* (1996), [1996] 5 S.C.C. 281 (India S.C.) (striking down certain amendments to the *Coastal Regulation Zone Notification 1991*).

[107] *People United for Better Living in Calcutta-Public v. State of West Bengal* (1992), [1993] A.I.R. Cal. 215 (Calcutta H.C.) (holding that law courts are to strike a balance between development and ecology with no compromise on each other).

[108] *Indian Council for Enviro-legal Action v. Union of India* (1996), [1996] 3 S.C.C. 212 (India S.C.) (holding companies absolutely liable for historic pollution of underground aquifers).

[109] *A.P. Pollution Control Board v. Prof. M.V. Nayudu I* (1999), [1999] A.I.R. S.C. 812 (India S.C.); *A.P. Pollution Control Board v. Prof. M.V. Nayudu II* (2000), [2001] 2 S.C.C. 62 (applying the principle of precaution and the new principle of burden of proof).

[110] "Rivers, forests, minerals and such other resources constitute a nation's natural wealth. These resources are not to be frittered away and exhausted by any one generation. Every generation owes a duty to all succeeding generations to develop and conserve the natural resources of the nation in the best possible way. It is in the interest of mankind. It is in the interest of the nation." *State of Tamil Nadu v. M.S. Hind Stone* (1981), A.I.R. 1981 S.C. 711 (India S.C.); *State of H.P. v. Ganesh Wood Products* (1995), [1995] 6 S.C.C. 363 (India S.C.) (relying on the principle of inter-generational equity to decide the issue of granting of permission to set up forest based industries).

[111] *M.C. Mehta v. Kamal Nath* (1996), [1997] 1 S.C.C. 388 (India S.C.) (incorporating the doctrine of public trust into Indian environmental jurisprudence); *Hinch Lal Tiwari v. Kamala Devi* (2001), [2001] 6 S.C.C. 496 (India S.C.) (holding that the material resources of the community such as forests, tanks, ponds, hillocks, and mountains need to be protected for a proper and healthy environment to enable people to enjoy a quality life, which is the essence of article 21).

[112] See generally M.K. Ramesh, "Environmental Justice: Courts and Beyond" (2002) 3:1 Indian J. Envtl. L. 20. See also *M.C. Mehta v. Kamal Nath* (2000), [2000] 6 S.C.C. 213 (India S.C.) (even though a pollution fine under the *Water Act* can be imposed only after a finding of guilt, the court has power under the law of torts to require the polluter to pay damages, including exemplary damages).

[113] For instance see *Suo Motu v. State of Rajasthan* (2004), [2005] A.I.R. Raj. 82 (Rajasthan H.C.) (guidelines to improve the facilities in Jaipur city as part of enforcing the right to live in a clean city).

[114] *Vellore Citizens' Welfare Forum v. Union of India* (1996), [1996] 5 S.C.C. 647 (India S.C.) [*Vellore Citizens' Welfare Forum*] (after laying down elaborate guidelines to discipline the tanneries, the Supreme Court directed the Chief Justice of the Madras High Court to constitute a special 'Green Bench' to monitor the guidelines).

the *parens patriae* jurisdiction to provide relief to the victims of mass disasters,[115] issuance of continuing mandamus,[116] constituting authorities,[117] special committees[118] and expert opinions,[119] and issuance of orders and directions[120] were some of the procedural innovations developed to further the cause. By innovating and employing existing tools of justice delivery imaginatively, India's green judges exhibited 'Marshall-like resourcefulness', which has become the hallmark of judicial intervention in the environment.[121]

Through these actions, the writ courts often stretched the extent of its legitimate confines. However, by crafting innovative environmental remedies that introduced efficiency into a system of environmental governance, which at best could be described as being tardy, lax, inefficient and corrupt, it weathered criticism of having overstepped the *lakshman-rekha*[122] thereby becoming the final

---

[115] See generally *Diwan*, *supra* note 100.

[116] Merely issuing a mandamus directing agencies to perform their task was found to be futile. Therefore, writ courts in India issue directions from time-to-time keeping the matter pending and requiring agencies to report progress to ensure better compliance. For e.g., see *T.N. Godavarman Thirumulpad v. Union of India*, [1995] Writ Petition (Civil) 202 (India S.C.), online: The Judgments Information System <http://www.courtnic.nic.in/courtnicsc.asp> [*Forest case*] (popularly known as the *Forest case*, the Supreme Court has been entertaining this continuing mandamus since 1995, rendering more than two hundred orders, enabling the court to have the final word on forest management). For further details see *Forest Case Update*, online: Forest Case Update <http://www.forestcaseindia.org/>.

[117] *S. Jagannath v. Union of India* (1986), [1997] 2 S.C.C. 87 (India S.C.) [*Jagannath*] (directing the central government to constitute an authority to implement the polluter pays and precautionary principles in the context of increased commercial aquaculture in coastal areas).

[118] *Rural Litigation and Entitlement Kendra, Dehradun v. State of U.P.* (1988), [1988] A.I.R. S.C. 2187 (India S.C.) (appointing an expert committee to evaluate the environmental impact of limestone quarrying operations in the region and to supervise the execution of court orders).

[119] *Dr. Shivarao Shantaram Wagle v. Union of India* (1988), [1988] 2 S.C.C. 115 (India S.C.) (appointing a committee to look into whether imported Irish butter was contaminated by radioactivity). See also Tony George Puthucherril, "Harnessing the Atom: Strengthening the Regulatory Board for Nuclear Safety in India Based on the Canadian Experience" (2008) 26:4 J. Energy & Nat'l Res. L. 553 at 578.

[120] *M.C. Mehta v. Union of India* (1988), [1988] 1 S.C.C. 471 (India S.C.) (issuing a series of directions calling upon the Kanpur Municipality to take steps to prevent the pollution of the river Ganga).

[121] See Tony George Puthucherril & Lekshmi Vijayabalan, "Judiciary in Quest for Gender Justice to Women" (2006) 4 Sri Lanka L.C.L. Rev. 3 at 16.

[122] Corruption is so perverse that while environmental justice is being delivered in the courts, the executive, which has to enforce the orders, does not do so. In fact, environmental enforcement has changed from being a part of the problem to becoming the problem itself. There is a need to incorporate the accountability principle into Indian environmental jurisprudence. M.C. Mehta, "The Accountability Principle: Legal Solutions to Break Corruption's Impact on India's Environment" Book Excerpt of *In the Public Interest* by M.C. Mehta, (2006) 21 J. Envtl. L. & Litig. 141 at 143, 147 (WLeC). See also *T.N. Godavarman Thirumulpad v. Ashok Khot* (2006), [2006] 5 S.C.J. 662 (India S.C.) (a former principal secretary to the Department of Forest, State of Maharashtra and the former minister in-charge of the Department of Forest was held guilty for contempt and sentenced to one month imprisonment for wilfully disobeying the orders of the court by granting permission to certain saw mills).

hope to secure environmental justice. The ever-increasing number of public interest litigations that are being filed in the Supreme Court and in the High Courts over every conceivable environmental problem bear testimony to this faith.[123]

B. *The Supreme Court of India, Hazardous Wastes Management and Shipbreaking*

To prevent India from becoming the toxic capital of the world, since 1995, the Supreme Court has been constantly monitoring the dumping of hazardous wastes into the country and has issued periodic orders in a continuing mandamus.[124] Popularly known as the *Hazardous Waste Management case*,[125] the court has utilised this opportunity to also ensure that scrapping of ships is both eco-friendly and sensitive to human rights.

Within two years of entertaining this case, the Supreme Court appointed a High Powered Committee (HPC) to study all aspects relating to the management of hazardous wastes.[126] Among its 14 terms of reference, the 14th related to decontamination of ships before being exported to India for breaking.[127] The HPC conducted a detailed study and filed a comprehensive report on hazardous wastes management that was accepted by the Supreme Court. The suggestions on shipbreaking in this report can be described as the first instalment of the judicially prescribed framework for sustainable ship recycling in India.[128]

1. *The Supreme Court Guidelines on Sustainable Ship Recycling*

In articulating the Indian version of a sustainable ship recycling model, the Supreme Court observed:

> We are not suggesting discontinuing of ship breaking activity but it deserves to be strictly and properly regulated. When the ship arrives at a port for breaking, the concerned authorities have to be vigilant about the hazardous waste which may be generated if appropriate timely action by various agencies, in particular, Maritime Board and the SPCB are not

[123] See generally Tony George Puthucherril, *One Step Forward, Two Steps Back: Constitutionalism and Environmental Jurisprudence in India,* online: OSGOODE Graduate Law Students' Association <http://www.osgoode.yorku.ca/glsa/2008conference/documents/Tony%20George.pdf>.

[124] *Research Foundation for Science, Technology and Natural Resources Policy v. Union of India*, [1995] Writ Petition (Civil) No. 657 (India S.C.), online: The Judgments Information System <http://www.courtnic.nic.in/courtnicsc.asp> [*Research Foundation*].

[125] *Ibid.*

[126] See India, Ministry of Environment & Forests, *Report of the High Powered Committee on Management of Hazardous Wastes*, vol. 1, at 1–15, online: Ministry of Environment & Forests <http://envfor.nic.in/cpcb/hpcreport/>.

[127] *Ibid.* at 8.

[128] *Ibid.* at 251–53.

> taken ... The ship breaking operation referred to ... cannot be permitted to be continued without strictly adhering to all precautionary principles, CPCB guidelines and taking the requisite safeguards which have been dealt with extensively in the report of HPC which include the aspect of the working conditions of the workmen.[129]

By permitting continuance of this industry subject to strict and proper regulation, the court was in effect striking a balance between the needs of development and environmental protection, thereby furthering the cause of sustainable development. This philosophy is also reflected in the HPC Guidelines on shipbreaking which operate at four levels: (1) they mandate certain important procedures that have to be complied with when importing obsolete ships, (2) they seek to secure environmental protection, (3) they provide for worker safety, and (4) they envisage certain general requirements.

The Guidelines, which apply to shipbreaking activities in all coastal states,[130] mandate that prior to the arrival of a ship, there has to be "proper consent" from the concerned authority or the state maritime board to the effect that it does not contain any hazardous wastes or radioactive substances.[131] It also requires shipowners to ensure that prior to the breaking, the ship is properly decontaminated,[132] and that a complete inventory of hazardous wastes on board is produced without which permission for breaking is not to be granted.[133]

To address environmental concerns, the Guidelines emphasise the need to maintain ambient air and noise levels and ensure effective waste disposal. Elaborating on measures to achieve these objectives, the Guidelines provide that the pollution control board continuously monitor air and noise levels and set up adequate infrastructure for wastes disposal.[134] It also bans the burning of any material on the beach[135] and requires classification of wastes into hazardous and non-hazardous, and determination of their quantities, which is to be brought

---

129 *Research Foundation for Science, Technology and Natural Resources Policy v. Union of India* (14 October 2003), Writ Petition (Civil) 657 of 1995 (India S.C.), online: The Judgments Information System <http://www.courtnic.nic.in/courtnicsc.asp> [*Research Foundation* (14 October 2003)].

130 *Ibid.*, guideline 18.

131 *Ibid.*, guideline 1. In appropriate cases, particularly in relation to radioactive substances, the AERB is to be consulted.

132 *Ibid.*, guideline 2. The fact that there has been decontamination has to be ensured by the SPCBs.

133 *Ibid.*, guideline 13.

134 The Guidelines provide that the pollution control boards are to install proper equipment and infrastructure for first level inspection of hazardous material, radioactive substances, etc. *Ibid.*, guideline 10. Also, the SPCBs together with the SMBs, are to prepare land fill sites and incinerators as per the guidelines of the CPCB and after obtaining its prior approval. *Ibid.*, guideline 15. Individual shipbreaking industries are also to have facilities for waste disposal in an environmentally sound manner. In fact, if the units lacked authorisation under Rule 5 of the *Hazardous Wastes (Management and Handling) Rules 2003*, they were to close shop. *Ibid.*, guidelines 5, 8.

135 *Ibid.*, guideline 7.

to the notice of the concerned authority.[136] Regarding waste disposal, the Guidelines require the adoption and use of technologies that can ensure an effective destruction efficiency of 99.9 percent.[137] Also, a generation of persistent organic pollutants is to be banned and there is to be a complete containment of all gaseous, liquid and solid residues.[138] The disposed materials are to be stored at a specified place earmarked for this purpose.[139] Special care is to be taken while handling asbestos wastes,[140] and its final disposal is to be subject to authorisation by the pollution control board.[141] Waste oil, sludge, paints, chips, etc., are to be carefully removed from the ship and taken to areas outside the beach for "immediate safe disposal".[142]

For worker protection, the Guidelines require that the Gujarat SPCB ensure that the GMB comply with the requirements of the *Gujarat Maritime Board (Prevention of Fire and Accidents for Safety and Welfare of Workers and Protection of Environment during Shipbreaking Activities) Regulations 2000*, and file compliance reports with the court.[143] The notification issued by the GMB on gas-free-for-hot-work is to have mandatory effect, and no ship is to be given beaching permission unless this certificate is produced.[144] Explosions are to be sternly dealt with.[145]

General requirements include the stipulation that India participate (at both the central and state levels) in international meetings on shipbreaking organised by the IMO and the Basel Convention's Technical Working Group with a clear man date to ensure decontamination of ships prior to their export to India.[146] The Guidelines also provide for the creation of administrative structures. They also direct the central government to establish an inter-ministerial committee comprising of the Ministries of Surface Transport, Steel, Labour and Environment and representatives from the shipbreaking industry, labour and environment organisations.[147] Perhaps the most important feature of this order is the constitution of a monitoring committee to oversee the timely implementation of the Guidelines on waste management, including further directions of the Supreme Court and the specific Guidelines on shipbreaking.[148]

---

136 *Ibid.*, guideline 3.
137 *Ibid.*, guideline 4.
138 *Ibid.*
139 *Ibid.*
140 *Ibid.*
141 *Ibid.*
142 *Ibid.*, guideline 6.
143 *Ibid.*, guideline 11.
144 *Ibid.*, guideline 12 .
145 *Ibid.*
146 *Ibid.*, guideline 16.
147 *Ibid.*, guideline 14.
148 For the composition of the Supreme Court Monitoring Committee, see India, Supreme Court Monitoring Committee on Management of Hazardous Wastes (For the period: November 2003 to

Although these Guidelines fall short of being a complete code, their importance cannot be underestimated. The Guidelines, which seek to provide a sustainable framework for the operation of shipbreaking, were introduced at a time when opposition to the industry was high. The court did not completely shut its doors on this industry; it did, however, stress the need for strict and proper regulation. However, vagueness in language and frequent use of general terminology rather than elaboration on specifics has left room for conflicting interpretations of its measures, severely affecting its efficacy. As these Guidelines were intended to fill in the legislative vacuum in regulation of this industry, it would have served its long-term sustainable growth had the principles been more specific.

Specific Guidelines can be criticised on several levels. The most controversial seems to be Guideline 2, which mandates that the ship be decontaminated "properly" by the owner prior to its breaking, and Guideline 13, which requires the shipowner to provide a complete inventory of hazardous wastes on board the ship. Both provisions embody the precautionary approach. However, neither provides any guidance as to what "properly decontaminated" is, nor have any acceptable standards for decontamination been provided. Also, Guideline 13 does not prescribe what an inventory of hazardous substances should contain. Although it is doubtful whether any ship would be carrying hazardous wastes or radioactive substances as cargo while requesting beaching and breaking facilities, Guideline 1 provides guidance to meet such an eventuality. However, an offshore inspection to determine the nature of any cargo is unlikely to be foolproof as it is an arduous task to access and inspect all compartments of an obsolete ship when it is on the high seas.

From the worker's perspective, the Guidelines also leave room for improvement. They merely lay down two stipulations and do not provide for worker training, compulsory and proper maintenance of rolls, improvement of work and living conditions, better wage rates, etc. Finally, the Guidelines though intended to govern shipbreaking activities in all coastal states, seem to be Alang-centric. It is doubtful whether other shipbreaking states in India have laws similar to those in Gujarat state, raising doubts as to whether these Guidelines could foster the development of a composite legal regime on shipbreaking for the whole country.

Despite these flaws, these judicial guidelines do contain several important principles for sustainable shipbreaking. However, the ship recycling industry in India views them as being 'unworkable', constituting a major hindrance to the free trade practices that characterise the working of this industry. As will be seen later

October 2006), *Report of the Supreme Court Monitoring Committee on Management of Hazardous Wastes* Vol. 1 (Chairman: Dr. G. Thyagarajan, New Delhi: Ministry of Environment & Forests, 2006) [unpublished] [on file with the author].

in this chapter, concerted attempts by the ship recycling industry, the bureaucracy and a player that one would least expect to be in this list, the judiciary itself, have diluted the rigour of many of these requirements.[149]

2. *Strengthening the Guidelines on Sustainable Shipbreaking: One Step Forward, Two Steps Back*

The initial set of Guidelines, established by virtue of the 2003 order, was in force for nearly four years. In the meantime, two major toxic ships, the 'Riky' and the 'Clemenceau', arrived in Indian waters. Even though the Supreme Court missed an opportunity to make a final pronouncement on the issues raised by the 'Clemenceau', given that "the problem of arrival of toxic ships was [a] recurring one", the court directed the Union of India to constitute yet another committee, the Committee of Technical Experts (TEC), to inquire into the adequacy of the infrastructural facilities at Alang and to determine whether shipbreaking was being conducted in an environment-friendly manner.[150] The TEC was also directed to examine the pre-conditions that the recycler had to satisfy prior to dismantling and were to suggest measures to control the environmental impacts from the asbestos dust generated during the breaking process.[151]

The TEC submitted its report on 30 August 2006. However, the report was not considered until the court became cognisant of the arrival of the third major toxic ship, the 'Blue Lady', in 2007. By that time, conditions in the shipbreaking

---

[149] For instance, the Supreme Court Monitoring Committee in its Majority Report on the 'Clemenceau' afforded an interpretation that seemed to nullify Guidelines 2 and 13:

1. The ship should be property decontaminated by the ship owner prior to the breaking. This should be ensured by the SPCBs.
2. A complete inventory of hazardous waste on board of a ship should be made mandatory for the ship owner. And no breaking permission should be granted without such an inventory. This inventory should also be submitted by the GMB to concerned SPCBs to ensure safe disposal of hazardous and toxics waste.
3. *Proper de-contamination under para 1 above is with reference to hazardous cargo*, if any, carried by the ship. Like-wise, *hazardous waste in para 2 above is with reference to waste cargo that may be carried by the ship and not with reference to materials used in ship construction*. In fact, if proper de-contamination under para 1 is taken to include materials used in ship construction, ships would not be able to reach dismantling yards under their own propulsion and possibly, not even be able to float, which would rule out towing as well. *Such an interpretation is clearly unwarranted*.

India, Supreme Court Monitoring Committee on Hazardous Wastes, *Final Report on Clemenceau: Majority Report* (Chairman: Dr. G. Thyagarajan, New Delhi: Ministry of Environment & Forests, 2006) [unpublished] at 2 [on file with the author] [*SCMC Majority Report*] [emphasis added].

[150] For the text of the order of the Supreme Court of India dated 17 February 2006, see *Research Foundation for Science and Technology and Natural Resource Policy v. Union of India* (2007), [2007] S.C. 890 ¶ 4 (India S.C.) (indlaw.com) [*Blue Lady case*].

[151] *Research Foundation for Science and Technology and Natural Resource Policy v. Union of India*, [2007] Writ Petition (Civil) No. 657 (India S.C.), online: The Judgments Information System <http://www.courtnic.nic.in/courtnicsc.asp> (issuing order dated 12 March 2007).

industry in India had changed considerably, and Alang was virtually on its death bed due to the strict conditions imposed by the court through its 2003 order. The "all pervasive"[152] report of the TEC was finally endorsed by the Supreme Court *via* its order on 6 September 2007. The recommendations in this report, together with the 2003 Guidelines and existing statutes on the subject, jointly form the legal framework on ship recycling in India.

In accepting the recommendations of the TEC, the Supreme Court held thus:

> It is desirable that the Government of India shall formulate a comprehensive Code incorporating the recommendations and the same has to be operative until the concerned Statutes are amended to be in line with the recommendations. Until the Code comes into play, *the recommendations shall be operative by virtue of this order.*"[153]

Thus, these judicial guidelines are intended to be only a stop-gap arrangement to govern the industry until the government promulgates comprehensive legislation on shipbreaking.

As far as the major recommendations of the TEC as elaborated in the Supreme Court order are concerned, it mainly focused on ship dismantling, specifying processes for anchoring, beaching and breaking. For anchoring, the recommendations require that well in advance of a ship's arrival, the owner or the recycler has to submit certain documents[154] for a desk review by the State Maritime Board (SMB) in consultation with the concerned SPCB and the Customs Department.[155] In case of refusal by any of these three agencies, the shipowner can appeal or apply for a review.[156] If permission is granted, the SMB can issue instructions for safe anchorage.[157]

With regard to beaching, the recycler is to submit certain documents as required by annex I of the Ship Recycling Regulation.[158] In case of oil tankers, the recycler is also to submit a gas-free-for-hot-working certificate.[159] To verify these submissions, the ship can be inspected while in anchorage by the Customs Department,

---

152 *Blue Lady case, supra* note 150 ¶ 6.

153 *Research Foundation for Science and Technology and Natural Resource Policy v. Union of India* (2007), [2007] S.C. 1484 ¶ 9 (India S.C.) (indlaw.com) [*Research Foundation for Science*] [emphasis added].

154 The documentation includes details regarding the name of the ship, IMO identification number, the flag, call sign, name of the master of the ship and nationality, list of crew members, the GRT/ NRT/LDT of the ship with supporting documents, and an assessment of hazardous wastes/hazardous substances in relation to the structure of the ship, and on board as far as practicable by referring to the ship's, drawings, technical specifications, ship's stores, manifest, in consultation with the shipbuilder, equipment manufacturers and others as appropriate. In the case of ships of special concern, in addition to identification and marking of all areas containing hazardous wastes/hazardous substances, quantification of such wastes/substances is also necessary. *Ibid.*, ¶ 4.

155 *Ibid.*

156 *Ibid.*

157 *Ibid.*

158 *Ibid.*

159 *Ibid.*

the SPCB, the Explosives Department and the Atomic Energy Regulatory Board (AERB).[160] If necessary, these agencies can take representative samples for verification.[161] On the basis of the clearance granted by these authorities, the SMB grants beaching permission[162] and thereafter, the recycler takes charge of the ship. If permission is refused, the shipowner is again entitled to both review and an appeal.[163]

In addition to the documents required under the Ship Recycling Regulation, for the process of breaking, the recycler has to submit a ship recycling plan consisting of a recycling facility management plan[164] and a ship specific dismantling plan.[165]

This Supreme Court order refers only to the recommendations by the TEC in relation to the processes of anchoring, beaching and breaking, and to certain measures regarding ships of concern.[166] But as the court accepted the TEC Report in its totality without any reservations, one can conclude that the recommendations on "environmental monitoring",[167] "management of occupational safety and health issues",[168] "handling of hazardous materials and hazardous wastes",[169] and "worker's welfare issues"[170] as detailed in the Report though not in the Supreme Court order, also apply to shipbreaking operations.[171]

3. *Toxic Ships and Progressive Dilution of the Guidelines*

It is clear that the Supreme Court succeeded in establishing a comprehensive framework for ship recycling in two instalments, with the first order of 2003

---

160 *Ibid.*

161 *Ibid.*

162 *Ibid.*

163 *Ibid.*

164 This should have as its goal "minimisation and ultimately elimination of the adverse effects on human health and environment caused by ship recycling". The plan is to put in place certain systems. *Ibid.*

165 This plan should address all the three phases of recycling – preparation, dismantling and waste stream management. It should include details about the ship, a fair assessment of the hazardous wastes; shipbreaking schedules and the sequence of work; operational work procedures; availability of material handling equipment and personal protection equipments; plan for removal of oil and cleaning of tanks; hazardous waste handling and disposal plan; "gas-free-for-hot-work" certificate; identification and marking of all non-breathable spaces and places containing or likely to contain hazardous substances; and confirmation that ballast water has been exchanged on the high seas. Since asbestos is the major area of concern, the plan is to include a scheme for the removal of asbestos, and asbestos containing materials (ACMs), both on board and on the shore, and also arrangements for its handling, treatment and disposal. Also, it should provide for systems and procedures to document and for keeping track of all hazardous wastes generated during recycling, as well as hazardous substances found on board the ship, and their transport to the disposal facility or registered recycling facility. *Ibid.*

166 *Ibid.*, ¶ 3.

167 *TEC Report*, *supra* note 3 at 36–41.

168 *Ibid.* at 24–29.

169 *Ibid.* at 30–35.

170 *Ibid.* at 46–50.

171 *Research Foundation for Science*, *supra* note 153 ¶ 6, 9.

providing the general principles and the second order in 2007 delving into specific aspects of ship dismantling. However, since the 2003 Guidelines were handed down, there has been a slump in shipbreaking activities in India. The loss of business at Alang has proven to be a boon for the industry in Bangladesh. Not surprisingly, the industry and government officials have made concerted efforts to dilute the Guidelines on one pretext or the other. Surprisingly, even the Supreme Court watered down some of the mandates as required under its earlier 2003 order. During the course of the 2007 proceedings, it was brought to the attention of the Supreme Court by the petitioner that there was to be an "additional precaution for decontamination".[172] It was suggested that before a ship leaves the foreign country, it should obtain a certificate that it was "totally decontaminated".[173] Turning this plea down due to the many "practical difficulties" involved, the court observed that, "(the) authorities in India can without the certificate at the stage of anchorage verify and come to a conclusion that if the ship is contaminated the same is to be sent back".[174] As mentioned earlier, the primary issue in shipbreaking relates to the amount of in-built hazardous wastes present in its structure.[175] By permitting ships not subjected to decontamination to enter Indian waters and leaving the matter to be decided according to the discretion of government officials, the Supreme Court practically weakened the rigour of its earlier Guidelines that stressed on "proper decontamination" by the shipowner.[176]

The first major success in diluting the 2003 Guidelines was achieved within a span of one and a half years, when the Danish ship the 'Riky' anchored in Indian waters in April 2005 for dismantling. Further dilution of the Guidelines also occurred with controversies surrounding the dismantling of the French aircraft carrier 'Clemenceau' in 2005 and the 'Blue Lady' in 2006. Each of these examples raises additional legal issues and is examined in turn below.

i. *The 'Riky'*

A ferry of 1950s vintage, Kong Frederik IX *alias* 'Riky', after years of service and changing through several hands, was finally sold to a company infamous for exporting toxic ships to India and Bangladesh for scrapping.[177] The former,

---

[172] *Ibid.*, ¶ 7.

[173] *Ibid.*

[174] *Ibid.*

[175] *Comments on the Indian Committee Inspection Report on the Hazardous Materials onboard the SS Blue Lady*, (31 July 2006), online: Basel Action Network <http://www.ban.org/Library/NGO_Platform_Critique_on_TC_Inspection_Report_Final.pdf> (pointing out the flaws in the inspection report on 'Blue Lady').

[176] *Ibid.* See also India, Gujarat Pollution Control Board, *Inspection Report of 'Blue Lady'* (India: Gujarat Pollution Control Board, 2006), online: GPCB Office <http://www.gpcb.gov.in/PDF/blue lady_main_report.pdf>.

[177] Ea Krogstrup & Karen King Nash Arleth, *Ship Scrapping – a Floating Scenario* (International Development Studies Final Thesis, University of Roskilde, Denmark, 2006) at 22, online: RUDAR

as well as the new owners, produced necessary documentation before the Danish authorities claiming that the ship was to be put in service in the Middle East as a ro-ro cargo vessel.[178] Accordingly, the ship was classified as non-waste, allowing it to leave Denmark. Instead of sailing to the Middle East, it headed directly to India for recycling.[179] On realising that they were duped, the Danish Environment Minister alerted her Indian counterpart:

> … I believe our interests are joint – and I call on you to co-operate in this case by denying the ship to be dismantled in India – and refer the ship to be returned to Denmark in order to be stripped of the hazardous waste. By this we can send a strong signal that neither India nor Denmark will accept export of environmental problems that could be solved locally, and that we – as governments – will not accept this kind of foul play which results in lasting damage to the environment.[180]

However, the Indian Minister refused to heed the Danish request and replied:

> We have determined that the ship cannot be classified as 'Wastes' within the scope of Art. (*sic*) 2.1 of the Basel Convention … According to the Gujarat Maritime Board, Gujarat Pollution Control Board, and the Central Pollution Control Board who have inspected the Vessel, there is no objectionable hazardous material on the ship. Accordingly the said ship now called 'Riky' has been beached on 23rd of April 2005 at Alang Gujarat. There are only in built insulation material which is part of the structure of all ships. As per Indian Laws and our position under the Basel Convention and the IMO, the ship has the requisite permission for beaching. I would like to assure you that India has adequate capacity to ensure environmentally sound disposal of the said ship.[181]

Diametrically opposite to the stance of the Indian Minister, the Chairman of the Supreme Court Monitoring Committee on Hazardous Wastes (SCMC) observed:

> The ship changing its name and arriving at Indian shore illegally clearly demonstrates its intention to cheat and deceive. Its arrival is in gross violation of the directives on ship

<http://rudar.ruc.dk/bitstream/1800/1521/1/Thefinalmaster.pdf>. See also Recycling of Ships: Proposal for an Enforceable Legally Binding System for the Recycling of Ships, Submitted by Denmark, IMO/MEPC 53/3/7, 13 May 2005, ¶ 3 (KR-CON) [Proposal for an Enforceable Legally Binding System] (noting that even though the ship was docked in Denmark, it was registered in St. Vincent and the Grenadines).

178 See also *ibid.*, ¶ 4.

179 In Denmark, the administration and management of wastes is the responsibility of local municipalities. This rendered the municipality of Korsør in a central position: it was the stakeholder with authority to prevent the ship from leaving Denmark. Since circumstantial evidence pointed in the direction of scrapping, the municipality required the former owners, Danish KFIX A/S, to document that the ship really was intended for further use. KFIX provided the authorities with documents stating that the ship was to continue operating in the Persian Gulf. Krogstrup & Arleth, *supra* note 177 at 22–24.

180 Letter from Connie Hedegaard, Danish Minister for the Environment to A. Raja, Minister for Environment and Forests, India (15 April 2005), online: Greenpeace Nederland <http://www.greenpeaceweb.org/shipbreak/frederik.pdf>.

181 Basel Action Network, "Legal Analysis of Letter from Mr. Raja to Ms. Hedegaard regarding the legal application of the Basel Convention to the Kong Frederik IX (aka Riky) (3 May 2005)", online: Basel Action Network <http://www.ban.org/Library/BAN_analysis_Ricky.pdf>.

> breaking … of the Honourable Supreme Court of India … The ship should not have been allowed to enter Indian territorial waters at all. The letter from the Danish Minister for the Environment cautioned India loudly and clearly. If the ship is considered hazardous by Denmark, the Basel Convention requires India also to treat it as such. I am surprised that an illegal arrival is being treated as acceptable for ship breaking when it has not been earlier decontaminated to remove asbestos, waste oils, paint chips and other hazardous materials. This will set a wrong precedent and encourage not only more such wicked incidents but defeat the very purpose of the Apex Court's directives on ship breaking … Riky must be mercilessly driven out of Indian sovereign territory without any further loss of time.[182]

Subsequently, the SCMC reversed its position on the 'Riky' and permitted the dismantling of the ferry.[183]

What caused this sudden change in stance, one may never know. The 'Riky' is a story of deceit and fraud at high levels of governance.[184] A plain reading of the extracts from the official correspondence reproduced above reveals several inconsistencies in the manner in which Indian regulators approached the issue and their failure to comply with relevant laws.[185] Though the matter was eventually brought

[182] Letter from Gopalakrishna Thyagarajan to Bhanujan (24 May 2005), online: Greenpeace Nederland <http://www.greenpeaceweb.org/shipbreak/drthya.pdf>.

[183] See also V. Venkatesan, "Pressures of monitoring" *Frontline [of India]* 22:25 (3 December 2005), online: Frontline <http://www.flonnet.com/fl2225/stories/20051216002904700.htm> [Venkatesan, "Pressures of monitoring"]. Quoting from a telephone interview with the Chairman:

> … [t]he ship was apparently carrying a small amount of asbestos, which cannot be removed, which is necessary to propel the ship. We are not concerned with the entry of the ship, which is within the jurisdiction of the Gujarat Maritime Board and the Customs. The legality of the entry of the ship is something that we are not concerned with. Riky was inspected, and we found nothing on the ship. Asbestos as in-built material is different from asbestos as cargo.

Ibid. See also Praful Bidwai "Confused and ambivalent" *Frontline [of India]* 22:11 (21 May 2005), online: Frontline <http://www.hinduonnet.com/fline/fl2211/stories/20050603005811600.htm> (observing that, "No country, leave alone India, has the technology to handle such toxic waste safely by manually breaking up a ship … "Riky" will go down as a dark chapter in India's environmental history").

[184] It is interesting to note that the flag under which the ship sailed into Indian waters is also not clear. While the customs division in Gujarat point out that 'Riky' was flying under the North Korean flag, the petitioner who brought the matter before the Supreme Court alleges that the ship sailed under the flag of Roxa, a non-existent country. For further details see Gopal Krishna, "The scrapping of Riky" *India Together* (16 July 2008), online: India Together <http://www.indiatogether.org/2006/mar/envriky.htm> [Krishna, "Riky"]; Interview of Claude Alvares by Lyla Bavadam [n.d. 2006?] in Cover Story, "We get the waste and pay for it" *Frontline [of India]* 23:2 (28 January 2006), online: Frontline <http://www.flonnet.com/fl2302/stories/20060210006601900.htm> (as a member of the Supreme Court Monitoring Committee on Hazardous Wastes, Alvares opined, "[w]e got a report saying the asbestos waste generated on demolition was 222 kg. But the Danish authorities claimed it had 17 tonnes of ACM on board. We would like to re-examine this aspect now").

[185] There are several discrepancies in the permission granted for dismantling the 'Riky'. Firstly, since 'Riky' arrived in India without Denmark's authorisation, it could not have carried Form 7. Even though the SCMC inspection team observed that the shipowners could not provide a detailed inventory of in-built materials, a post facto inspection conducted after the beaching granted permission to break down the ship. *Ibid.*

to the notice of the Supreme Court, the ship had since been broken down.[186] The 'Riky' episode raises questions about obligations under three legal frameworks, namely, the European Union Waste Management Law, the Basel Convention, and the Indian Supreme Court Guidelines on shipbreaking.

One of the most controversial issues raised by 'Riky' relates to when a ship can be categorised as 'waste'. Under the European Union waste management framework,[187] waste is defined as anything that the holder discards or intends to discard.[188] This definition has been the subject of intense controversy as shipowners claim that obsolete ships are actually not being discarded, rather they are being recycled. As long as the ship is seaworthy, and the owner has not declared his/her intention to dispose of it, the ship cannot be categorised as waste.[189] This poses problems in applying the European Union waste management framework. As illustrated by 'Riky', a shipowner can by-pass this law by refraining or falsely declaring an intention not to dispose the ship until it leaves European waters.[190]

Under article 2(1) of the Basel Convention, 'waste' is defined as substances or objects which are disposed of or are intended or required to be disposed of by the provisions of national law. Even though the Government of India determined that the ship could not be classified as hazardous waste within the scope of article 2(1), under article 1(b), if waste is defined as, or considered to be, hazardous by the domestic legislation of the party of export, import or transit, then it could be deemed as such for the purposes of the Basel Convention.

In 'Riky', Denmark (country of export) had already determined the ship to be hazardous, which implies that India acted contrary to the Basel Convention.[191] Any transboundary movement of hazardous or other waste under article 9 of the Basel Convention without notification to all concerned states or without consent, or with consent obtained through falsification, misrepresentation or fraud, is deemed to be illegal traffic. In this case, there was evidence to show falsification and misrepresentation regarding the bona fides of 'Riky', which was supposedly going to be used as a ro-ro cargo ship in the Middle East. Under the Basel

---

[186] *Ibid.*

[187] EC, *Council Regulation 259/93 of 1 February 1993 on the supervision and control of shipments of waste within, into and out of the European Community,* [1993] O.J. L 30/1 [*WSR 259/93*]; EC, *Regulation (EC) 1013/2006 of the European Parliament and of the Council of 14 June 2006 on shipments of waste*, [2006] O.J. L 190/1 [*WSR 1013/2006*].

[188] *Ibid.*

[189] U.K., H.C., "Environment, Food and Rural Affairs Committee on Dismantling Defunct Ships in the UK", HC 834 (18th Report of Session 2003–2004) 11, online: United Kingdom Parliament <http://www.publications.parliament.uk/pa/cm200304/cmselect/cmenvfru/834/834.pdf>.

[190] *Ibid.*

[191] Proposal for an Enforceable Legally Binding System, *supra* note 177 ¶ 8 (noting that "Denmark is of the opinion that the export of this ship should be regarded as an export of hazardous waste, and … falls under the Basel Convention … [A]s the exporting State, [Denmark] has a responsibility to re-import the ship. India does not share this opinion").

Convention, if there is an illegal traffic of waste, the state of export has to ensure that the waste is properly taken care of either by re-import or by environmentally-sound disposal. Although Denmark asked to have the 'Riky' returned to its shores, knowing well that India's capability to ensure the environmentally-sound disposal of the ship was suspect, India turned down the request.[192]

The Supreme Court Guidelines on shipbreaking have, as one of its central tenets, the principle of "proper decontamination" of the ship by the owner prior to the breaking.[193] In 'Riky', there was no record to suggest any prior decontamination, yet the ship was permitted to be broken down at Alang, in violation of the Supreme Court order, by the Monitoring Committee appointed by the Supreme Court to oversee the implementation of its Guidelines.[194] 'Riky' highlights the inconsistencies and the ineffectiveness of the present legal regime for shipbreaking.

ii. *The 'Clemenceau'*

Like 'Riky', the dismantling of the 'Clemenceau' was not without controversy and raised several legal issues. Initially, the contract to dismantle the 'Clemenceau', the French aircraft-carrier, was won by a Spanish company. Following frustration over attempts to take the ship to Spain and later to Turkey, this contract fell through.[195] The contract was re-awarded to the next highest bidder, Ship Decommissioning Industries Company (SDIC) and *Eckhardt Marine Gmbh*.[196] Under the initial proposal, the decontamination was to be carried out in Greece with the final demolition in India.[197] As Greece declined entry, a new contract had to be re-negotiated under which there was to be partial asbestos removal in Toulon, France, before its journey to India.[198] In the meantime, there was a futile legal battle in the French courts by anti-asbestos groups alleging that the export would violate both the Basel Convention and the European Union legislation on this subject. Finally, on 31 December 2005, the carrier left Toulon for Alang under tow.[199]

As early as January 2005, the SCMC began to receive complaints on the possible arrival of the 'Clemenceau' at Alang. It issued directions to the Gujarat SPCB to ensure the decontamination of the ship in France and the disposal of the

---

[192] V. Venkatesan, "Breaking rules" *Frontline [of India]* 22:25 (3 December 2005), online: Frontline <http://www.flonnet.com/fl2225/stories/20051216002304500.htm>.

[193] See *Research Foundation* (14 October 2003), *supra* note 129.

[194] Venkatesan, "Pressures of monitoring", *supra* note 183.

[195] India, Supreme Court Monitoring Committee on Hazardous Wastes, *Final Report on Clemenceau: Minority Report* (Chairman: Dr. G. Thyagarajan, New Delhi: Ministry of Environment & Forests, 2006) [unpublished] ¶ 15 [on file with the author] [*SCMC Minority Report*].

[196] *Ibid.*

[197] *Ibid.*

[198] See Vaiju Naravane, "A shipload of trouble from France" *The Hindu [of India]* (3 January 2006), online: The Hindu <http://www.thehindu.com/>.

[199] Laurie Kazan-Allen, *Le Clemenceau: Action and Reaction*, online: International Ban Asbestos Secretariat <http://www.btinternet.com/~ibas/lka_le_clemenceau.htm>.

asbestos generated in an environmentally-sound manner.[200] A sub-committee of the SCMC also visited Alang and forwarded a series of recommendations in relation to 'Clemenceau', which were accepted by the SCMC.[201] Once the warship began its journey from France, events began to move rapidly in India. As there were contradictory media reports on the exact quantity of asbestos and other hazardous materials on board, the SCMC called upon the French company, Technopure, which carried out the decontamination of the ship, to provide further details. It became obvious that only 70 tonnes of asbestos had been removed, leaving behind nearly 500 tonnes.[202] This was in sharp contrast to the French government's assessment of 45 tonnes as the balance.[203] Consequently, the SCMC placed the matter before the Supreme Court with the recommendation that 'Clemenceau' not be allowed into Indian waters until its final report was issued.[204]

The final report was divided. The majority, endorsed by seven out of the ten members, favoured the arrival of the 'Clemenceau', subject to certain conditions.[205] The minority, supported by the remaining three members including the Chairman, vehemently opposed its arrival.[206]

The divergent opinions and the fact that 'Clemenceau' was a warship prompted the Supreme Court, once again, to order the formation of yet another committee,

---

200 *SCMC Minority Report, supra* note 195.

201 It recommended that the company provide additional documentation relating to decontamination in France, in particular an independent third-party audit verifying certification from the French authorities that the ship had been decontaminated and that it did not violate the Basel Convention. In addition, the Union Ministry of Environment and Forests was to procure from the French Embassy documents relating to the ship including an official statement from the French government that hazardous materials, including asbestos, had been removed up to 98 percent and that the balance would be recovered at Alang under guidance and co-ordination from the SDIC. See generally India, Supreme Court Monitoring Committee on Hazardous Wastes, *First Report of the Monitoring Committee on the Entry of the French Warship "Le Celmenceau" for Breaking at Alang in India with Recommendations* (Chairman: Dr. G. Thyagarajan, New Delhi: Ministry of Environment & Forests, 2006) [unpublished] [on file with the author] [*SCMC First Report*].

202 It seems that the French government disputed the extent of work done by Technopure claiming that the company had removed 106 tonnes of asbestos. Technopure did not complete the assignment and the work was assigned to another company, Prestosid, which claimed to have removed 8.5 tonnes of asbestos. V. Venkatesan, "Deadly Vessel" *Frontline [of India]* 23:2 (28 January 2006), online: Frontline <http://www.flonnet.com/fl2302/stories/20060210008300400.htm> [Venkatesan, "Deadly Vessel"]. See also *SCMC First Report, ibid.*

203 *Comments on Statement of French Government on Clemenceau by the Basel Action Network on Behalf of the Greater Coalition Demanding Return of Clemenceau to France for Decontamination* (24 January 2006) 6, online: Basel Action Network <http://www.ban.org/Library/statement_of_french_government_on_clemenceau.pdf> [*Comments on Statement of French Government*].

204 See also *SCMC First Report, supra* note 201. *Research Foundation for Science, Technology and Natural Resources Policy v. Union of India* (16 January 2006), Writ Petition (Civil) 657 of 1995 (India S.C.), online: The Judgments Information System <http://www.courtnic.nic.in/courtnicsc.asp>. See also Zubair Ahmed, "Stay out, India tells toxic ship" *BBC News* (6 January 2006), online: BBC News <http://news.bbc.co.uk/2/hi/south_asia/4588922.stm>.

205 See generally *SCMC Majority Report, supra* note 149.

206 See generally *SCMC Minority Report, supra* note 195.

this time comprised of experts from "the field of construction/demolition of warships".[207] By the same order, the court also permitted the importer of the hull to file the bill of entry with the customs authorities. This set the ball in motion for the arrival of the 'Clemenceau'.[208] For the first time since the court began to dictate the shipbreaking policy, it threatened to unleash its contempt powers, ironically to punish those holding public demonstrations and writing articles on issues surrounding the dispute.[209] The 'Clemenceau' drama abruptly ended on 15 February 2006 when the Council of State in France suspended the export of the ship forcing the French President to order its return to France.[210]

Like the 'Riky', the 'Clemenceau' also raises legal questions involving the Basel Convention, the European Union waste management framework (WSR 259/1993), and the Supreme Court Guidelines on Ship Recycling.[211] In the written statement filed on behalf of the Government of France before the SCMC, three major arguments were put forward to buttress the claim that the breaking of this warship did not violate national and international law.[212] These are analysed below.

One of the most crucial questions raised by the 'Clemenceau' was whether this state-owned warship could be classified as hazardous waste under the Basel Convention. France contended that 'Clemenceau' was weaponry and hence not waste. However, the French stance contradicts article 1 of the Basel Convention, which offers no exemption to warships.[213] A related aspect is that of waste determination under the Basel Convention. Waste subject to transboundary movement can become hazardous waste depending on the determination to be made by any one of the parties namely, the state of export, import or transit, in accordance with their respective national legislations. France and India, both parties to the Basel Convention, claimed that 'Clemenceau' was not hazardous waste. As the ship had to cross the Suez Canal to reach India, Egypt (also a party to the Basel Convention) became the state of transit. Egypt also determined 'Clemenceau' was non-hazardous,[214] permitting it to cross the Suez Canal. Thus, the Basel

---

[207] *Research Foundation for Science, Technology and Natural Resources Policy v. Union of India* (13 February 2006), Writ Petition (Civil) 657 of 1995 (India S.C.), online: The Judgments Information System <http://www.courtnic.nic.in/courtnicsc.asp>.

[208] *Ibid.*

[209] *Ibid.* See also "India media ban over 'toxic' ship" *BBC News* (13 February 2006), online: BBC News <http://news.bbc.co.uk/2/hi/south_asia/4708530.stm>.

[210] See Gopal Krishna, "French apex court rules, Clemenceau recalled" *India Together* (16 July 2008), online: India Together <http://www.indiatogether.org/2006/feb/env-shipretn.htm>.

[211] "Basel Non-Compliance Notification Report" *Basel Action Network* (31 January 2006), online: Basel Action Network <http://www.ban.org/notifications/BNN2006_1.pdf>.

[212] See generally *Comments on Statement of French Government, supra* note 203.

[213] *Basel Convention on the Control of Transboundary Movements of Hazardous Wastes and their Disposal*, 22 March 1989, 1673 U.N.T.S 126, 28 I.L.M. 657 (entered into force 5 May 1992) [*Basel Convention*].

[214] Heba Saleh, "Egypt grants 'toxic' ship passage" *BBC News* (15 January 2006), online: BBC News <http://news.bbc.co.uk/2/hi/south_asia/4615662.stm>.

Convention was inapplicable. The 'Clemenceau' incident demonstrates the inconsistencies in the international legal regime on shipbreaking.

A further question raised by the 'Clemenceau' was whether the presence of in-built hazardous substances in the ship structure could render the ship as waste.[215] In this case, 'Clemenceau' was not carrying asbestos as cargo; rather, asbestos was an integral part of its structure, a common feature of all ships of that vintage. Because of this, France held the view that the ship could not be classified as waste.[216]

Although not an authoritative precedent, the opinion of the Council of State in *Upperton Ltd., with offices in Mauritius v. The Minister of Housing, Spatial Planning and the Environment*,[217] involving an interpretation of the WSR 259/1993, does not support the French argument. In this case, the respondent's action of preventing the appellant from letting the ship 'Sandrein' sail from the Netherlands to India by applying "executive coercion" was raised in appeal before the Council of State.[218] One of the major issues related to whether 'Sandrein' constituted waste or not. According to the appellant, the ship was to be treated as a green list substance under WSR 259/1993 as it contained no hazardous cargo residues.[219] Moreover, the phrase "other materials arising from the operation of the vessel" could not be construed to include asbestos that was used in the ship's construction.

Article 1(3) of the WSR 259/1993, *inter alia*, states that the shipment of waste described in its annex II, and intended exclusively for recovery, are excluded from its operational ambit. annex II contains the "green list" of wastes and its heading reads as follows:

> regardless of whether or not wastes are included on this list, they may not be removed as green waste substance if they are contaminated by other materials to an extent which: a) increases the risks associated with the waste sufficiently to render it appropriate for

[215] India has consistently held the stance that merely because a negligible quantity of hazardous materials is present on board a ship, the scrap ship as a whole cannot be considered as a hazardous waste. Recycling of Ships: Comments on Document MEPC 48/3, Submitted by India, IMO/MEPC 48/3/2, 9 August 2002, ¶ 2.1 (KR-CON). But see Recycling of Ships: Comments on the Report of the Correspondence Group, Submitted by Greenpeace International, IMO/MEPC 47/3/5, 11 January 2002, ¶ 7 (KR-CON) (even in a single ship these hazardous wastes may be considerable in quantity posing significant threats to health and the environment, even though when compared to the total mass of a single ship the amount may appear to be 'minor'. Moreover, accumulation of such wastes from many ships multiplies the problem).

[216] *Comments on Statement of French Government*, *supra* note 203 at 11.

[217] (19 June 2002), LJN-No.: AE4310 Case No.: 200105168/2 (Council of State, The Hague), reproduced in Recycling of Ships: Decision Concerning the Export of the "Sandrien" for Ship Recycling, Submitted by Greenpeace International, IMO/MEPC 48/INF.11, 10 July 2002, annex (KR-CON).

[218] *Ibid.*, ¶ 1.

[219] *Ibid.*, ¶ 2.2.

inclusion in the amber or red lists; or b) prevents the recovery of the waste in an environmentally sound manner.[220]

In the present case, there was evidence to prove that the ship's hull contained considerable quantities of asbestos, to the extent that the risks attached to the waste qualified the ship to be included on the red list. Accordingly, the Council of State ordered that the ship could not be transported.[221]

Another argument advanced by France was that 'Clemenceau' was not intended to be disposed; rather, it was to be sold as steel after removal of the residual asbestos.[222] Again, the decision of the Administrative Jurisdiction Division of the Council of State in *Stichting Greenpeace Nederland, a foundation that has its seat in Amsterdam v. the State Secretary for Housing, Spatial Planning and the Environment*,[223] though not authoritative, is relevant.

The primary issue in this case was whether the scrapping of 'Otapan', a decommissioned oil tanker, was a disposal operation or recycling. The 'Otapan', which arrived in Amsterdam port, originally belonged to Navimin, a corporate body under Mexican law. Sometime in 2005 Navimin was declared bankrupt, and Basilisk acquired ownership over the 'Otapan' following a ruling by the Mexican courts. Thereafter, Basilisk gave notice of the proposed removal of the ship to Turkey for scrapping. This was objected to by the defendant on the grounds that there was no financial guarantee for shipment. Upon Basilisk providing the requisite bank guarantee, the defendant authorised the shipping of the 'Otapan'.[224] The claimants objected to this, arguing that the scrapping of the ship was wrongly classified in the notification as a recovery operation as the ship contained asbestos and other hazardous substances that had to be removed before it could be broken down.

In the meantime, the 'Otapan', which had already left for Turkey, was disallowed entry as it was found to have more asbestos than what was indicated in the notification.[225] Therefore, the defendant decided to bring the ship back to the Netherlands and, once the quantity of asbestos was brought in line with the notification, the 'Otapan' was to be shipped back. Accordingly, the defendant argued that since the 'Otapan' had returned to the Netherlands, no interest would be served by assessing the merits of the case, since a fresh notification was now required for the shipment. However, the court proceeded to decide the matter and determine whether scrapping the ship was a disposal operation or recycling.

---

220 *Ibid.*, ¶ 2.2.4.

221 *Ibid.*, ¶ 2.2.5.

222 *Comments on Statement of French Government*, *supra* note 203 at 12.

223 (21 February 2007), Case No. 200606331/1 (Council of State, Administrative Jurisdiction Division), online: Basel Convention: Dismantling of Ships – Relevant Caselaw <http://www.basel.int /ships/relevcaselaw.html>.

224 *Ibid.*, ¶ 2.1.

225 *Ibid.*, ¶ 2.3.1.

In its finding, the Council of State referred to a declaratory judgment by the court of Justice of the European Communities. This judgement held that where a waste treatment process comprises of several distinct stages, to classify it either as disposal or recovery, "only the first operation that the waste is to undergo subsequent to shipment"[226] is to be considered. The Council of State accordingly held:

> … [T]he treatment of the waste specified in the notification is a process consisting of several stages. The scrapping plan states that the first part of the operation is to remove the asbestos in the ship. It follows from the plan that the removal of the asbestos is necessary in order to enable the ship to be scrapped. The plan also emphasises that no other work on the ship, for example the removal of liquids and gases, should start until the asbestos has been removed. In view of the above, the Division considers that the waste treatment process is wrongly classified on the notification form as a recovery operation … Nor is this altered by the fact that substances obtained from the dismantling of the ship will subsequently be transferred … to the metalworking industry for recycling, which could possibly be classified as a recovery operation. In these circumstances, it must be concluded that the defendant wrongly failed to object to the proposed shipment on the ground of an incorrect classification on the notification form.[227]

In this case, as the object of the shipment of 'Otapan' to Turkey was for disposal which was prohibited by article 4(1) of the WSR 259/1993, the notification was quashed.

Finally, analysing the 'Clemenceau' episode from the perspective of the Supreme Court Guidelines, it is clear that throughout the controversy, France failed to provide a complete inventory of the hazardous wastes on board the vessel. France could not even provide a true picture of the extent of decontamination that it carried out. Thus, it is clear that France did not satisfy the requirements as mandated by the Supreme Court Guidelines.[228]

iii. *The 'Blue Lady'*

From its launching in 1962 as the 'SS France', to its final journey for scrapping at Alang as the 'Blue Lady', the story of this classic ocean liner, one of the 50 most toxic ships in the world, has been eventful.[229] A transatlantic passenger ship, the

---

[226] *Ibid.*, ¶ 2.5.1.

[227] *Ibid.*, ¶ 2.5.2.

[228] It is interesting to note that the decontamination related only to asbestos while other deadly substances such as PCBs were omitted in the contract granted for removal. Interestingly, while France claimed that 45 tonnes of asbestos containing products remained, constituting less than 0.2 percent of the total weight of the ship, which is nearly 22,000 tonnes, the SCMC was referring to removal of 98 percent of the total asbestos. If only 45 tonnes of asbestos were decontaminated, this would constitute about 28 percent of the total quantity of asbestos on the ship. Venkatesan, "Deadly Vessel", *supra* note 202.

[229] The longest passenger ship of her time, the 'Blue Lady' is a steam turbine driven ship with a power and rating of 30,000 KW and 40,760 HP respectively. The ship is about 291 metres long and 34 metres wide and has a capacity of about 45,886 MT. It also has 16 floors, 1,400 rooms for

'SS France' was decommissioned following the oil crisis in the 1970s. Later, it was sold to the Norwegian Cruise Line Company, which refurbished it, and in its new *avatar* as 'SS Norway', the ship revolutionised the cruise industry. Its odyssey came to an end in May 2003, when a boiler explosion claiming several lives severely crippled the ship. 'SS Norway' was towed to Germany from where it was confined under the Basel Convention because of the heavy presence of asbestos.

Like the 'Riky', the German authorities were assured that the ship was to go to Malaysia for repairs for later use as a floating hotel.[230] Finally, 'SS Norway' was allowed to leave under tow, reaching Malaysia in 2005. Thereafter, efforts were made to dismantle it in Bangladesh. However, Bangladesh refused entry due to the huge amounts of hazardous wastes in the ship.[231]

In April 2006, the Indian media carried news that a ship by the name, 'SS Blue Lady', registered as a barge under the Bahamas flag, was sold to a shipbreaker in Alang.[232] Subsequently, an application was filed before the Supreme Court to ensure that in case the 'Blue Lady' arrived at Alang, all relevant laws in relation to shipbreaking would be complied with.[233] On 5 June 2006, the apex court allowed the ship to drop anchor in Indian territorial waters. Later, it was also permitted to be beached.[234]

As mentioned earlier, the Supreme Court, by its order dated 6 September 2007, accepted the recommendations of the TEC in relation to shipbreaking. By doing so, it seems that the court was in fact setting the stage for its subsequent and more important order delivered just five days thereafter, *i.e.,* on 11 September 2007, which granted permission to dismantle the 'Blue Lady' at Alang.[235] It should be noted that earlier on the TEC was also directed by the Supreme Court to look into the issue of the anchoring, beaching and dismantling of the 'Blue Lady'.[236]

---

accommodation, in addition to restaurants, cinemas, health clubs, and shopping complexes. *Blue Lady case, supra* note 150 ¶ 2.

230 See generally Bavadam, *supra* note 33.

231 Kalpana Sharma, "Breaking ships need not break lives" *The Hindu [of India]* (21 August 2006), online: The Hindu <http://www.thehindu.com/>.

232 See Mahesh Langa, "Blue Lady: End of an era" *Express India* (3 August 2006), online: expressindia <http://cities.expressindia.com/fullstory.php?newsid=195268>. See also *Blue Lady case, supra* note 150 ¶ 2.

233 Gopal Krishna, "Will the Blue Lady do a Le Clemenceau?" *India Together* (14 July 2008), online: India Together <http://www.indiatogether.org/2007/feb/env-bluelady.htm>.

234 *Research Foundation for Science, Technology and Natural Resources Policy v. Union of India* (5 June 2006), Writ Petition (Civil) 657 of 1995 (India S.C.), online: The Judgments Information System <http://www.courtnic.nic.in/courtnicsc.asp> [*Research Foundation* (5 June 2006)]. Gopal Krishna, "Blue Lady seduces SC on Environment Day" *Basel Action Network* (6 June 2006), online: Basel Action Network <http://www.ban.org/Library/Editorials/060606_environment_day.html>.

235 See generally *Blue Lady case, supra* note 150.

236 *Research Foundation for Science, Technology and Natural Resources Policy v. Union of India* (12 March 2007), Writ Petition (Civil) 657 of 1995 (India S.C.), online: The Judgments Information System <http://www.courtnic.nic.in/courtnicsc.asp>.

In another report dated 10 May 2007, the TEC recommended the granting of permission for the dismantling.[237] This was accepted by the Supreme Court through its order dated 11 September 2007, since India had by now, arguably, a legal regime on shipbreaking in place, even though it was only six days old.

The permission to dismantle the ship was a boost for the withering shipbreaking industry in India, particularly since Bangladesh, the market leader with the dubious reputation for accepting all kinds of toxic ships, had refused permission to dismantle this ship. However, the appropriateness of the *Blue Lady dictum* has been the subject of intense debate among environmentalists and human right activists.[238] Analysis of this judgement reveals several inconsistencies in the judicial reasoning, its adverse impact on the sustainable development of this industry, and the broader implications it has for constitutional and human rights jurisprudence in India.

The *Blue Lady* judgement highlights that the breaking of the 'Blue Lady' would yield about 41,000 tonnes of steel, which could significantly reduce the pressure for more mining and, in the process, generate employment for nearly 700 people.[239] It seems that these are the major factors that weighed on the mind of the *Blue Lady* court when it permitted the dismantling of the ship. Even though a commendable approach to resource conservation and an endemic unemployment problem, the court ignored the fact that these 700 shipbreakers were being put to great risk. Breaking of ships is one of the most dangerous operations, particularly in countries such as India where there is little infrastructure for safe dismantling. Moreover, as the work force is unorganised, their vulnerability to abuse is also high.

The TEC Report relied on by the court to establish systems for safe ship recycling, classifies its recommendations (including those relating to worker safety) into four categories: immediate, short-, mid-, and long-term.[240] The time duration

---

[237] See generally India, *Report of the Technical Experts Committee on Dismantling Plan of the Ship 'Blue Lady', Beached at Alang (Gujarat)*, (Chairman: Dr. Prodipto Ghosh, New Delhi: Ministry of Environment & Forests, 10 May 2007) [on file with the author].

[238] See NGO Platform on Shipbreaking, Press Release, "Human Rights and Environmental Groups Condemn India for Callous Disregard for its Poorest Labourers and the Law after Shipbreaking Ruling" (12 September 2007), online: Basel Action Network <http://www.ban.org/ban_news/2007/070912_shipbreaking_ruling.html>. Quoting Ingvild Jenssen:

> [t]he Blue Lady ruling … makes a mockery of the Indian judicial system and shows it has no respect for their own rulings, and international law, but likewise has officially condemned the ship breaking workers to death by accident or from occupational disease such as asbestosis and cancer. This ruling sends an unmistakable signal that India does not care about the welfare of its poorest most desperate workers. *Ibid.*

[239] *Blue Lady case, supra* note 150 ¶ 11.

[240] *Research Foundation for Science, supra* note 153 ¶ 6. *TEC Report, supra* note 3 at 51 (the time required for implementing immediate recommendations is three months; for short-term, it is six months; medium, it is two years; and, long-term it is five years).

needed to operationalise these measures ranges between three months for the immediate to five years for the long-term.[241] For example, the appointment of a safety officer, which falls under the category of short-term recommendation and one vital to ensure worker safety, would require at least six months to put into effect. By the time a safety officer was appointed for Alang, the 'Blue Lady' would already have reached the re-rolling mills.[242] Given the generally poor state of law enforcement in India, one can safely conclude that these 700 workers would not in any way benefit from the safeguards contemplated by the TEC Report. The judgement is also silent on monitoring, as it does not call for any periodic compliance or progress reports from the concerned authorities to assess the extent of enforcement of laws.

The "foolproof"[243] TEC Report outlines several principles in relation to asbestos and PCB removal and hazardous waste management; however, the changes envisaged therein are only cosmetic. Shipbreaking at Alang is performed in the inter-tidal zone, creating substantial marine pollution. Neither the report nor the judgement suggests the need to upgrade the infrastructure at Alang by investing in alternative methods to beaching. Rather, the judgement misleads one to believe that the suggestions in the TEC Report are the solution to all the environmental ills at Alang.[244]

The judgement does little to resolve questions surrounding the mysterious disappearance of radioactive substances on board the 'Blue Lady'. It mentions that on the basis of the apprehension expressed by the petitioner regarding the presence of radioactive materials on board the 'Blue Lady', the AERB and the GMB inspected the ship and confirmed their presence.[245] Subsequently, these authorities certified that there were no more radioactive materials on board.[246] There was evidence to show that the ship carried nearly 5,500 smoke detectors containing radioactive material.[247] This finding merits an inquiry as to where the radioactive substances disappeared. Were the radioactive wastes removed before the court granted its permission to dismantle the ship? Who removed them? Were the rules

---

[241] *Ibid.*

[242] *Ibid.* at 52.

[243] *Blue Lady case, supra* note 150 ¶ 14 (describing the Report as "foolproof").

[244] See also U.K., Department for Environment, Food and Rural Affairs, *Overview of Ship Recycling in the UK: Guidance* (London: Crown Copyright, Department for Environment, Food and Rural Affairs, 2007) at 14, online: Defra <http://www.defra.gov.uk/environment/waste/strategy/pdf/shiprecycle-strategy-overview.pdf> (noting that in the United Kingdom, beaching of ships would not be afforded the necessary permission).

[245] *Blue Lady case, supra* note 150 ¶ 6.

[246] *Ibid.*

[247] "Blue Lady's fate uncertain as activists to challenge SC" *The Economic Times [of India]* (18 September 2007) (Prowler); "Radioactive material aboard Blue Lady" *The Times of India* (30 July 2007) (Prowler) (reporting that Tom Haugen, the erstwhile project manager of the ship, had written to the Committee of Technical Experts set up by the Supreme Court claiming that the ship has radioactive material – Americium 241 – on board in at least 5,500 fire detection points).

relating to the handling and disposal of such substances followed? Was the AERB involved in the removal of these substances? By settling for incomplete information on such a critical issue, the court seems to have compromised human and environmental safety.

One of the most glaring inconsistencies between the Supreme Court judgement in the *Blue Lady* of 2007 and its earlier order dated 5 June 2006 relates to granting permission for beaching the 'Blue Lady'. By relying on the TEC Report, the *Blue Lady* judgement affirms that beaching is an "irreversible process".[248] If that is true, why did the court in its earlier order dated 5 June 2006 allow the beaching of the 'Blue Lady' at Alang?

In its 2006 order, the court observed as follows:

> ... [T]he Committee appointed by this Court ... has made three recommendations in respect of anchorage, beaching and dismantling. The conditions stipulated ... appear to be prima facie reasonable. Let action in terms of said recommendations be taken. This shall, however, not confer any *equity* on the owners of the ship which is sought to be put on anchorage, beaching and dismantling.[249]

Thus, it is clear that the fact of beaching was not to confer any equity on the owners of the ship. However, did the *Blue Lady* court in disregard of this earlier order, confer equity or was the court misled to do so?

The TEC was apprised by the recycler that the ship could not be safely anchored on the high seas and that holding the ship with the help of two tugs would cost approximately USD30,000 per day.[250] Moreover, in light of the fast approaching monsoons and limited provisions on board, the TEC adopted a lenient stance and recommended its anchoring.[251] Since anchoring the ship for long in monsoons was not advisable, the TEC also recommended beaching subject to the concerned authorities certifying that "the ship was not having hazardous materials of a kind or quantity that could not be safely removed, handled and disposed of at Alang".[252]

Even though the TEC was aware that beaching was an irreversible process,[253] it is doubtful whether the court was informed of this important aspect when the TEC placed its recommendation on the issue of beaching of the 'Blue Lady', which led to the order of 5 June 2006.[254] This is critical since once the ship was

---

248 *Blue Lady case, supra* note 150 ¶ 6, 14.

249 *Research Foundation* (5 June 2006), *supra* note 234 [emphasis added].

250 *Proceedings of Third Meeting of the Committee of Technical Experts on Management of Hazardous Waste Relating to Ship Breaking* (20 May 2006) at annex A-2, page 1, in the Affidavit dated 31 August 2006 filed by Dr. V. Rajagopalan, Member Convenor, Committee of Technical Experts on Ship Breaking in *Research Foundation* [on file with the author].

251 *Ibid.* at 3.

252 *Ibid.*

253 *Ibid.*

254 *Research Foundation* (5 June 2006), *supra* note 234.

permitted by the court to be beached, with no possibility to refloat the ship due to its poor condition, the fate of the 'Blue Lady' had already been determined; a case of *fiat accompli* for the shipowners.

Perhaps the greatest drawback of the *Blue Lady dictum* is that it reverses the sustainable development jurisprudence that the Supreme Court and the High Courts' assiduously built over the years through introduction of new principles.[255] In determining sustainable development *vis-à-vis* ship dismantling, the *Blue Lady* court introduced the principle of proportionality based on the concept of balance. Justice S.H. Kapadia, writing for the court, explains thus:

> In an emergent economy, the principle of proportionality based on the concept of balance is important. It provides level playing field to different stakeholders ... When we apply the principle of sustainable development, we need to keep in mind the concept of development on one hand and the concepts like generation of revenue, employment and public interest on the other hand. This is where the principle of proportionality comes in.[256]

In articulating this concept, the court relied on the keynote address on "Global Constitutionalism"[257] delivered by Lord Goldsmith, former Attorney General of the United Kingdom, the exposition of Justice Arijit Pasayat in the *K.M. Chinnappa Applicant in T.N. Godavarman Thirumalpad v. Union of India*[258] (*Kudremukh Iron Ore Company's case*), and on Amartya Sen's treatise, 'Development as Freedom'. While the *Blue Lady* court makes only a casual mention to Sen's work, specific references to the other sources are provided.[259]

Lord Goldsmith developed his thesis in the context of the "plenty of bad times"[260] in the aftermath of 9/11. In responding to the challenges post-9/11, he noted that though "grave and real", these "do not justify a response that would throw away or undermine the [basic] fundamental freedoms and values which terrorists seek to destroy".[261] He put forward three principles to guide the formulation

---

[255] *Essar Oil Ltd. v. Halar Utkarsh Samiti* (2004), [2004] 2 S.C.C. 392 (India S.C.) (there cannot be *a priori* presumption that laying pipelines through ecologically sensitive areas will lead to destruction of wild life); *Jagannath, supra* note 117 (rejecting the dollar based argument forwarded to justify shrimp farming, it was held that damage to ecology was higher than the earnings from the sale of coastal aquaculture products); *M.C. Mehta (Taj Trapezium Matter) v. Union of India* (1996), [1997] 2 S.C.C. 353 (India S.C.) (directing re-location of 292 industries situated nearby the Taj Mahal); *M.C. Mehta v. Union of India* (2002), [2002] 4 S.C.C. 356 (India S.C.) (defeating the attempts of the executive to frustrate earlier court orders directing introduction of compressed natural gas as the alternate fuel for buses plying in New Delhi city).

[256] *Blue Lady case, supra* note 150 ¶ 11.

[257] Rt. Hon. Lord Goldsmith QC, "Keynote Address" (Remarks delivered at the Stanford Law Review Symposium in Global Constitutionalism at the Stanford Constitutional Law Center, 16 February 2007), (2007) 59 Stan. L. Rev. 1155 (WLeC).

[258] *K.M. Chinnappa Applicant in T.N. Godavarman Thirumalpad v. Union of India*, [2002] S.C. 1386 (indlaw.com), [2002] 10 S.C.C. 606 [*Kudremukh Iron Ore Company's case*].

[259] See *Blue Lady case, supra* note 150.

[260] Goldsmith QC, *supra* note 257.

[261] *Ibid.* at 1171.

of a response, namely, respect for the rule of law, commitment to fundamental values and freedoms, and the principle of proportionality.[262] He observed that even though under the *European Convention on Human Rights, 1950* the concept of balance is one of its key themes, some rights are absolute and there can be no compromise on them[263] (e.g., the prohibition on torture and the right to a fair trial). However, certain other rights may be subject to adjustment or derogation in the interests of others. In such cases, however, the action has to be proportionate "going no further than is necessary to meet the threat".[264]

In elucidating the concept of balance under proportionality, the *Blue Lady* court also referred to one of its earlier decisions in *Kudremukh Iron Ore Company's case*.[265] Justice Arijit Pasayat (who was also on the *Blue Lady* court along with Justice S.H. Kapadia) therein observed:

> It cannot be disputed that no development is possible without some adverse effect on the ecology and environment, and the projects of public utility cannot be abandoned and it is necessary to adjust the interest of the people as well as the necessity to maintain the environment. A balance has to be struck between the two interests. Where the commercial venture or enterprise would bring in results which are far more useful for the people, difficulty of a small number of people has to be bypassed. The comparative hardships have to be balanced and the convenience and benefit to a larger section of the people has to get primacy over comparatively lesser hardship.[266]

The primary dispute in this case related to the renewal of a mining lease granted in and around a national park situated in the Western Ghats, one of 18 recognised hotspots for biodiversity conservation in the world.[267] A Central Empowered Committee (CEC), which was appointed by the court to consider the exceedingly rich biodiversity of the area, recommended that the company wind up its operations within five years, or on the exhaustion of the oxidised weathered secondary ore, whichever was earlier.[268] The company objected to this, claiming an existing legal right to renewal. It contended that any order for discontinuance would affect its foreign exchange earning capacity, in addition to rendering a large number of employees jobless.[269] Rejecting these arguments and in light of relevant provisions under the *Forest Conservation Act 1980*, the court concluded that there was no basis to vary the opinion of the CEC.[270]

---

[262] *Ibid.* at 1171, 1172.
[263] *Ibid.* at 1172.
[264] *Ibid.*
[265] *Kudremukh Iron Ore Company's case, supra* note 258.
[266] *Ibid.*, ¶ 36.
[267] *Ibid.*, ¶ 1.
[268] *Ibid.*, ¶ 4.
[269] *Ibid.*, ¶ 7.
[270] *Ibid.*, ¶ 49.

In deciding the "seminal issue" in *Kudremukh Iron Ore Company's case* as to whether the approach should be "dollar friendly" or "eco-friendly",[271] Justice Pasayat elaborately discussed the need to protect the environment.[272] He expounded on "the concept of balance under the principle of proportionality applicable in sustainable development" (though the judgment does not mention the development of this new doctrine) in paragraph 36. Thereafter, he emphasises the importance of environmental impact assessment reports.[273] A combined reading of both paragraphs reveals that what the court intended, even in cases where the difficulty of a small number of people had to be bypassed, was not a *carte blanche* where their rights could be violated, rather there had to be proper environmental impact assessments. The *Blue Lady* court does not appear to have considered this vital aspect while developing the principle of balance under proportionality.

Lord Goldsmith's thesis and Justice Pasayat's exposition disclose entirely different settings and the manner in which the "balance" under proportionality was envisaged and developed. By merely lifting the concept, not giving due attention to its context, the *Blue Lady* court seems to have misapplied it. Proportionality envisages that any administrative authority, while exercising a discretionary power, ought to maintain a sense of proportion between the goals and the means employed for achieving them, so that an action impinges on individual rights only to the minimum extent possible.[274] This is the balance that is sought.

For development to be sustainable, conflicting interests must be balanced.[275] It is not the case that one group is to be made worse off, particularly when they have a limited voice in the decision-making process.[276] By disproportionately

---

[271] *Ibid.*, ¶ 11.

[272] *Ibid.*, ¶ 42.

[273] *Ibid.*, ¶ 37.

[274] See generally *Union of India v. G. Ganayutham* (1997), [1997] 7 S.C.C. 463 (India S.C.) (expounding the doctrine of proportionality).

[275] For instance see *Narmada Bachao Andolan v. Union of India* (2000), [2000] 10 S.C.C. 664 (India S.C.). The people in Gujarat State had only one solution to their water woes and that was the construction of the Sardar Sarovar dam which, unfortunately, would have adverse consequences on another set of people (as their houses and agricultural lands would be submerged) who belonged to disadvantaged groups, mostly from the neighbouring State of Madhya Pradesh. To mitigate the hardships, the oustees were provided with alternative lands, sites and compensation with the objective of ensuring that rehabilitation in new locations would render them better off than they were. *Ibid.*

[276] *Rural Litigation and Entitlement Kendra Dehradun v. State of Uttar Pradesh* (1985), [1985] (2) S.C.C. 431 (India S.C.). A letter to the Supreme Court dated 2 July 1983 by Rural Litigation and Entitlement Kendra, Dehradun, together with affidavits detailing unauthorised and illegal mining operations carried on in the Mussoorie Hills, which was adversely affecting the local ecology, was treated as a writ petition. In striking a balance, the court observed thus:

> … the lessees of lime-stone quarries which have been directed to be closed down permanently … would be thrown out of [a] business in which they have invested large sums of money and

applying the concept of balance under the principle of proportionality, the *Blue Lady* court seems to have reduced sustainable development to an oxymoron where developmental concerns can override legitimate environmental and human rights considerations. This approach runs counter to the accepted jurisprudence on sustainable development as developed by Indian courts in earlier cases as well as the norms of international environmental law.

Finally, the *Blue Lady* ruling reflects the far-reaching implications a reshaped economic order has had for the development of constitutional and human rights jurisprudence in a globalised India. The fall of socialism in Eastern Europe, the dominance of neo-individualism and the free market economy, the rise of the World Trade Organization (WTO), and the wave of economic liberalisation gripping Asian societies have led to re-assertion of Nozick's libertarian theory of justice.[277] This theory extols the virtues of a free market economy in which there is no place for the deprived, the unskilled, and no scope for state-sponsored welfare in favour of the marginalised sections of the society.[278] Instead, the main function of economic policy is to ensure that the right of a minority to accumulate profits at the highest rate possible is safeguarded (euphemistically referred to as 'growth'). Development can ensue only if this growth is sustained.[279] Once this is ensured, other members of society stand to benefit from the associated spin-offs (the trickle-down effect).[280] Freedom can be achieved and enjoyed only if such growth occurs. Accordingly, state expenditures are to be directed towards creating an enabling climate for growth. Resources should not be 'wasted' on providing public services which, in any case, can be offered more efficiently by private players.[281] Every effort is to be made to ensure that the invisible hand of the market is nurtured and developed.[282]

> expanded considerable time and effort. This would undoubtedly cause hardship to them, but it is a price that has to be paid for protecting and safeguarding the right of the people to live in a healthy environment with minimal disturbance of ecological balance and without avoidable hazard to them and to their cattle, homes and agricultural land and undue affectation of air, water and environment. *Ibid.*, ¶ 12.

Nevertheless to mitigate the hardship, the court directed that whenever any other area in Uttar Pradesh was thrown open for grants for lime-stone or dolomite quarrying, the displaced lessees, if found fit and eligible, were to be afforded priority.

277 Robert Nozick, *Anarchy, State and Utopia*, 1st ed. (New York: Basic Books, 1974).

278 Mark S. Stein, "Nozick: A Utilitarian Reformulation", Essay, (1997) 18 N. Ill. U.L. Rev. 339 (WLeC). Nozick is willing to grant human beings only those rights conferred by the law of nature, namely, the rights to life, health, liberty, and possessions. John D. Hodson, "Nozick, Libertarianism and Rights" (1977) 19 Ari. L. Rev. 212 (HeinOnline).

279 Firoze Manji, Book Review of *Development as Freedom* by Amartya Sen, online: Fahamu <http://www.fahamu.org/downloads/DevelopmentasFreedom.pdf>.

280 *Ibid.*

281 *Ibid.*

282 World Bank, *World Development Report, 1997: The State in a Changing World* (New York: Oxford University Press, 1997) at 61.

In contrast, John Rawls and his theory of "justice as fairness" seek to tone down the excesses of the market economy.[283] Rawls' ideals are more in conformity with Indian constitutional values.[284] In particular, his "difference principle" allows inequalities in income and wealth provided that the conditions of those at the bottom are also improved.[285] A similar stance is adopted by Amartya Sen in his "Development as Freedom" which highlights the well-being of humans as both the goal and the means for development.[286] For Sen, development is a process of expanding freedoms.[287] Poverty is seen as deprivation of basic capabilities rather than merely as lowness of incomes, the standard criteria to identify poverty.[288] The ultimate goal of development is to enhance people's freedom and their capabilities. Five freedoms are instrumental to meeting this goal and are consequently vital to overall economic development: political freedoms, economic facilities, social opportunities, transparency guarantees, and protective security.[289] By moving away from the traditional theories of "development" that are identified merely in terms of economic growth, Sen rejects approaches to development where enhancement of people's capabilities and freedoms are placed on the "backburner while jumpstarting the economy".[290]

Nozick's philosophy, which stands in contrast to the socialist values as enshrined in the Indian Constitution, has tremendously impacted the re-shaping of state policies in India in this era of globalisation.[291] For instance, the Indian Constitution, through its preambular statement[292] and certain directive principles,[293] makes it clear that the object of the Indian polity is to secure social and

---

[283] See generally John Rawls, *A Theory of Justice*, 1st ed. (Oxford: Oxford University Press, 1999). See also John Rawls, *Justice as Fairness*: *A Restatement*, Ering Kelly (Massachusetts: Harvard University Press, 2001).

[284] Vinod Dixit, "Justice as Fairness – A Restatement" (2005) 1 J. Nat'l Judicial Academy 396 at 399.

[285] See generally Rex Martin, "Rawls's New Theory of Justice" (1994) 69 Chicago-Kent L. Rev. 737 (WLeC); David Elkins, "Responding to Rawls: Toward a Consistent and Supportable Theory of Distributive Justice" (2007) 21 BYU J. Pub. L. 267; Xiaobing Xu & George Wilson, "On Conflict of Human Rights" (2006) 5 Pierce L. Rev. 32 (WLeC).

[286] See generally Amartya Sen, *Development as Freedom* (Oxford: Oxford University Press, 1999) (arguing for a development model that enhances people's capabilities in contra-distinction to an approach that focuses solely on income enhancement). See also Martha Nussbaum, "Human Rights and Human Capabilities" (2007) 20 Harv. Hum. Rts. J. 21 (WLeC) (noting that the elaboration of the "Human Development Approach," or the "Capability Approach," has been "the most important theoretical development in human rights during the past two decades" and has been operationalised in the Human Development Reports of the United Nations Development Programme).

[287] Sen, *ibid.* at 3.

[288] *Ibid.* at 87.

[289] *Ibid.* at 38.

[290] Maggi Carfield, "Enhancing Poor People's Capabilities through the Rule of Law: Creating an Access to Justice Index", Notes, (2005) 83 Wash. U.L.Q. 339 at 352 (HeinOnline).

[291] For instance see generally *BALCO v. Union of India* (2001), [2002] 2 S.C.C. 333 (India S.C.).

[292] See *Indian Constitution, supra* note 90 Preamble.

[293] For instance see *ibid.*, arts. 38, 39 (commanding the state to secure a social order for the promotion of welfare of the people, in particular, of women, children, workers, and the disabled).

economic justice for its citizens. In this scheme, the judiciary, as the ultimate guarantor of the right to human dignity, has a unique role to ensure basic rights of individuals. Through its power of judicial review, activism and creativity, the judiciary assumes a decisive role in influencing constitutional governance. It even has a prominent say in the control, utilisation and management of natural resources and of the economy.[294] As a result, the judiciary has generated a unique dignity-centric jurisprudence grounded on the principle of distributive justice.[295]

The new economic policy introduced in the early 1990s was a turning point in the development strategy that has affected the nature of constitutional governance in India.[296] "License raj"[297] disappeared and the private sector took over the engines of development. With India joining the WTO, sovereignty of the Indian state in economic matters began to progressively diminish.[298]

Globalisation initiated a process of sweeping economic change in India. Wealth is being created at unthinkable levels, but its asymmetric effects on the rich and the poor have resulted in a gulf that is widening with each passing day.[299] As the government withdraws itself, more or less, from its responsibility to ensure distributive justice,[300] and the most recent jurisprudence from the Supreme Court favouring the market forces, the situation is becoming more complex. The *Blue Lady* is the latest in this succession of rulings favouring market forces.[301]

---

[294] *Forest case*, *supra* note 116. For one of the most important decisions in this regard see *Bangalore Water Supply and Sewerage Board v. A. Rajappa* (1978), [1978] 2 S.C.C. 213 (India S.C.) [*Rajappa*]. This case widened the definition of the term industry under the Industrial Disputes Act, 1947 bringing 'workmen' engaged in most establishments under the Act. *Rajappa*, which held the field for two decades, enabled the moulding of a labour friendly jurisprudence based on socialist moorings. However, in the post-globalised world, *Rajappa* is fast losing relevance. *Ibid.*

[295] Jeremy Cooper, "Poverty and Constitutional Justice: The Indian Experience" (1993) 44 Mercer L. Rev. 611 at 624 (WLeC).

[296] Montek S. Ahluwalia, "Economic Reforms in India since 1991: Has Gradualism Worked?" (2002) 16:3 J. Econ. Persp. 67.

[297] Following independence, the founding fathers of the Indian Republic decided to have a planned economy, where the state was to command the heights of the economy. Consequently, an elaborate system of licences were established which severely curtailed the growth of private enterprises between 1947 and 1990. In 1990, the new economic policy was launched, ushering in a new era where there was de-regulation and dismantling of the license systems, etc.

[298] Professor N.R. Madhava Menon, "Role of the State and Judiciary in Advancing Equity and Equality in a Market Economy" (Address at the National Seminar on Equity & Equality in a Market Economy during the Golden Jubilee Celebrations of the Indian Law Institute, New Delhi, 5 August 2006) [unpublished].

[299] B.S. Chimni, "A Just World under Law: A View from the South" (2007) 22 Am. U. Int'l L. Rev. 199 at 211 (WLeC). See also UNDP, *Human Development Report 2005* (New York: United Nations Development Programme, 2005) at 30–31, online: Human Development Reports <http://hdr.undp.org/en/media/hdr05_complete.pdf>.

[300] Menon, *supra* note 298.

[301] *Hindustan Lever Employees' Union v. Hindustan Lever Limited* (1994), [1995] A.I.R. S.C. 470 (India S.C.) (amalgamation and allotment of more than 51 percent of shares to a foreign company

In effectuating the balance by affording primacy to the convenience and benefit to one section of the people who are economically, politically and socially well-off, ignoring the hardships caused to a disadvantaged minority, the court has set a precedent which, if followed in the future in similar or like environmental or human rights cases, could result in infraction of basic rights. It seems that the court is treading down the theoretical path set by Nozick, *i.e.,* even if the workers in the shipbreaking yards suffer, their suffering is perfectly justified as it brings in economic benefits.

It is interesting to note that the *Blue Lady* court referred to Amartya Sen's treatise to support its theory of balance. However, Sen was articulating an entirely different concept of development, one that is based on enhancement of human capabilities. One can examine the *Blue Lady* judgement from the perspective of the five freedoms identified by Sen to reveal whether shipbreaking actually enhances the capabilities of workers. First, from the standpoint of "political freedoms",[302] shipbreaking is an unorganised industry with workers having no voice in the business. No measures have been adopted to ensure that the work force moves away from contract work and are conferred status through unionisation or other means. Thus, shipbreaking does not meet the political freedom test.

Regarding the second element, "economic facilities",[303] apart from the meagre wages provided to the workers, there is no redistribution of benefits from the upper end (consisting of shipowners, the scrapping entrepreneurs, and governments) to the workers. Rather, shipbreaking poses serious health hazards and debilitates people, often with fatal consequences. The economic benefits that a worker accumulates during his (being a male-dominated industry) productive years by breaking ships often are erased by dealing with the diseases the worker

was upheld even though it would be prejudicial to the worker's interest); *T.K. Rangarajan v. Government of Tamil Nadu* (2003), [2003] A.I.R. S.C. 3032 (India S.C.) (holding that government employees do not have a moral, fundamental, statutory, or equitable right to strike); *Steel Authority of India Ltd. v. National Union Waterfront Workers* (2001), [2001] 7 S.C.C. 1 (India S.C.) (holding that on a notification under section 10 of the *Contract Labour (Regulation & Abolition) Act* there could not be any automatic absorption of the employees); *State of U.P. v. Jai Bir Singh* (2005), [2005] 5 S.C.C. 1 (India S.C.) (observing that, "a worker oriented approach in construing the definition of industry unmindful of the interests of the employer or the owner of the industry and the public would be a one sided approach" and therefore the matter of revisiting *Rajappa* was placed before the Chief Justice of India to constitute a larger bench).

[302] Professor Sen defines political freedoms as the "... opportunities that people have to determine who should govern and on what principles ... to scrutinise and criticise authorities, to have freedom of political expression and an uncensored press, to enjoy the freedom to choose between different political parties ..." Sen, *supra* note 286 at 38.

[303] Economic facilities have been defined as the "opportunities that individuals respectively enjoy to utilise economic resources for the purpose of consumption, or production or exchange." *Ibid.* at 38–39.

contracts after years of work in the unhealthy ship recycling yards. The mounting health bills and the eventual loss of the breadwinning shipbreaker economically ruin an entire family. Other able-bodied family members are forced to replace the lost one leaving school if necessary. The cycle of poverty repeats itself.[304]

As far as "social opportunities"[305] are concerned, the shipbreaking yards have substandard living conditions. Since the accommodation is generally in the vicinity of the scrap yards, toxins are omnipresent. The sanitary conditions are unhygienic, and there is lack of access to potable water and adequate and emergency health care.

The ship recycling yards generally function in an atmosphere of secrecy, violating "transparency guarantees".[306] Situated in far flung areas that are difficult to access, coupled with the migratory nature of the work force, monitoring the recycling yards becomes a daunting task. There is always a dichotomy between official casualty figures and those put forward by environmental non-governmental organisations. Information is often entangled in bureaucratic red tape.

The final element, "protective security",[307] also fails to be met in the recycling yards. A dangerous occupation for shipbreakers, shipbreaking also affects the local population who has no protection from the chemicals which contaminate their water, the soil, and local fisheries

The *Blue Lady* judgement, in highlighting some of the overall economic benefits which the scrapping industry provides, equates development in terms of increased industrialisation. By doing so, the court relied on an outmoded

---

[304] For an overview of the troubles plaguing rural India see generally P. Sainath, *Everybody Loves a Good Drought: Stories from India's Poorest Districts* (London: Headline Book Publishing, 1998); P. Sainath, "The decade of our discontent" *The Hindu [of India]* (9 August 2007), online: The Hindu <http://www.thehindu.com/> (noting that even after sixty years since independence, the state has not been able to solve the structural inequalities in rural India); P. Sainath, "17,060 farmer suicides in one year" *The Hindu [of India]* (31 January 2008), online: The Hindu <http://www.thehindu.com/>; P. Sainath, "One farmer's suicide every 30 minutes" *The Hindu [of India]* (15 November 2007), online: The Hindu <http://www.thehindu.com/> (noting that one Indian farmer committed suicide every 32 minutes between 1997 and 2005, and that since 2002, this has become one suicide every 30 minutes).

[305] It refers to the "arrangements that society makes for education, health care and so on, which influences the individual's substantive freedom to live better." Sen, *supra* note 286 at 39.

[306] This "deals with the need for openness that people can expect: the freedom to deal with one another under guarantees of disclosure and lucidity."*Ibid.*

[307] Protective security seeks to "… provide a social safety net for preventing the affected population from being reduced to abject misery, and in some cases even starvation and death." *Ibid.* at 40. It is to be noted that an application was filed in the Supreme Court by the village head on behalf of the 30,000 villagers and 12 *panchayats* (a unit of local self government) situated in the close vicinity of the shipbreaking yard at Alang against the health and livelihood impacts of shipbreaking when the matter of the *Blue Lady* was being considered by the court. This application is yet to be heard even though the *Blue Lady* matter has been disposed. See *Justice K.G. Balakrishnan, Chief Justice of India*, online: Mediavigil <http://mediavigil.blogspot.com/2007_08_01_archive.html>.

development paradigm rather than looking to the social context and to broader issues of capability enhancement, poverty reduction, and distributive justice.[308]

4. *Justice to the Junkyards: A Mirage?*

Despite the limitations of courts as institutions to delve into areas of high science to craft solutions for human suffering and environmental degradation, credit should go to the Supreme Court of India for having taken up these challenges. Given the legislative vacuum, the court had to intervene in this area to discipline the shipbreaking industry. In the process, it laid down a robust albeit imperfect regulatory framework. However, since the judiciary began to interfere in shipbreaking, the legislature and the executive have treated the matter as *sub-judice*, enabling them to ignore the problems created by the industry and leaving the *onus* solely on the judiciary to find solutions. Unfortunately, the involvement of the judiciary has had a negative impact on the development of a sustainable model for ship recycling in India. For one, the power of the courts to choose among two or more alternatives (judicial discretion) has led to confusion and allowed vested interests to violate applicable laws with impunity. As well, the judiciary's response has been sporadic, depending on applications that are moved before it when elite toxic ships enter Indian waters. There are several other shortcomings in this judiciary sponsored model. For instance, in developing its Guidelines, the Supreme Court relied on the expertise of two committees that studied the issues. However, as revealed by the 'Blue Lady' beaching episode (where the court it seems was not rightly apprised of the non-possibility to refloat the 'Blue Lady'), relying blindly on technical committees can pose further problems. The muddled body of Guidelines has also resulted in inconsistencies in an area that requires standardised norms of regulation if it is to operate sustainably. Even though the 2003 Supreme Court Guidelines contained prescriptions that could facilitate the sustainable development of this industry, the court failed to achieve its goals

---

[308] However Amartya Sen's vision of development as being enhancement of human capabilities is slowly catching on in Indian jurisprudence. *N.D. Jayal v. Union of India* (2003), [2004] A.I.R. S.C. 867 (India S.C.) holding that:

> The right to development cannot be treated as a mere right to economic betterment or cannot be limited to as misnomer to simple construction activities. The right to development encompasses much more than economic well being, and includes within its definition the guarantee of fundamental human rights. The 'development' is not related only to the growth of GNP … The right to development includes the whole spectrum of civil, cultural, economic, political and social process, for the improvement of people's well being and realisation of their full potential. It is an integral part of human right. *Ibid.*, ¶ 24.

See also *Proposed ILO Declaration on Social Justice for a Fair Globalization*, online: International Labour Organization <http://www.ilo.org/wcmsp5/groups/public/—ed_norm/—relconf/documents/meetingdocument/wcms_094042.pdf>. Even though the Declaration acknowledges the benefits of globalisation, it calls for renewed efforts to implement decent work policies. The Declaration was adopted by the International Labour Conference at its 97th session held at Geneva on June 2008.

because it did not ensure compliance by the major players to its Guidelines. This failure is all the more striking when one considers that it had its own creation, the SCMC, to monitor compliance. The 'Riky' and the 'Clemenceau' fiascos are prime examples of these weaknesses and the subsequent consequences. At every stage, concerted efforts were made by the votaries of this industry to dilute the Guidelines and the Supreme Court did not respond. In the 2007 Guidelines, the court went a step further to accept the "all pervasive" report of the TEC. However, in the process, it watered down the rigour of the 2003 Guidelines, particularly in relation to the requirement for prior decontamination. This meant that toxic ships of all hues and sizes could enter Indian waters. The Supreme Court has also never applied the polluter pays principle to hold the ship recyclers or shipowners liable for the environmental vandalism that is being perpetuated at Alang. This is surprising given that the court has come down heavily on polluters in the past for other pollution incidents.[309]

The *Blue Lady* pronouncement also raises questions regarding the court's ability to balance the conflicting interests of environmental protection and developmental requirements as it introduced proportionality into sustainable development decision-making. More often than not, the adverse impacts of developmental projects and environmental degradation fall disproportionately on the poor. In a democratic polity grounded on the rule of law, where individual protection flows from a guaranteed bill of rights, one segment of the population cannot be made to suffer a disproportionate burden to benefit another, particularly when that group belongs to a disadvantaged section of the general population. In such situations, the courts should attempt to ensure that none is made worse off, if not better. Given the sorry state of dismantling practices at Alang, the courts must ensure that the workers, the environment, and the local communities are not the ultimate sufferers because of the shipbreaking industry.

Clearly, the courts may not always be the best forum to deal with regulatory development for an industry such as shipbreaking. We must consider alternative paths for developing appropriate legal regulatory frameworks. In any case, the judicial guidelines should be seen as a stop-gap arrangement, and the final solution should emanate from the legislature.

Before concluding this discussion, it is relevant to note that in a "National Workshop on the Development of the International Convention for the Safe and Environmentally Sound Recycling of Ships" (held at Mumbai, India in January 2008) which brought together officials from the IMO, the European Union, ILO, the Basel Convention Secretariat and other international experts,[310] it emerged that

---

[309] *Vellore Citizens' Welfare Forum*, *supra* note 114.

[310] See generally National Workshop on International Convention for Safe and Environmentally Sound Recycling of Ships (7–10 January 2008), online: Gujarat Maritime Board <http://www.gmb ports.org/international_workshop.htm>.

"the new regime governing ship recycling in India was similar to the requirements of the draft text of the IMO Convention",[311] and that, "there was no serious impediments to India ratifying the new Convention".[312] This observation should be heartening for the industry as the present legal regime (despite all its imperfections as highlighted) can facilitate subscription to the upcoming international legal regime by India, ensuring continuity of business. However, India must invest resources to revamp and consolidate its existing legal framework. More importantly, India must develop its recycling capacity to ensure that the constitutional promises made to its disadvantaged and deprived, in this case, the workers and the local inhabitants at Alang are met and the industry operates in a sustainable manner.

IV. The Economics of Shifting Ship-Scrap Markets and Efficacy of National Legal Regimes

Contemporary history of the ship demolition industry reveals two distinct phases. The first phase, from 1994–2003, witnessed the exit of OECD countries and the emergence of India as the market leader. India continued to rule the market until 2003 (second phase), when the Supreme Court laid down its Guidelines for regulating the industry. Since then, the industry in India has been on the decline.

The economics of ship demolition reveal the reasons for these market shifts. The ship scrapping industry is profit-driven; market forces dictate its nature and growth. Generally, an owner of a ship has cash inflow from two sources: the freight market and the scrapping market.[313] Other factors being constant, a shipowner's decision to scrap a vessel is generally determined by weighing the advantages of keeping the ship against selling it as scrap. If the expected income from continued trading is less than the scrap value, the shipowner will decide to send the ship to the scrap yard.[314] To maximise profit, the shipowner chooses the scrapping location that offers the highest scrap price for the ship.

The recycler's decision to buy an obsolete ship is determined by the possibility of selling the steel and other reusable items that can be recovered from the ship, and more importantly, by the probable costs that s/he will incur while breaking the

---

[311] Recycling of Ships: National Workshop on the Development of the International Convention for the Safe and Environmentally Sound Recycling of Ships, Submitted by India, IMO/MEPC 57/3/5, 25 January 2008, ¶ 10 (KR-CON).

[312] *Ibid.*, ¶ 11.

[313] Katrine Vetaas Vedeler, *From Cradle to Grave – Value Chain Responsibility in the Ship Scrapping Industry* (Master Thesis, Norwegian School of Economics and Business Administration, 2006), online: BORA-NHH <http://bora.nhh.no/>.

[314] A determinant factor in arriving at a reasonable economic decision would be the age of the ship given that older ships have a lower future earning potential due to higher maintenance costs. Moreover, aging ships have to undergo regular surveys (the fifth survey, at the age of 25 years, tends to be expensive) to obtain classification and face rising insurance costs. *Ibid.* at 21–22.

ship.[315] The recycler will quote the lowest price or, in the alternative, ensure that the running costs are kept at a minimum so that the expected returns do not diminish. In the event of escalating scrap prices, the only possibility then is for the recycler to keep running costs as low as possible. One way to reduce costs is to not comply with legal requirements relating to occupational safety and environmental protection.[316] Accordingly, the ship scrapping industry, which has considerable political and economic clout, ensures that governments regulate it as little as possible.

Currently, any regulation of the shipbreaking industry emanates locally. While some countries want to implement labour and environmental standards, their overall economic situation often forces them to look the other way. More importantly, the moment a regulatory regime is enforced, costs increase. This makes it difficult for the recycler to offer better prices for the obsolete tonnage, inducing the trade to migrate to other jurisdictions where controls are minimal and profits are greater. Thus, a municipal legal regime may not provide the appropriate pathway for the sustainable growth of this highly dynamic and global industry. This is precisely what has happened at Alang. With the Supreme Court of India insisting on higher standards by its 2003 order, the industry migrated to Bangladesh. It was just a matter of time before progressive dilution of the domestic regulatory regime began in India with the aim of re-attracting the redundant tonnage and rendering the industry once again competitive. This accentuates the need for a strong international legal regulatory regime to lay down uniform standards that major shipbreaking nations can subscribe to in order to ensure responsible and sustainable ship scrapping.

V. Conclusion

The major issues that affect sustainability in the shipbreaking industry relate to environmental protection and occupational safety and health in the scrap yards. To this list, one must add another important element, namely, the viability of

[315] EC, Directorate-General Energy & Transport, *Oil Tanker Phase Out and the Ship Scrapping Industry: A Study on the Implications of the Accelerated Phase out Scheme of Single Hull Tankers Proposed by the EU for the World Ship Scrapping and Recycling Industry* (EC, 2004) at 45, online: European Commission <http://ec.europa.eu/transport/maritime/safety/doc/prestige/2004_06_scrapping_study_en.pdf> (the costs incurred are determined by the price of scrap steel; the running costs, which involves taxes, salaries, rent and other charges; fluctuations in exchange rate; enforcement of labour and environmental standards; etc.); EC, Directorate General Environment, *Ship Dismantling and Pre-cleaning of Ships: Final Report* (EC, 2007) at 122, online: European Commission <http://ec.europa.eu/environment/waste/ships/pdf/ship_dismantling_report.pdf> (the main drivers in the scrapping industry are freight rates, which determine when to scrap; labour costs, which determine where to scrap; and steel prices, which determine the size of the shipowner's profit).

[316] See Martin Stopford, *Maritime Economics* 2d ed. (London: Routledge 1997) at 484–87. See also France, Interdepartmental Committee on the Dismantling of Civilian and Military End-of-Life Ships, *Le rapport de la Mission Interministérielle portant sur le Démantèlement des Navires civils et militaires en fin de vie*, at annex I (27 March 2007, Chair: Xavier de la Gorce), online: SGMer <http://www.sgmer.gouv.fr/IMG/pdf/Annex_1_Economic_factors.pdf>.

national legal regimes to ensure the sustainable growth of this industry. Even though legal solutions to the problems posed by ship scrapping can emanate nationally, the international underpinnings of the ship scrapping industry can undermine these domestic efforts. This is an issue of concern for developing countries as they may end up losing an economic opportunity solely because of the measures that they initiated to discipline their scrap yards. Left to themselves, national regulators may be unable to come up with workable solutions for the sustainable growth and development of this industry. One hopes that the *Hong Kong International Convention for the Safe and Environmentally Sound Recycling of Ships, 2009* will facilitate the development of effective national legal regimes that can serve as concise roadmaps for the sustainable growth of the ship recycling industry.

Chapter 4

Contemporary International Law and Ship Recycling

I. Introduction

Shipping was initially dependant on the rules laid down by powerful maritime states.[1] As the movement of people, goods, cargo, and freight *via* ships grew, the shipping industry also developed at a rapid pace, necessitating greater cooperation between the different states to meet the myriad of challenges that emerged. The realisation that global solutions were appropriate for shipping found concrete expression with the establishment of the International Maritime Organization (IMO) as a specialised agency of the United Nations with the obligation to regulate and support shipping services engaged in international maritime trade.[2]

An integral component of the shipping business, ship recycling is also a global industry. While there are very few binding standards at the international level that directly deal with shipbreaking, several international hazardous waste management instruments do impact on the ship recycling industry. In particular, the *Basel Convention on the Control of Transboundary Movements of Hazardous Wastes and their Disposal, 1989*, the primary international legal hazardous waste management law, as well as the Basel Ban and the Technical Guidelines for the Environmentally Sound Management of the Full and Partial Dismantling of Ships, govern the transboundary movement of ships for recycling. Elements of the 1982 *United Nations Convention on the Law of the Sea* (LOS Convention) are also relevant to shipbreaking. The International Labour Organization (ILO) has established standards to protect workers for occupational safety and health. Several IMO conventions on shipping and maritime safety are also apposite to shipbreaking. These include the *Convention on the Prevention of Marine Pollution by Dumping of Wastes and Other Matter, 1972* (London Convention)

---

[1] See generally Edgar Gold, *Maritime Transport: The Evolution of International Marine Policy and Shipping Law* (Massachusetts: Lexington Books, 1981). Dividing the historical development of maritime law into five distinct periods – the first beginning from pre-history to the early Mediterranean period; the second from the years 1000 to 1400 when Codifications began to appear; the third which included the years 1400 to 1700 which saw European dominance (Spain, Portugal, France, the Netherlands, the Hansa and England); the fourth, including the years 1700 to 1840, where maritime power was concentrated in England, the Netherlands and France; and finally the century from 1840 to 1940 when England was the predominant maritime power. Edgar Gold, *International Maritime Law: Basic Principles Lecture Notes and Materials* (Faculty of Law, World Maritime University, Sweden, 30 June 1986) at 22.

[2] Edgar Gold, Aldo Chircop & Hugh Kindred, *Maritime Law* (Toronto: Irwin Law, 2003) at 77.

and its 1996 Protocol, the *Nairobi International Convention on the Removal of Wrecks, 2007*, the *International Convention on the Control of Harmful Anti-fouling Systems on Ships, 2001*, the *International Convention for the Control and Management of Ships' Ballast Water and Sediments, 2004*, and the *International Convention for the Prevention of Pollution from Ships, 1973* and its *Protocol of 1978*. Perhaps the most important among the IMO instruments having a direct bearing on shipbreaking prior to the recent adoption of the *Hong Kong International Convention for the Safe and Environmentally Sound Recycling of Ships, 2009* (Ship Recycling Convention) are the IMO Guidelines on Ship Recycling, 2003 (IMOGSR).

This chapter maps out some of the major features of these international legal instruments with a view to understand what scope these have in terms of their relevance to managing the ship recycling industry at an international level. It also reviews the adequacy of this framework to meet the challenges posed by the operation of this industry.

II. Ship Dismantling and International Hazardous Waste Management Law

The *Basel Convention on the Control of Transboundary Movements of Hazardous Wastes and their Disposal, 1989*[3] is the foremost international legal instrument that recognises and provides for measures to contain the growing threat to human health and to the environment due to the transboundary movement of hazardous wastes. These wastes are transferred primarily from the developed to the developing world,[4] following "the path of least resistance."[5] Sponsored by the United

---

3 22 March 1989, 1673 U.N.T.S. 126, 28 I.L.M. 657 (entered into force 5 May 1992) [*Basel Convention*]. See also the *United Nations Environment Programme: Stockholm Convention on Persistent Organic Pollutants*, 22 May 2001, 40 I.L.M. 532 (entered into force on 17 May 2004). Mindful of the precautionary approach, its objective is to protect human health and the environment from persistent organic pollutants. *Ibid.*, art. 1. It specifically limits importation and exportation of PCBs unless environmentally sound disposal and use are provided for. *Ibid.*, art. 3(2). Even though it does not specifically deal with the ship scrapping issue, this convention imposes the duty on states to consider persistence, bioaccumulation, potential for long-range environmental transport, and adverse effects when (if ever) it reassesses a chemical already in use. *Ibid.*, art. 3(4), annex D. Furthermore, the convention recognises the need for developed countries to render "timely and appropriate technical assistance" to developing countries and parties with economies in transition. *Ibid.*, art. 12. See also *Convention on the Prior Informed Consent Procedure for Certain Hazardous Chemicals and Pesticides in International Trade*, 11 September 1998, 38 I.L.M. 1 (entered into force 24 February 2004).

4 Hazardous waste treatment facilities in the industrialised nations are filled to capacity, and due to rising environmental awareness, fewer new facilities are being built. This has caused the cost of disposal in the United States to skyrocket to more than USD2,000 per tonne in some areas, whereas developing nations charge as little as USD20 per tonne. Diane Kilcoyne, "The Basel Convention: Will It Curtail Hazardous Waste Exports" (1992) 16:2 Environs Envtl. L. & Pol'y J. 47 (HeinOnline).

5 See generally Katharina Kummer, *The Basel Convention and Related Legal Rules*, Professor Ian Brownlie, International Management of Hazardous Wastes, Oxford Monographs in International Law

Nations Environment Programme, the convention entered into force in 1992. To date, the Basel Convention has over 172 parties.[6]

As stated in its Preamble, the Basel Convention aims to "protect, by strict control, human health and the environment against the adverse effects which may result from the generation and management of hazardous wastes[7] and other wastes".[8] This goal is to be achieved in two ways. First, the Basel Convention provides for enhanced control over the "transboundary movement"[9] of hazardous and other wastes as an incentive for environmentally sound management (ESM).[10] Second, the parties are obliged to adopt "appropriate measures" to minimise the generation of hazardous and other wastes and ensure the establishment of adequate disposal facilities within the generating state.[11] It is anticipated

---

(New York: Oxford University Press, 1995). The 'Khian Sea' incident exposed the inadequacies of international legal controls on the matter. In 1986, the 'Khian Sea', owned by a Bahamian company, loaded 28 million pounds of toxic incinerator ash from Philadelphia. The ship unsuccessfully tried to offload its cargo in the following countries, namely, the Bahamas, the Dominican Republic, Honduras, Costa Rica, Guinea Bissau, and the Cape Verde Islands. In 1987, it offloaded 28 million pounds of ash in Haiti. On realising the true nature of the cargo, it was ordered to leave. Thereafter, it sailed to Senegal, Morocco, Yugoslavia, Sri Lanka, and Singapore looking for a place to dump its toxic load. Meanwhile, the ship changed its name to 'Felicia' and thereafter to 'Pelacano'. Its ownership and registration also changed (first from Liberia to the Bahamas and later to Honduras) in an attempt to cloak its true identity. Finally, somewhere in the Indian Ocean, between Singapore and Sri Lanka, the ash disappeared. *Additional Case Studies*: What a Long, Strange Trip it Has Been, online: Principles of Environmental Science <http://highered.mcgrawhill.com/sites/0072919833/student_view0/chapter13/additional_case_studies.html>. Michael Tsimplis, "Liability and Compensation in the International Transport of Hazardous Wastes by Sea: The 1999 Protocol to the Basel Convention" (2001) 16 Int'l J. Mar. & Coast. L. 295 at 297–98 (HeinOnline).

[6] *Parties to the Basel Convention*, online: Basel Convention <http://www.basel.int/ratif/convention.htm>. See also Pierre Portas, "The Basel Convention, Back to the Future", Commentary, (2006) 6:3 S.D.L. & Pol'y 38 (HeinOnline).

[7] It defines wastes as "substances or objects which are disposed of or are intended to be disposed of or are required to be disposed of by the provisions of national law." *Basel Convention*, *supra* note 3 art. 2(1). A waste is hazardous if it is listed under annex I unless it does not possess any of the characteristics enumerated in annex III namely, flammability, explosivity, toxicity, and eco-toxicity. For wastes that are not covered by this clause, they can still be treated as hazardous wastes, if they are so considered by the national law of the party of export, import or transit. *Ibid.*, art. 1(1). Radioactive wastes and wastes arising from the normal operation of a ship (governed by MARPOL 73/78) are excluded from its ambit. *Ibid.*, art. 1(3)–(4). Annexes I and II of the Basel Convention (subject to the criteria mentioned in annex III) sets out the wastes that are regulated by it. Annexes VIII and IX clarify the specific wastes that are covered by the convention.

[8] Wastes that belong to any category contained in annex II that are subject to transboundary movement shall be "other wastes". *Ibid.*, art. 1(2). These generally include household wastes and their residuals.

[9] *Ibid.*, art. 2(3).

[10] See *ibid.*, Preamble.

[11] *Ibid.*, art. 4(2)(a)–(b). See also *ibid.*, art. 10(2)(c) (imposing obligations on parties to cooperate, subject to their national laws, to develop and implement new environmentally sound low-waste technologies, and to work towards improving existing technologies with a view to eliminating, within practical limits, the generation of hazardous and other wastes and achieving more effective and efficient methods of ensuring their management in an environmentally sound manner). See also *ibid.*, art. 11 (empowering parties to enter into "bilateral, multilateral, or regional agreements or arrangements" with parties or non-parties for the transboundary movement of hazardous or other wastes, provided

that these measures will reduce the frequency of the transboundary movement of wastes.[12]

Throughout the convention text, ESM appears as the touchstone to determine whether or not a transboundary movement of wastes or its disposal conforms to the Basel Convention stipulations.[13] As defined in article 2(8), ESM requires the adoption of practical steps to ensure that hazardous or other wastes are managed in a manner that will not jeopardise human health and the environment. A series of technical guidelines developed in relation to the management of specific waste streams implement this concept. In the context of ship de-construction, ESM is elucidated in the Technical Guidelines on the Environmentally Sound Management of the Full and Partial Dismantling of Ships, 2002 (TGSD).[14]

As far as the other features of the Basel Convention are concerned, it envisages five natural or legal persons[15] that can be involved in a transport of hazardous wastes.[16] The first is the "generator" whose activity produced the wastes or, in cases where this entity is not known, the person who has possession or control over the wastes.[17] The "exporter" is the individual or entity who arranges for the export of hazardous wastes and is under the jurisdiction of the exporting state.[18] The "carrier" is the performing carrier.[19] The "importer" is any person under the jurisdiction of the importing state who imports the hazardous wastes.[20] The final player is the "disposer", the individual or entity that receives the hazardous cargo for disposal.[21] These entities operate in a broad framework where the main drivers are the states of export,[22] import[23] and transit.[24]

---

that, such agreements or arrangements do not derogate from the ESM requirements, taking into account the interests of developing countries). Z. Lipman, "Transboundary Movement of Hazardous Waste: Environmental Justice Issues for Developing Countries" [1999] Acta Jur. 266 at 275 (HeinOnline).

[12] Tsimplis, *supra* note 5 at 299.

[13] Muthu S. Sundram, "Basel Convention on Transboundary Movement of Hazardous Wastes: Total Ban Amendment" (1997) 9 Pace Int'l L. Rev. 1 at 13 (WLeC) (referring to ESM as the soul of the *Basel Convention*).

[14] See also Recycling of Ships: Control System of the Basel Convention on the Control of Transboundary Movements of Hazardous Wastes and their Disposal, Submitted by the Secretariat of the Basel Convention, IMO/MEPC 53/INF.19, 27 May 2005, ¶ 12 (KR-CON) [Control System of the Basel Convention].

[15] *Basel Convention, supra* note 3 art. 2(14) (defining person to mean any natural or legal person).

[16] Tsimplis, *supra* note 5 at 300.

[17] *Basel Convention, supra* note 3 art. 2(18).

[18] *Ibid.*, art. 2(15).

[19] *Ibid.*, art. 2(17).

[20] *Ibid.*, art. 2(16).

[21] *Ibid.*, art. 2(19).

[22] "State of export" is the party from which the transboundary movement of hazardous wastes or other wastes is planned to be initiated or is initiated. *Ibid.*, art. 2(10).

[23] "State of import" is the party to which a transboundary movement of hazardous or other wastes is planned or takes place for the purpose of disposal therein or for the purpose of loading prior to disposal in an area not under the national jurisdiction of any state. *Ibid.*, art. 2(11).

[24] It means any state, other than the export or import state through which the movement of hazardous or other wastes is planned or takes place. *Ibid.*, art. 2(12).

The Basel Convention prohibits parties from exporting or importing hazardous or other wastes to or from non-party states.[25] To this end, it establishes a system of controls on the export and import of hazardous waste.[26] In order to prevent illegal trade[27] in hazardous and other wastes, the Basel Convention affirms the "sovereign right"[28] of a party to unilaterally prohibit the import of hazardous wastes, and there is a corresponding obligation on other parties to respect this right.[29] Exports of hazardous wastes are permitted only in cases where the exporting state lacks the technical expertise and capacity for sound disposal or in situations where the waste is required as a raw material for recycling or for recovery in the state of import.[30] In all cases, the duty to ensure ESM of the hazardous waste lies solely on the state of export and cannot be transferred to the importing or transit state.[31] As well, the exporting party is also required to prohibit an export, if it believes that the wastes will not be managed in accordance with ESM principles in the intended destination.[32]

To enable an importing state to assess the nature of the risks involved in an import,[33] the convention provides for elaborate control measures based on the principle of prior informed consent.[34] Accordingly, the Basel Convention requires certain information to be provided to the importing state in order to enable it to make an informed decision, namely, information on the wastes,[35] proposed methods of transportation[36] and disposal,[37] and evidence of a contract between the parties ensuring environmentally-sound recycling.[38] The prior informed consent of the transit state irrespective of its party/non-party status is also required.[39]

To initiate a transboundary movement of hazardous wastes, the convention contemplates two documents. The first is the notification document by which the

---

[25] *Ibid.*, art. 4(2)(5).

[26] For instance, see *ibid.* art. 4(5) (stating that a party is not to permit hazardous or other wastes to be exported to a non-party or to be imported from a non-party).

[27] In cases of non-compliance with the notification and consent requirements; or where the consent was obtained through falsification, misrepresentation or fraud; or if there is discrepancy between the material particulars in the documents and the transport of the hazardous wastes or where it is a case of "deliberate disposal", the movement is considered to be an illegal traffic. *Ibid.*, art. 9(1). This is a criminal offence under the convention and parties are called upon to initiate steps to introduce appropriate domestic laws to prevent and punish the same. *Ibid.*, art. 9(5).

[28] *Ibid.*, Preamble.

[29] *Ibid.*, art. 4(1)(b).

[30] *Ibid.*, art. 4(9).

[31] *Ibid.*, art. 4(10).

[32] *Ibid.*, art. 4(2)(e).

[33] Sundram, *supra* note 13 at 15 (terming ESM as the soul of the Basel Convention).

[34] *Basel Convention, supra* note 3 art. 6. See also Control System of the Basel Convention, *supra* note 14 ¶ 15.

[35] *Basel Convention, ibid.,* annex V A, items 13–17.

[36] *Ibid.*, annex V A, item 11.

[37] *Ibid.*, annex V A, item 18.

[38] *Ibid.*, annex V A, item 21.

[39] *Ibid.*, arts. 6(4), 7.

exporting states notify or, alternatively, ensure that the generator/exporter notifies the competent authority about the proposed export and obtains the written consent of both the importing and transit states[40] before transporting the wastes.[41] The second is the "movement document",[42] which contains information on 14 items,[43] namely, declarations by the generator/exporter that the information provided is correct,[44] assurances that none of the states involved object to the transport,[45] and an overall description of the waste.[46]

Hazardous and other wastes intended for transboundary movement should be packaged, labelled, and transported in accordance with generally-accepted international standards.[47] As may be required by the states of import or transit, the transboundary movement must be covered by insurance, bond or other guarantee.[48] The Basel Convention also establishes rules and procedures on matters relating to liability and compensation for damage during transportation and disposal.[49] Another requirement is that only persons authorised by national law can transport or dispose of hazardous or other wastes,[50] which in effect implies that recycling facilities are to be authorised.[51] Additional provisions deal with re-importing the hazardous waste in cases where the disposal contract stands frustrated.[52] In such cases, the state of export has to ensure that the exporter takes the wastes back if alternative arrangements for its environmentally sound disposal cannot be made.

The Basel Convention is managed by a secretariat[53] that assists parties in identifying illegal traffic,[54] provides assistance to states in cases of emergency,[55] and

---

40 *Ibid.*, art. 6(4). Even a transit state that is not a party to the convention has to be provided with the same notification and written consent is also to be obtained from it. *Ibid.*, art. 7.

41 *Ibid.*, art. 6(1), (3) (in fact, for the export to commence, two conditions need to be fulfilled. Firstly, the state of import and the states of transit will have to give their written consent to the transport of wastes and secondly, the state of import has to confirm the existence of a contract between the exporter and the disposer specifying ESM of the wastes). See also Tsimplis, *supra* note 5 at 301.

42 *Basel Convention*, *supra* note 3 art. 4(7)(c).

43 *Ibid.*, annex V B.

44 *Ibid.*, annex V B, item 12.

45 *Ibid.*, annex V B, item 13.

46 *Ibid.*, annex V B, item 8.

47 *Ibid.*, art. 7(b).

48 *Ibid.*, art. 6(11).

49 *Ibid.*, art. 2. The Fifth Conference of the Parties to the Basel Convention adopted the *Basel Protocol on Liability and Compensation for Damage Resulting from Transboundary Movements of Hazardous Wastes and Their Disposal, 1999.*

50 *Ibid.*, art. 4(7)(a).

51 EU Comments Received Pursuant to OEWG6 Decisions, "Assessment on Ship Dismantling with Particular Reference to the Levels of Control and Enforcement Established by the Basel Convention and the Expected Level of Control and Enforcement to be provided by the Draft Ship Recycling Convention in their Entirety" ¶ 8, online: Basel Convention <http://www.basel.int/ships/commentsOEWG6/oewg6.html> [EU Comments].

52 *Basel Convention*, *supra* note 3 art. 8.

53 *Ibid.*, art. 16.

54 *Ibid.*, art. 16(i).

55 *Ibid.*, art. 16(j).

receives and conveys information.[56] Most importantly, the convention includes provisions for the Conference of Parties (COP)[57] to "review and evaluate the effective implementation" of its terms. The COP represents the governing body of the Basel Convention[58] and consists of all state parties.

The COP has played a vital role in adjusting the Basel Convention tools to ship dismantling. The first COP meeting considered measures needed to assist states to fulfil their responsibilities with respect to protection and preservation of the marine environment.[59] In 1994, at the second meeting of the COP, an immediate ban on the export of hazardous wastes from the Organisation for Economic Co-operation and Development (OECD) to non-OECD countries was introduced.[60] However, this decision was not incorporated into the Basel Convention text, and there were doubts regarding its legal nature.[61] The matter was sorted out in 1995, at the third meeting of the COP, wherein the ban was formally incorporated into the convention as an amendment.[62] Accordingly, Decision III/1 bans exports of hazardous wastes for final disposal and recycling from what are known as annex VII countries (Basel Convention parties that are members of the European Union, OECD, and Liechtenstein) to non-annex VII countries (all other parties to the convention).[63] Even though it does not use the 'OECD/non-OECD' distinction, in essence, the decision (known as the BAN amendment) prohibits the movement of wastes from OECD to non-OECD countries. This amendment requires ratification by three-fourths of the Basel parties[64] to come into effect. Thus far, the Ban amendment has not come into force due to insufficient ratifications.[65]

Ship dismantling began to figure prominently in the work of the COP only from its fifth meeting onwards. At this meeting, the Technical Working Group was called upon to work with the secretariat of the Basel Convention to collaborate

---

56 *Ibid.*, art. 16(g).

57 *Ibid.*, art. 15. In addition, the COP has to promote harmonisation of appropriate policies, strategies and measures for minimising harm to human health and the environment by hazardous and other wastes; consider and adopt amendments to this convention, taking into consideration, *inter alia*, available scientific, technical, economic and environmental information; consider and undertake any additional action that may be required for achieving the purposes of this convention; consider and adopt protocols as required; and establish such subsidiary bodies as are deemed necessary for the implementation of this convention.

58 For further information, see *The Conference of the Parties*, online: Basel Convention <http://www.basel.int/meetings/cop/cop-0.html>.

59 *Basel Convention*, *supra* note 3 art. 15(4).

60 *Ban Amendment to the Basel Convention on the Control of Transboundary Movements of Hazardous Wastes and their Disposal Geneva, 22 September 1995*, online: Basel Convention <http://www.basel.int/ratif/ban-alpha.htm>.

61 *Ibid.*

62 *Ibid.*

63 *Ibid.*

64 *Ibid.*

65 Sundram, *supra* note 13 at 19–24 (outlining some of the difficulties which states encounter in ratifying the BAN Amendment).

with the IMO on the issues of full and partial dismantling of ships and to prepare guidelines.[66] Thereafter, at its sixth meeting (Geneva, 9–13 December 2002), the TGSD were adopted.[67] The sixth meeting also called upon the Open-Ended Working Group (OEWG) to present recommendations on the legal issues concerning the full and partial dismantling of ships.[68]

The seventh COP meeting dealt with three important issues: (1) ESM in ship dismantling;[69] (2) the Joint Working Group (JWG) of the ILO, the IMO and the Basel Convention on Ship Scrapping;[70] and (3) abandonment of ships.[71]

The eighth COP also considered three important matters related to ship breaking. The first again related to ESM in ship dismantling.[72] Welcoming the steps taken by the IMO for developing the Ship Recycling Convention, it called upon the IMO to ensure that this convention establishes an equivalent level of controls as contemplated by the Basel Convention. In doing so, however, duplication of regulatory instruments having the same objectives was to be avoided.[73] The second aspect related to the third meeting of the JWG of the ILO, the IMO and the Basel Convention on ship scrapping.[74] Finally, the COP examined the issue of abandonment of ships, urging parties to apply the *Convention on the Prevention of Marine Pollution by Dumping of Wastes and Other Matter* (the London Convention and its 1996 Protocol).[75]

At the recently concluded ninth COP, held at Bali, Indonesia, 23–27 June 2008, in its Decision IX/30 on the dismantling of ships, the COP established a process to assess whether the Ship Recycling Convention establishes an equivalent level

---

[66] For the text of the *V/28. Dismantling of Ships* decision, see UNEP, *Report of the Fifth Meeting of the Conference of the Parties to the Basel Convention – Advance Copy,* 5th Mtg., UNEP/CHW.5/29, (10 December 1999) 55 at 56, online: Basel Convention <http://www.basel.int/meetings/cop/cop5/cop5reportfinal.pdf#v28>.

[67] UNEP, *Report of the Conference of the Parties to the Basel Convention on the Control of Transboundary Movements of Hazardous Wastes and their Disposal*, 6th Mtg., UNEP/CHW.6/40 (10 February 2003) 138, online: Basel Convention <http://www.basel.int/meetings/cop/cop6/english/Report40e.pdf#vi24>.

[68] *Ibid.*

[69] *VII/26. Environmentally Sound Management of Ship Dismantling*, online: Basel Convention <http://www.basel.int/ships/relev-decisions-cop-vii.pdf#VII_26> [*VII/26*].

[70] *VII/25. Joint Working Group of the International Labour Organization, the International Maritime Organization and the Basel Convention on Ship Scrapping*, online: Basel Convention <http://www.basel.int/ships/relev-decisions-cop-vii.pdf#VII_25>.

[71] *VII/27. Abandonment of Ships*, online: Basel Convention <http://www.basel.int/ships/relev-decisions-cop-vii.pdf#VII_26>.

[72] *VIII/11: Environmentally Sound Management of Ship Dismantling*, online: Basel Convention <http://www.basel.int/meetings/cop/cop8/docs/16e.pdf#viii11>.

[73] *Ibid.*

[74] *VIII/12 Joint Working Group of the International Labour Organization, the International Maritime Organization and the Basel Convention on Ship Scrapping*, online: Basel Convention <http://www.basel.int/meetings/cop/cop8/docs/16e.pdf#viii11>.

[75] *VIII/13: Abandonment of Ships on Land or In Ports*, online: Basel Convention <http://www.basel.int/meetings/cop/cop8/docs/16e.pdf#viii11>.

of control as that established under the Basel Convention. The COP also invited the seventh session of its Open-ended Working Group to carry out a preliminary assessment on whether the Ship Recycling Convention establishes an equivalent level of control and enforcement as under the Basel Convention.[76] The results of this assessment are to be transmitted to the COP in its tenth meeting.[77] Furthermore, as part of its mandate to foster international cooperation and technical assistance in respect of ESM in ship dismantling, the COP also welcomed the establishment of the Global Programme for Sustainable Ship Recycling. This programme seeks to improve working practices and environmental standards in the ship recycling yards of Asia.[78] The OEWG has also issued "decisions" dealing with different aspect of ship dismantling.

As mentioned earlier, ESM in the context of ship dismantling is explained in the Basel Convention Technical Guidelines for the Environmentally Sound Management of the Full and Partial Dismantling of Ships. Adopted in 2002, the TGSD[79] provides guidance to countries that desire to establish ship dismantling facilities.[80] The guidelines provide information and recommendations on procedures, processes and practices that needs to be implemented to achieve ESM at such facilities.[81] The TGSD qualify the definition of ESM in relation to hazardous and other wastes by recognising worker safety as being integral to a scheme on ESM.[82]

In implementing ESM in ship dismantling, the guidelines identify current practices in the dismantling facilities, which are compared to a model facility enabling identification of gaps and formulation of measures to fill-in these gaps.[83]

---

[76] *Recycling of Ships, Report of the Ninth Meeting of the Conference of the Parties to the Basel Convention*, Note by the Secretariat of the Basel Convention, IMO/MEPC 58/3/9, 29 July 2008 (KR-CON); Recycling of Ships: Appropriate Criteria to Carry out a Preliminary Assessment of Equivalency between the Basel Convention and the Hong Kong International Convention for the Safe and Environmentally Sound Recycling of Ships, 2009, Note by the Secretariat of the Basel Convention, IMO/MEPC 59/3/5, 8 May 2009 (KR-CON). See also UNEP, Conference of the Parties to the Basel Convention on the Control of Transboundary Movements of Hazardous Wastes and their Disposal, 9th Mtg., (23–27 June 2008) Environmentally Sound Management of Ship Dismantling and the Joint Working Group of the International Labour Organization, the International Maritime Organization and the Basel Convention on Ship Scrapping, Note by the Secretariat, UNEP/CHW.9/34*, 14 April 2008, online: Basel Convention <http://www.basel.int/meetings/cop/cop9/docs/34e.pdf>.

[77] UNEP, *ibid*.

[78] *Ibid*.

[79] UNEP, Conference of the Parties to the Basel Convention on the Control of Transboundary Movements of Hazardous Wastes and their Disposal, 6th Mtg., (9–13 December 2002) Technical Guidelines for the Environmentally Sound Management of the Full and Partial Dismantling of Ships, UNEP/CHW.6/23, 8 August 2002 at 7, online: Basel Convention <http://www.basel.int/meetings/cop/cop6/cop6_23e.pdf> [TGSD].

[80] See generally *ibid*.

[81] *Ibid*. at 7.

[82] *Ibid*.

[83] *Ibid*. at 8, 80–83.

Accordingly, the TGSD lays down best practices in relation to the "[d]esign, construction and operation of ship-dismantling facilities"[84] and "environmental control procedures at ship-dismantling facilities".[85] The TGSD also provides advice on achieving ESM practices.[86] Its major drawback is that it does not address measures to minimise the hazardous materials on board a ship prior to it being sent to a ship recycling facility.[87] Also, the TGSD is silent on the need to introduce the most environmentally sound method for recycling of ships, namely, dry docking. This silence undermines the prospect of ensuring effective ESM in ship dismantling.[88]

In short, there is a vast body of international hazardous waste management law that has relevance for ship dismantling. However, the shipping industry has consistently held the view that the Basel Convention tools are inappropriate to the business of dismantling ships.[89] To support this argument, the shipping industry relies mainly on the exception to article 1(4) of the Basel Convention which states: "[w]astes [that] derive from the normal operations of a ship, the discharge of which is covered by another international instrument, are excluded from the scope of this Convention". Ship wastes are primarily regulated at the international level by MARPOL 73/78. Under MARPOL 73/78, a ship is defined in article 2(4) as "a vessel of any type whatsoever, operating in the marine environment". Accordingly, wastes that are derived from the normal operation of ships that sail under their own power irrespective of whether they are on their first or last journey are excluded from the ambit of the Basel Convention.[90] Although certain components in the ship, particularly those that are used to sustain normal operations, may contain materials that possess some of the "Basel characteristics",[91] as long as "the ship as a unit" is subject to the MARPOL 73/78, it is not to

[84] *Ibid.* at 59–72.

[85] *Ibid.* at 40–57.

[86] *Ibid.* at 75–79.

[87] *Ibid.* at 8.

[88] *Comments on the Draft Technical Guidelines on Environmentally Sound Management for Full and Partial Dismantling of Ships*, (10 January 2001), at 2, online: Basel Action Network <http://www.ban.org/Library/BANComments%20ShipbreakingGuidelines2.PDF> (referring to the Guidelines as a bandage over a festering sore due to several issues being left unaddressed). See also Recycling of Ships: Comments on the Report of the Correspondence Group to MEPC 49, Submitted by India, IMO/MEPC 49/3/3, 13 May 2003, ¶ 2.2 (KR-CON) [Comments on the Report].

[89] Peter M. Swift, "How Can Ship Breaking Become a Sound Industry In its Own Right?" (PowerPoint presented in Shanghai, China, 3 March 2005) at slide 14, online: Intertanko <http://www.intertanko.com/templates/DocumentList.aspx?id=20528> (noting that a "ship" does not exhibit any of the elements necessary to characterise it as hazardous waste and is therefore not subject to the provisions of the Basel Convention).

[90] Recycling of Ships: The Application of the Basel Convention, Submitted by the International Chamber of Shipping (ICS) on behalf of the Industry Working Party on Ship Recycling, IMO/MEPC 52/INF.12, 5 August 2004, annex ¶ 2 (KR-CON).

[91] *Ibid.*, ¶ 8.

be considered as hazardous waste.[92] Moreover, the industry argues that any hazardous waste that emanates from the ship is due to the recycling operation and is therefore "created at the recycling facility."[93]

The legal soundness of this opinion is strongly disputed by Greenpeace International.

> [O]f all of the instruments currently in place that impact this issue, the Basel Convention is the only one that is a) legally binding, and b) is in a clear position to actually minimize the export of ships containing hazardous materials to developing countries, and thus is the only instrument well placed to quickly prevent more impoverished workers from being poisoned or otherwise killed from risks associated with hazardous wastes such as flammable materials, carcinogens, etc.[94]

Even though arguments by the shipping industry seem compelling, their line of reasoning is flawed. For one, the primary object of concern is not the wastes that are derived from the normal operation of the ship, rather it is the wastes that are in-built in its structure. Secondly, wastes are not created at the ship recycling facility. All that happens there is that the in-built wastes are exposed when the hulls are ripped open. Thus, it is clear that the Basel Convention tools are relevant to ship dismantling. Such a view received endorsement at the seventh meeting of the COP, which recognised a dual personality for ships. It was decided that even though a ship can become waste under article 2 of the Basel Convention, it can still continue to be a ship under other international rules.

The core principles related to ship dismantling in the Basel Convention include the polluter pays principle, the proximity principle, common but differentiated responsibilities and, most importantly, the benchmark for hazardous wastes disposal, namely, the ESM. Other rules in the convention seek to ensure minimisation of waste material generation[95] and their transboundary movement,[96] the proper labelling, packaging, and transport of hazardous wastes,[97] and securing the availability of adequate disposal facilities within national boundaries.[98] Further, state parties are obliged not to allow the export of the hazardous wastes if the exporting state has reason to believe that the waste will not be

---

[92] *Ibid.*

[93] *Ibid.*, ¶ 4.

[94] Recycling of Ships: Analysis of Inconsistencies between the IMO Guidelines on Ship Recycling and the Basel Convention, Submitted by Greenpeace International, IMO/MEPC 51/INF.11, 6 February 2004, ¶ 3 (KR-CON) [Analysis of Inconsistencies].

[95] *Basel Convention*, *supra* note 3 art. 4(2)(a). See also Recycling of Ships, Proposals for Mandatory Guidelines, Submitted by Greenpeace International, IMO/MEPC 52/3/2, 2 August 2004, ¶ 14 (KR-CON) [Proposals for Mandatory Guidelines] (noting that this obligation can be interpreted to include the development of green designs for shipbuilding).

[96] *Basel Convention*, *ibid.*, art. 4(2)(d).

[97] *Ibid.*, art. 4(7)(b).

[98] *Ibid.*, art. 4(2)(b).

managed in an environmentally sound manner.[99] One of the most important principles in the Basel Convention that can be applied to ship dismantling is the prior informed consent procedure.[100] In addition, this convention also requires parties to prohibit persons under their national jurisdiction from disposing of hazardous wastes unless they are authorised.[101] Though the "criteria and procedure" are not elaborated, article 4(2)(c) requires parties to ensure that the persons involved in the management of hazardous waste take the necessary steps to prevent pollution.

Each of these principles is relevant to ship dismantling and in applying some of them, the following picture emerges: firstly, state parties have to ensure that new ships are built with a minimum of hazardous substances on board; secondly, they must ensure that there is proper decontamination of ships prior to dismantling; and thirdly, the informed consent procedures require a contract between the parties and transactions are to be concluded *via* state parties rather than individual entities.

There are, nonetheless, certain practical difficulties in applying these general Basel Convention principles to the ship recycling industry. Some claim that the Basel Convention was never intended to apply to ships[102] and that the presence of certain hazardous materials in its structure should not necessarily imply that the entire ship is to be treated as hazardous waste. However, the major difficulty stems from the issue of identification of an export state in respect to ships that have reached their end-of-life stage. Under international maritime law, ships are governed by their flag state (the state of registration).[103] It is easy to register and de-register a ship regardless of the shipowner's citizenship or place of residence. In the case of flags of convenience and open registers, there may not be any genuine link between the flag state and the owner of the ship. Therefore, it becomes difficult to identify the export state. Consequently, application of the notification requirements of the Basel Convention to merchant shipping is difficult.[104] Even if it is possible to identify an export state, generally, that state will have very little control over the shipowner.[105] Despite these concerns, the Basel Convention remains a primary binding legal instrument prescribing rules that are relevant to environmentally sound ship dismantling.

---

99 *Ibid.*, art. 4(2)(e).

100 *Ibid.*, art. 6.

101 *Ibid.*, art. 4(7)(a).

102 *Ibid.*, art. 4(12).

103 *United Nations Convention on the Law of the Sea*, 10 December 1982, 21 I.L.M. 1261 (entered into force 1994) [*LOS Convention*], art. 92(1).

104 EU Comments, *supra* note 51 ¶ 38.

105 France, Interdepartmental Committee on the Dismantling of Civilian and Military End-of-Life Ships, *Le rapport de la Mission Interministérielle portant sur le Démantèlement des Navires civils et militaires en fin de vie*, ¶ 3.1 (27 March 2007, Chair: Xavier de la Gorce), online: SGMer <http://www.sgmer.gouv.fr/IMG/pdf/2007-06-18_-_Rapport_MIDN_english_version.pdf>.

III. Shipbreaking and the Law of the Sea

After 14 years of negotiations involving more than 150 countries, the LOS Convention was adopted in 1982, by the Third United Nations Conference on the Law of the Sea. Known as "a constitution for the oceans"[106] because of its comprehensive coverage of the concerns and challenges engendered by the global usage and evolving interests in the oceans, their resources and amenities, the LOS Convention deals with almost all aspects of ocean management. Even though it contains several provisions that are relevant to shipbreaking,[107] the most important relate to obligations on the part of the state to control pollution from land-based activities such as breaking ships and the right to innocent passage of end-of-life ships, particularly when they are being towed to the recycling yard.

International concern over land-based sources of marine pollution emerged in 1970s[108] following the 1972 Stockholm Conference.[109] However, rules of international law in this area continue to be limited, and the LOS Convention remains the most important international instrument that imposes obligations on state parties in respect of land-based sources of marine pollution.[110] Part XII of the LOS Convention, which epitomises the philosophy and jurisprudence of the

---

[106] Tommy T.B. Koh, "A Constitution for the Oceans" (Remarks at the Final Session of the Third United Nations Conference on the Law of the Sea, Montego Bay, 1–11 December 1982), online: United Nations <http://www.un.org/Depts/los/convention_agreements/texts/koh_english.pdf>.

[107] Relevant LOS Convention articles include those dealing with "use of technologies or introduction of alien or new species" (art. 196) which obliges states to prevent the discharge of ballast waters containing harmful pathogens and "measures relating to seaworthiness of vessels to avoid pollution" (art. 219) which enables states to detain obsolete leaking vessels in their ports before they embark on their journey to the recycling yards. Other relevant articles deal with measures to avoid pollution arising from maritime casualties (art. 221) relevant in the context of wreck removal; pollution by dumping (art. 210); enforcement with respect to pollution by dumping (art. 216); nationality of ships (art. 91); status of ships (art. 92); duties of flag state (art. 94); enforcement by flag states (art. 217); enforcement by port states (art. 218); enforcement by coastal states (art. 220); pollution from vessels (art. 211); scientific and technical assistance to developing states (art. 202); and preferential treatment for developing states (art. 203), *etc.* See *LOS Convention*, *supra* note 103.

[108] Daud Hassan, "International Conventions Relating to Land-Based Sources of Marine Pollution Control: Applications and Shortcomings" (2004) 16 Geo. Int'l Envtl. L. Rev. 657 at 661 (WLeC).

[109] For instance, see Principles 6 and 7, *Declaration of the United Nations Conference on the Human Environment*, 1972, 11 I.L.M. 1416 at 1418. See also *Convention for the Prevention of Marine Pollution from Land-Based Sources*, 21 February 1974, 13 I.L.M. 352. The UNEP Montreal Guidelines for the Protection of the Marine Environment against Pollution from Land Based Sources, 24 May 1985, UNEP/GC. Decision 13/8/II, online: UNEP <http://www.unep.org/law/PDF/UNEPEnv-LawGuide&PrincN07.pdf>; *UNEP Global Programme of Action for the Protection of the Marine Environment from Land-Based Activities*, online: UNEP <http://www.gpa.unep.org/index.html?ln=6>; *Washington Declaration on Protection of the Marine Environment from Land-based Activities*, 1 November 1995, online: UNEP <http://www.gpa.unep.org/documents/washington_declaration_english.pdf>.

[110] John H. Bates & Charles Benson, *Marine Environment Law*, looseleaf (London: Lloyd's of London Press Ltd., 1993) ¶ 8.1.

convention's environmental regime,[111] imposes on each state, consistent with its "sovereign right to exploit … natural resources pursuant to its own environmental policies"[112] the general obligation to "protect and preserve the marine environment".[113] When this duty is interpolated onto the shipbreaking scenario, states with shipbreaking facilities who are parties to the LOS Convention are legally obligated to adopt measures to ensure that the shipbreaking process does not contaminate the marine environment. Furthermore, article 194 of the LOS Convention calls on all parties either individually or jointly, to take all measures necessary to prevent, reduce and control pollution of the marine environment.[114] In particular, these measures should minimise to the fullest extent possible[115] "the release of toxic, harmful or noxious substances, especially those that are persistent, from land-based sources"[116] and from vessels.[117] The LOS Convention further obliges state parties to "adopt laws and regulations to prevent, reduce and control pollution of the marine environment from land-based sources … taking into account internationally agreed rules, standards and recommended practices and procedures".[118] The land-based activity of breaking ships introduces a host of toxic substances and energy into the marine environment. Unfortunately, the major shipbreaking nations (India, Pakistan and Bangladesh), even though are parties to the LOS Convention, have yet to enact specific measures to curb such pollution.[119]

Generally, ships despatched to the recycling yard tend to be in leaking condition, and it may happen that contaminated rusty ships may have to be towed through the oceans. For instance, in the action involving the export of ghost ships to the United Kingdom, it was contended that the tandem towing of these defunct ships containing toxic substances for over 45 days across the North Atlantic would pose serious threats to the marine environment of the states through which these ships were to pass.[120] This raises the question of how the LOS Convention

---

[111] Moira L. McConnell & Edgar Gold, "The Modern Law of the Sea: Framework for the Protection and Preservation of the Marine Environment?" (1991) 23 Case W. Res. J. Int'l L. 83 (WLeC); David M. Dzidzornu, "Coastal State Obligations and Powers Respecting EEZ Environmental Protection under Part XII of the UNCLOS: A Descriptive Analysis" (1997) 8 Colo. J. Int'l Envtl. L. & Pol'y 283 (WLeC).

[112] *LOS Convention*, *supra* note 103 art. 193.

[113] *Ibid.*, art. 192.

[114] 'Pollution of the marine environment' has been defined as "the introduction by man, directly or indirectly, of substances or energy into the marine environment, including estuaries, which results or is likely to result in such deleterious effects as harm to living resources and marine life, hazards to human health, hindrance to marine activities, including fishing and other legitimate uses of the sea, impairment of quality for use of seawater and reduction of amenities". *Ibid.*, art. 1(4).

[115] *Ibid.*, art. 194(3).

[116] *Ibid.*, art. 194(3(a).

[117] *Ibid.*, art. 194(3)(b).

[118] *Ibid.*, art. 207(1). See also McConnell & Gold, *supra* note 111 at 92–93.

[119] For further information on the marine pollution from land-based shipbreaking, see Chapter Two entitled, "The Global Business of Shipbreaking".

[120] See *Basel Action Network v. Maritime Administration*, 285 F.Supp.2d 58 (D.D.C., 2003).

provisions on innocent passage apply to end-of-life ships in transit to ship recycling yards.

Although, coastal states have been afforded full sovereignty within the twelve nautical mile territorial sea, normally measured from the baselines,[121] this sovereignty is subject to the right of innocent passage.[122] Even the Basel Convention, while affirming the sovereignty of states over their territorial sea, recognises the exercise by ships of their navigational rights and freedoms as provided under international law.[123] However, this navigational right under the LOS Convention is subject to the condition that it be "innocent". Innocence has been defined negatively to mean passage that is not "prejudicial to the peace, good order or security of the coastal state".[124] In clarifying these terms, article 19(2) of the LOS Convention enumerates a list of activities that can render passage non-innocent. This raises the question of whether the towing of old rusty ships that are contaminated with hazardous substances fall under any of the prohibited acts set out in the LOS Convention.

Under the LOS Convention "[p]assage of a foreign ship shall be considered to be prejudicial to the peace, good order or security of the coastal state if in the territorial sea it engages in … any act of wilful and serious pollution contrary to this Convention."[125] Even though parties know that towing of aged ships can endanger the marine environment, this by itself does not constitute "wilful and serious pollution", which generally extends to intentional acts like dumping.[126] Thus end-of-life ships do have the right to innocent passage. However, this right is not absolute. Coastal states can adopt laws and regulations to protect their interests[127] in the territorial sea, which foreign ships exercising their right of innocent passage will have to comply with.[128] For instance, a ship may not exercise the right of innocent passage in all parts of the territorial sea. Having regard to the safety of navigation, the coastal state can require foreign ships exercising the right of innocent passage through its territorial sea to use such sea lanes and traffic separation schemes as it designates.[129] Even though the LOS Convention does not specify whether the refusal of a ship to obey a coastal state law renders the passage non-innocent, the definition of non-innocent passage suggests that the violation of coastal state laws

---

[121] *LOS Convention*, *supra* note 103 art. 3.

[122] *Ibid.*, art. 17 (noting that in the territorial sea, "ships of all States, whether coastal or land-locked, enjoy the right of innocent passage").

[123] *Basel Convention*, *supra* note 3 art. 4(12).

[124] *LOS Convention*, *supra* note 103 art. 19(1).

[125] *Ibid.*, art. 19(2)(h).

[126] Elaine B. Weinstein, "The Impact of Regulation of Transport of Hazardous Waste on Freedom of Navigation" (1994) 9 Int'l J. Marine & Coast. L. 135 at 141 (HeinOnline).

[127] These include aspects relating to safety of navigation and the regulation of maritime traffic, preservation of the environment of the coastal state, and conservation of living resources of the sea. *LOS Convention*, *supra* note 103 art. 21(1)(a)–(h).

[128] *Ibid.*, art. 21(4).

[129] *Ibid.*, art. 22(1).

can cause the passage to become non-innocent,[130] thereby empowering the coastal state to take "necessary steps" to prevent the passage.[131] Nonetheless, the coastal state cannot adopt measures that have the practical effect of denying or impairing the right of innocent passage through its territorial sea.[132] Thus while end-of-life ships have the right of innocent passage through the territorial sea of coastal states, these states, taking into consideration the condition of the ship, can require such ships to transit only through designated sea lanes or impose terms and conditions under its national laws that the ship will have to respect.

Drafted at a time when shipbreaking was not an issue of international concern, the broad provisions of the LOS Convention continue to influence the developing law for sustainable shipbreaking. However, its provisions are insufficient to regulate the ship recycling business, particularly, in cases where ships are towed to the subcontinent. Article 23 of the LOS Convention states that certain ships, while carrying nuclear or other inherently dangerous or noxious substances, in exercising their right of innocent passage, should carry documents and observe "special precautionary measures". However, end-of-life ships cannot be brought under the convention's ambit under this article as they do not carry hazardous substances as cargo. Nonetheless the fact that these ships have in-built hazardous substances and are in poor physical condition suggests that provisions could be included in the Ship Recycling Convention or its Guidelines concerning the right of innocent passage of such ships subject to similar precautionary measures and documentation.

IV. Safe Shipbreaking and the International Labour Organization

One of the major factors that has affected the sustainability of shipbreaking is the lack of mechanisms – legal, administrative and infrastructural – to ensure occupational safety and health in the yards. This has led the ILO to examine the legal regime for occupational safety and health in relation to shipbreaking.[133] The ILO, the United Nations' specialised body that seeks to promote social justice and internationally recognised human and labour rights at the work place, has played a vital role in developing "international guidance" to protect occupational safety and health in the scrap yards.[134]

---

[130] William K. Agyebeng, "Theory in Search of Practice: The Right of Innocent Passage in the Territorial Sea" (2006) 39 Cornell Int'l L.J. 371 at 383 (HeinOnline).

[131] *LOS Convention, supra* note 103 art. 25(1).

[132] *Ibid.*, art. 24(1).

[133] For a discussion on the legal and administrative structures with regard to protection of labour in the scrap yards, see Chapter Three entitled, "Limitations of a National Response to Regulate the Global Shipbreaking Industry: A Study of the Indian Experience".

[134] Aage Bjørn Andersen, "An Issues Paper: Worker Safety in the Ship-breaking Industries" (2001) at 45 (ILO, Sectoral Activities Programme), online: International Labour Organization <http://www.ilo.org/public/english/dialogue/sector/papers/shpbreak/wp-167.pdf>.

Since its establishment in 1919, the ILO has strived to protect labour through maritime law. Initially, the ILO focused on the rights of seafarers,[135] extending its efforts to protect fishermen's rights[136] and those of dockworkers.[137] Although certain resolutions of the ILO Metal Trades Committee draw attention to the rights of shipbreakers as distinct from other workers, very little has been done to develop a culture of safe shipbreaking. In November 2000, the 279th session of the ILO's governing body took a decision that the ILO develop a compendium of best practices for shipbreaking tailored to the local conditions in the scrap yards.[138]

In 2003, as part of its "Safe Work" agenda, the ILO established the Safety and Health in Shipbreaking: Guidelines for Asian Countries and Turkey (ILO Guidelines). The ILO Guidelines provide direction to those who have the responsibility to ensure occupational safety and health in the shipbreaking yards. Though not legally binding, these Guidelines seek to gradually transform the present informal nature of the activity into a more formal economy.[139] They lay down

[135] *Maritime Labour Conventions and Recommendations including Standards Relating to Fishing, Dock Work and Inland Navigation*, 4th ed. (N.p., 1998) at 2.

[136] For instance, see Hours of Work (Fishing) Recommendation, 1920 (No. 7) and *Fishermen's Competency Certificates Convention, 1966 (No. 125)*. *Ibid.* at 169, 176.

[137] For instance, see *Dock Work Convention, 1973 (No. 137)* and *Occupational Safety and Health (Dock Work) Convention, 1979 (No. 152)*. *Ibid.* at 201, 208.

[138] Paul J. Bailey, "Is There a Decent Way to Break Up Ships?" (2000) Discussion Paper (ILO, Sectoral Activities Programme), online: International Labour Organization <http://www.ilo.org/public/english/dialogue/sector/papers/shpbreak/index.htm#N_1_>. The following ILO conventions and recommendations on occupational safety and health may have relevance to shipbreaking: *Guarding of Machinery Convention, 1963 (No.119)* and Guarding of Machinery Recommendation, 1963 (No.118), 1963; *Maximum Weight Convention, 1967 (No.127)* and Maximum Weight Recommendation, 1967 (No.128); *Occupational Cancer Convention, 1974 (No.139)* and Occupational Cancer Recommendation, 1974 (No.147); *Working Environment (Air Pollution, Noise and Vibration) Convention, 1977 (No.148)* and Working Environment (Air Pollution, Noise and Vibration) Recommendation, 1977 (No.156); *Occupational Safety and Health Convention, 1981* (No.155) and Occupational Safety and Health Recommendations, 1981 (No.164); *Occupational Health Services Convention, 1985* (No.161) and Occupational Health Services Recommendation, 1985 (No.171); *Asbestos Convention, 1986* (No.162) and Asbestos Recommendation, 1986 (No.172); *Chemicals Convention, 1990* (No.170) and Chemicals Recommendation, 1990 (No.177). See *International Labour Standards*, online: International Labour Organization <http://www.ilo.org/ilolex/english/conventions.pdf>; *International Labour Standards*, online: International Labour Organization <http://www.ilo.org/ilolex/english/recommendations.pdf>. There are several other ILO codes of practices on occupational safety and health that can influence shipbreaking, namely, Ambient Factors in the Workplace, 2001; Guidelines on Occupational Safety and Health Management Systems, 2001; Recording and Notification of Occupational Accidents and Diseases, 1995; Safety in the Use of Chemicals at Work, 1993; Technical and Ethical Guidelines for Workers' Health Surveillance, 1998; Safety in the Use of Asbestos, 1984; Occupational Safety and Health in the Iron and Steel Industry, 1983; Occupational Exposure to Airborne Substances Harmful to Health, 1980; Protection of Workers against Noise and Vibration in the Working Environment, 1977; and Safety and Health in Ship Building and Ship Repairing, 1974. See *Safe Work Codes of Practice*, online: International Labour Organization <http://www.ilo.org/public/english/protection/safework/cops/english/index.htm>.

[139] See generally Recycling of Ships: Safety and Health in Shipbreaking: Guidelines for Asian Countries and Turkey, Note by the International Labour Organization (ILO), IMO/MEPC 51/INF.4, 23 January 2004 (KR-CON) [Safety and Health in Shipbreaking, Note].

certain fundamental propositions that encapsulate the labour dimension of shipbreaking:

1. Shipbreaking is one of the most hazardous occupations.[140]
2. Shipbreaking is hazardous waste management.[141]
3. Shipbreaking is not always covered by labour law and social protection.[142]
4. Shipbreaking locations make enforcement of laws and regulations difficult.[143]

Divided into two parts, the ILO Guidelines establish parameters for sound national frameworks for responsible shipbreaking[144] and specify measures for improved safety and health.[145] In developing national laws and regulations, the ILO Guidelines require that the legal framework should be "adequate and appropriate to the specific type of shipbreaking facilities and worker's status in employment."[146] They also provide for the establishment of administrative structures,[147] management systems for occupational safety and health,[148] reporting systems,[149] responsibilities of employers,[150] the rights and duties of workers,[151] and the rights and responsibilities of contractors.[152] Regarding safe shipbreaking operations,[153] the ILO Guidelines call upon shipbreakers to promote a "safety first" culture and segment the shipbreaking process.[154] By doing so, it becomes easier to identify hazardous tasks whereby the breaking can be done in a controlled manner to enhance worker safety and health.[155] The ILO Guidelines elaborate on operational planning,[156] preventive and protective measures,[157] management of hazardous substances,[158] measures against physical, biological and psychosocial hazards,[159]

---

140 Safety and Health in Shipbreaking: Guidelines for Asian Countries and Turkey, ILO, Interregional Tripartite Meeting of Experts on Safety and Health in Shipbreaking for Selected Asian Countries and Turkey, Bangkok, 7–14 October 2003, MESHS/2003/1, ¶ 2.3.1, online: International Labour Organization <http://www.ilo.org/public/english/standards/relm/gb/docs/gb289/pdf/meshs-1.pdf> [ILO Guidelines].

141 *Ibid.*, ¶ 2.3.2.

142 *Ibid.*, ¶ 2.3.3.

143 *Ibid.*, ¶ 2.3.4.

144 *Ibid.* at 6–22.

145 *Ibid.* at 23–72. See generally Safety and Health in Shipbreaking, Note, *supra* note 139.

146 ILO Guidelines, *supra* note 140 ¶ 3.2.2.

147 *Ibid.*, ¶ 3.1(responsibilities and duties of competent authorities), 3.3 (duties of labour inspectorates).

148 *Ibid.*, ¶ 4.

149 *Ibid.*, ¶ 5.

150 *Ibid.*, ¶ 3.4.

151 *Ibid.*, ¶ 3.6, 3.5.

152 *Ibid.*, ¶ 3.8.

153 *Ibid.* at 23–72.

154 *Ibid.* at 23.

155 *Ibid.*, ¶ 7.1.1.

156 *Ibid.*, ¶ 7.

157 *Ibid.*, ¶ 8.

158 *Ibid.*, ¶ 9.

159 *Ibid.*, ¶ 10–12.

safety requirements in respect of tools, machines and equipment,[160] personal protective equipment and clothing,[161] special protection,[162] competence and training,[163] emergency preparedness,[164] and welfare measures.[165]

The ILO Guidelines constitute "the most notable work" done by this international body in the area of shipbreaking.[166] However, given the poor conditions under which most recycling facilities operate, implementation of the ILO Guidelines in the near future may prove difficult. Nevertheless, the ILO Guidelines are an important element in the quest to ensure the sustainable development of this industry.

V. International Maritime Law and Ship Recycling

Even though the IMO has conventions that govern almost all facets of shipping, long conspicuous by its absence were rules and regulations on shipbreaking, a land-based activity. However, several IMO instruments that secure its mandate to ensure "safe, secure and efficient shipping on clean oceans" have relevance to shipbreaking. In analysing these instruments, it is useful to consider whether or not these international maritime law instruments are appropriate for dealing with some of the issues posed by shipbreaking.

A. *Dumping Obsolete Ships: The London Convention and its 1996 Protocol*

A study on shipbreaking cannot remain oblivious to the related issues of dumping, sinking, and abandonment of ships for disposal and their associated legal frameworks. There is always the possibility that shipowners may simply choose to abandon or dump their ships rather than have them recycled. The dumping of obsolete ships and offshore installations is on the rise.[167] A form of marine

---

160 *Ibid.*, ¶ 13.

161 *Ibid.*, ¶ 15.

162 *Ibid.*, ¶ 17.

163 *Ibid.*, ¶ 14.

164 *Ibid.*, ¶ 16.

165 *Ibid.*, ¶ 18.

166 France, Interdepartmental Committee on the Dismantling of Civilian and Military End-of-Life Ships, *Le rapport de la Mission Interministérielle portant sur le Démantèlement des Navires civils et militaires en fin de vie*, annex VI at 2 (27 March 2007, Chair: Xavier de la Gorce), online: SGMer <http://www.sgmer.gouv.fr/IMG/pdf/Annex_6_ILO.pdf>.

167 The controversy over 'Brent Spar' brought this issue to public attention. 'Brent Spar' was an offshore loading and storage facility situated at the Brent field in the North Sea. Commissioned in 1976, it was taken out of service in 1991. Shell UK submitted a proposal for the disposal of the installation at sea to the competent licensing authority in the United Kingdom. It was approved, and a licence to dump was granted. Pursuant to the Oslo Convention, the United Kingdom notified the other parties, none of whom raised any objections even though Greenpeace raised questions regarding its disposal. Shortly thereafter, Greenpeace representatives boarded the 'Brent Spar' in a

pollution,[168] dumping of ships is governed primarily by the *Convention on the Prevention of Marine Pollution by Dumping of Wastes and Other Matter, 1972* (London Convention) and its 1996 Protocol.[169]

In force since 1975, the London Convention was concluded to protect the marine environment from human activities.[170] It calls upon contracting parties "to take all practicable steps to prevent the pollution of the sea by dumping of wastes and other matter."[171] Dumping was defined to include, among other things, "any deliberate disposal at sea of vessels ... platforms or other man-made structures."[172] It defines sea to mean all marine waters other than the internal waters of states.[173] Under the London Convention, dumping is permitted, subject to obtaining a prior "special permit", if the waste or other matter is enumerated in annex II, or to obtaining a "prior general permit" in all other cases of wastes or matter.[174] Dumping of wastes and materials listed under annex I is prohibited.[175]

---

non-violent protest and attracted extensive media coverage. Eventually, Shell abandoned the disposal operation and had it dismantled on land. Louise de La Fayette, "New Developments in the Disposal of Offshore Installations" (1999) 14 Int'l J. Mar. & Coast. L. 523 at 525 (HeinOnline). See also Phillip Abraham, "Offshore Oil and Gas Facility Decommissioning in Nova Scotia and Newfoundland" (2001) 24 Dal. L.J. 333 at 335 (HeinOnline); Guidelines and Standards for the Removal of Offshore Installations and Structures on the Continental Shelf and in the Exclusive Economic Zone, IMO Assembly Res. A.672(16), adopted on 19 October 1989 (KR-CON).

[168] Aldo Chircop, "Law of the Sea and International Environmental Law Considerations for Places of Refuge for Ships in Need of Assistance", in Aldo Chircop & Olof Linden, eds., *Places of Refuge for Ships: Emerging Concerns of a Maritime Custom* (Boston: Martinus Nijhoff Publishers, 2006) 231 at 254–55.

[169] 29 December 1972, 1046 U.N.T.S. 120, 11 I.L.M. 1291 (entered into force 30 August 1975) [*London Convention*], online: International Maritime Organization <http://www.imo.org/includes/blastDataOnly.asp/data_id%3D16925/LC1972.pdf> replaced by *1996 Protocol to the Convention on the Prevention of Marine Pollution by Dumping of Wastes and Other Matter, 1972 and Resolutions Adopted by the Special Meeting,* 7 November 1996, 36 I.L.M. 1 (entered into force 11 March 2008).

[170] IMO, *The London Convention and Protocol: Their Role and Contribution to Protection of the Marine Environment*, online: International Maritime Organization <http://www.imo.org/includes/blastDataOnly.asp/data_id%3D21278/LC-LPbrochure.pdf>; Status of the London Convention and Protocol: Report of the Secretary-General on the Status of the London Convention 1972, IMO/LC 30/2, 2 October 2008, online: International Maritime Organization <http://www.imo.org/includes/blastDataOnly.asp/data_id%3D23854/2.pdf> (as at 2 October 2008, 85 governments have ratified or acceded to the Convention). Hassan, *supra* note108 at 663.

[171] *London Convention*, *supra* note 169 art. I.

[172] *Ibid.*, art. III (1)(a)(ii).

[173] *London Convention*, *supra* note 169 art. III(3). Internal waters are defined as the waters on the landward side of the baseline from which the breadth of the territorial sea is measured. *LOS Convention*, *supra* note 103 art. 8(1).

[174] *London Convention*, *ibid.*, art. IV(1).

[175] *Ibid.* See also Chircop, *supra* note 168 at 225. See also Louise de La Fayette, "The London Convention 1972: Preparing for the Future" (1998) 13 Int'l J. Mar. & Coast. L. 515 at 517 (HeinOnline).

In 1996, a Protocol to the London Convention, based on the precautionary principle[176] and the polluter pays approach,[177] was adopted. This Protocol has replaced the original convention.[178] It requires contracting parties "to prevent, reduce and where practicable eliminate pollution caused by [the] dumping or incineration at sea of wastes or other matter".[179] Like the London Convention, the 1996 Protocol also defines dumping to include "any deliberate disposal into the sea of vessels ... platforms or other man-made structures."[180] In addition, dumping is also defined as "any abandonment or toppling at site of platforms or other man-made structures at sea, for the sole purpose of deliberate disposal." The definition of sea under the Protocol follows the London Convention and it means "all marine waters other than the internal waters of States ..."[181] The striking difference between the original convention and the 1996 Protocol lies in the adoption by the latter of a "reverse listing approach". This means that the dumping of all materials is prohibited unless they are listed in its annex. Even if listed, such materials can be dumped only if there is compliance with certain procedures.[182] The 1996 Protocol also obliges parties to adopt appropriate measures in accordance with international law to prevent and, if necessary, punish acts contrary to its provisions.[183]

Both the London Convention and the 1996 Protocol permit the disposal of obsolete ships at sea.[184] However, the dumping operation has to comply with the

176 *Protocol to the Convention on the Prevention of Marine Pollution by Dumping of Wastes and Other Matter, 1972 and Resolutions Adopted by the Special Meeting,* 7 November 1996, 36 I.L.M. 1 (entered into force 11 March 2008) [*Protocol to London Convention*], art. 3(1).

177 *Ibid.*, art. 3(2).

178 The Protocol enters into force on the 30th day following the date on which 26 states have ratified it, among which there has to be at least 15 contracting parties to the London Convention. *Ibid.*, art. 25. As on 22 January 2009, 37 states have ratified or acceded to the Protocol, of which 31 are contracting parties to the London Convention. Thus, the Protocol is now in force. For further details, see *1996 Protocol to the London Convention 1972: Overview of Contracting* Parties, online: International Maritime Organization <http://www.imo.org/includes/blastDataOnly.asp/data_id%3D25138/PROTPART.1st.doc>. Jon M. Van Dyke, "Sharing Ocean Resources – In a Time of Scarcity and Selfishness", in Harry N. Scheiber, ed., *Law of the Sea: The Common Heritage and Emerging Challenges* (The Hague: Martinus Nijhoff Publishers) 3 at 7–8 (characterising the "transformation" of the *London Convention* as "one of the most impressive success stories of the Nineties". The name of this convention was also changed from London Dumping Convention to London Convention, in order not to give an impression that the convention authorised "dumping").

179 *Protocol to London Convention, supra* note 176 art. 2.

180 *Ibid.*, art. 1(4)(1)(2).

181 *Ibid.*, art. 1(7).

182 *Ibid.*, art. 4(1).

183 *Ibid.*, art. 10(2).

184 *Ibid.*, annex 1(1)(4) (permits the dumping of "vessels and platforms or other man-made structures at sea"); IMO Guidelines on Ship Recycling: Res. A.962(23) Adopted on 5 December 2003 (Agenda Item 19), A 23/Res.962, 4 March 2004, s. 9.7 (KR-CON) [IMOGSR].

"Specific Guidelines for Assessment of Vessels"[185] (SGAV), which permits dumping of vessels for the purpose of creating artificial reefs, for marine habitat enhancement and for recreational diving sites. The SGAV sets out factors that need to be addressed when a state approves the disposal of a vessel at sea. It emphasises that prior to sea disposal being determined as the preferred mode, other alternatives need to be evaluated.[186] The SGAV identifies several disposal options, ranging from re-use of the vessel or its parts to recycling. There has to be a comparative risk assessment of each of these methods,[187] taking into account potential environmental[188] and health impacts,[189] technical and practical feasibility,[190] and economic factors.[191] There is also to be an "impact hypothesis", a concise statement of the expected consequences of the disposal option which provides a basis to decide whether or not to approve or reject the proposed disposal option.[192] If the permitting authority determines that opportunities exist to re-use or recycle the waste without causing undue risks to human health or to the environment, or that it does not involve disproportionate costs, the permit to dump can be refused.[193]

In disposing of a vessel at sea, a "pollution prevention plan"[194] is to be developed. Certain "best environmental practices", the primary objective of which is to ensure that wastes that can potentially pollute the marine environment have been removed to the maximum extent possible, are to be followed.[195] Once cleaned, the next step is to identify a proper "dump-site at sea".[196] Factors to be considered in selecting a site include the physical and biological characteristics of the seabed and its surrounding areas, potential implications of the vessel's presence on amenities in the area under consideration, economic and operational feasibility, and size of the dump-site and its capacity.[197]

Another related issue is that of abandonment of a ship for disposal. Once shipowners find that a ship is no longer useful and that it needs recycling, or that the

---

[185] *Specific Guidelines for Assessment of Vessels, 1997*, online: International Maritime Organization <http://www.imo.org/includes/blastDataOnly.asp/data_id%3D17023/4-Vessels.pdf>. This document is an "authoritative interpretation" meeting the requirement of both the London Convention and the 1996 Protocol by the 19th Consultative Meeting of Contracting Parties to the London Convention.

[186] *Ibid.*, ¶ 1.5.

[187] *Ibid.*, ¶ 3.3.

[188] *Ibid.*, ¶ 3.3.1.

[189] *Ibid.*, ¶ 3.3.2.

[190] *Ibid.*, ¶. 3.3.3.

[191] *Ibid.*, ¶ 3.3.4.

[192] *Ibid.*, ¶ 7.1.

[193] *Ibid.*, ¶ 3.2. See also *ibid.*, ¶ 9.

[194] *Ibid.*, ¶ 4.1.

[195] *Ibid.*, ¶ 5.

[196] *Ibid.*, ¶ 6.1.

[197] *Ibid.*, ¶ 6.2, 6.5–6.6.

ship has become a "constructive total loss" or is "detained for safety reasons by a port state control inspector,"[198] the shipowner may choose to abandon the ship to avoid hassles and expenses.[199] Any act of abandonment of a ship for disposal (*i.e.* taking the crew off and simply letting it go) constitutes an uncontrolled dumping and is covered by the 1996 Protocol.[200] In cases where the abandonment is in internal waters, the matter will have to be determined according to the national laws of the state concerned.[201] Even though "internal waters" have been excluded under the Protocol, states have the discretion to opt-in and to extend the application of the 1996 Protocol in respect of dumping or incineration to their internal waters.[202] In cases where a ship is abandoned consequent to a maritime casualty, the legal position is different and falls under the terms of the *Nairobi International Convention on the Removal of Wrecks, 2007*.[203]

B. *Recycling Shipwrecks and the Wreck Removal Convention*

The issue of shipwrecks and their removal may intersect with ship recycling.[204] The IMO reports that abandoned wrecks, numbering nearly 1,300 worldwide, pose considerable threats to safe navigation and to the ecological integrity of the marine and coastal environments.[205] To illustrate this point, reference is made to the shipwreck 'Tricolor'.

On 14 December 2002, the car carrier 'Tricolor' collided with the container ship 'Kariba' when both ships were about to enter the English Channel.[206] The 'Tricolor' sank and its wreck settled in the middle of a busy sea lane in French waters, close to the exclusive economic zones of Belgium and the

---

198 Chircop, *supra* note 168 at 261.

199 *Ibid.*

200 Consideration of the Work Programmes of the Pertinent Bodies of ILO, IMO and the Conference of Parties to the Basel Convention on the Issue of Ship Recycling: Abandonment of ships, Note by the IMO Secretariat, ILO/IMO/BC WG 1/2/2, 17 January 2005, online: Basel Convention <http://www.basel.int/ships/abandonment.html> [Work Programmes of Pertinent Bodies].

201 The term 'sea' has been defined to mean "all marine waters other than the internal waters …". *Protocol to London Convention, supra* note 176 art. 1(7). See also *LOS Convention, supra* note 103 art. 25(1) ("[i]n the case of ships proceeding to internal waters …, the coastal State also has the right to take the necessary steps to prevent any breach of the conditions to which admission of those ships to internal waters or such a call is subject").

202 *Protocol to London Convention, ibid.*, art. 7(2).

203 See Work Programmes of Pertinent Bodies, *supra* note 200.

204 Gold, Chircop & Kindred, *supra* note 2 at 623.

205 *Nairobi International Convention on the Removal of Wrecks, 2007*, online: International Maritime Organization <http://194.196.162.45/Legal/mainframe.asp?topic_id=1319>. Valerie J. Brown, "Warfare: Iraq's Toxic Shipwrecks" *Environmental Health Perspectives* 113:4 (April 2005) at A230 (JSTOR) (pointing out to the dangers posed by "Iraq's toxic shipwrecks", numbering nearly 260, that clog local waters).

206 Francis Kerckhof, Patrick Roose & Jan Haelters "The Tricolor Incident: From Collision to Environmental Disaster" (2004) 6:3 Atlantic Sea Birds 85, online: Royal Netherlands Institute for Sea Research <http://www.nioz.nl/public/mee/birds/0272.pdf>.

United Kingdom.[207] Two days later, the German coaster 'Nicola' ran into the wreck of the 'Tricolor'. However, as 'Nicola' was empty at the time of the collision, there was only minor damage.[208] On 1 January 2003, despite adoption of safety measures and mariners being alerted, the Turkish tanker, 'Vicky', carrying gasoline and several tonnes of heavy fuel on board, ran into the 'Tricolor'.[209] Not only did the collision damage the 'Vicky', but it further damaged the wreck. Both vessels began to leak hydrocarbons, contaminating the marine environment and resulting in thousands of oiled seabirds being washed ashore.[210]

Removing wrecks is expensive, and it may happen that shipowners attempt to evade responsibility for their removal.[211] At the international level, the *Nairobi International Convention on the Removal of Wrecks, 2007* (WRC)[212] provides uniform international rules and procedures for prompt and effective removal of wrecks and payment of compensation for the costs involved. The objective of the WRC is to ensure that wrecks do not pose any hazards to navigation or to the marine environment.[213] Wrecks[214] in the convention area[215] and, in certain cases, those situated within the territory of a state, including its territorial sea,[216] are brought within the regulatory ambit of the WRC. The WRC also specifies the standard which state parties are to consider while initiating measures[217] to effect wreck removal, namely that it should be "proportionate to the hazard."[218]

---

207 *Ibid.* at 86.

208 *Ibid.* at 87.

209 *Ibid.*

210 *Ibid.*

211 In November 1997, a Belize general cargo ship, "An Tai", arrived in Port Klang, Malaysia, from China to discharge its cargo. However, the ship broke its hull and sank. Its owner was a Chinese company, 'An Tai Navigation Enterprise', which had insurance with Ocean Marine Mutual Insurance Association Ltd. After sinking, the vessel discharged fuel oil, severely polluting the harbour. Although the Marine Department ordered removal of the wreck, the owner claimed inability to do so due to a lack of financial resources. The insurer also raised reservations on the cover provided. Finally, the Malaysian authorities had to remove the wreck at their own expense. Draft Convention on Wreck Removal, Experience of An Tai incident, Submitted by Malaysia, IMO/LEGAL COMMITTEE LEG 83/5/2, 14 September 2001 (KR-CON).

212 For the text, see 46 I.L.M. 697 (open for signature since 19 November 2007, not yet in force) [*WRC*].

213 *Ibid.*, Preamble.

214 Under the convention, a wreck is the result of a maritime casualty, and means a sunken or stranded ship; or any part of a sunken or stranded ship, including any object that is or has been on board such a ship; or any object that is lost at sea from a ship and that is stranded, sunken or adrift at sea; or a ship that is about, or may reasonably be expected, to sink or to strand, where effective measures to assist the ship or any property in danger are not already being taken. *Ibid.*, art. 1(4).

215 Convention area is the exclusive economic zone of a state party established in accordance with international law or, if a state party has not established such a zone, an area beyond and adjacent to the territorial sea of that state determined by that state in accordance with international law and extending not more than 200 nautical miles from the baselines from which the breadth of its territorial sea is measured. *Ibid.*, art. 1(1).

216 *Ibid.*, art. 3(2).

217 *Ibid.*, art. 2(1).

218 *Ibid.*, art. 2(2).

In the event of a maritime casualty resulting in a wreck, a state party is to require that the master and the operator of a ship flying its flag report[219] the matter immediately to the affected state,[220] which is to make a determination as to whether the wreck poses a hazard or not.[221] If a determination is made that the wreck constitutes a hazard, then the affected state is to immediately inform the matter to the ship's registry and the registered owner, setting out a reasonable deadline within which the registered owner is to remove the wreck. In cases that warrant immediate action, the affected state, after informing the state where the ship is registered and the registered owner, can remove the wreck.[222] In such cases, the registered owner is liable for the costs involved.[223]

Wreck removal is integral to a holistic understanding of the dynamics of ship-breaking. However, the WRC does not specify any steps that need to be adopted to ensure the safe disposal of a recovered wreck.[224] This is yet another matter that the IMO will have to consider while giving final shape to the ship recycling regime.

C. *The Anti-Fouling Systems Convention and Ship Recycling*

Sea life such as algae, barnacles and molluscs attach to the hull and quickly cover parts of the ship that are submerged below the water line. This leads to increased drag, which affects the hydrodynamics of the ship, increasing fuel consumption and costs.[225] Such ships also unwittingly act as vectors facilitating the spread of aquatic organisms and pathogens from one ecosystem to another. To combat this menace, anti-fouling paints have been used to coat the bottoms of ships.[226]

---

[219] Such reports are to provide relevant information in order for the affected state to determine whether the wreck poses a hazard or not. *Ibid.*, art. 5(2).

[220] *Ibid.*, art. 5(1).

[221] *Ibid.*, art. 6.

[222] *Ibid.*, art. 9(8).

[223] *Ibid.*, art. 10. A registered owner of a ship of 300 GT and above is to maintain insurance or other financial security to cover liability under this convention. *Ibid.*, art. 12.

[224] The majority of wrecks are found in inshore waters and the territorial sea, considerably enhancing the interest of coastal states in ensuring their removal. Moreover, the WRC is applicable mainly to the exclusive economic zone even though states can apply it to their territorial waters as well. Accordingly, wreck removal is normally carried out by applying the domestic law of the jurisdiction where the wreck lies. For an overview of the Canadian position, see Gold, Chircop & Kindred, *supra* note 2 at 620.

[225] Gary Conner, "Scraping the Hull: Ridding your organization of barnacles" (6 April 2004), online: thefabricator.com <http://www.thefabricator.com/CEOs/CEOs_Article.cfm?ID=877> (pointing out that fuel costs due to fouling could be as high as USD1.4 billion a year). *The Oxford Encyclopedia of Maritime History, e-reference ed.*, John J. Hattendorf ed. (USA: Oxford University Press, 2007) "Fouling" (noting that in 1905, the Russian navy suffered arguably the most decisive naval defeat in history at Tsushima in the Korea Strait; one of the reasons for this being that the Russian ships were so heavily fouled that they could not manoeuvre effectively).

[226] Vessel bottoms that are not protected by anti-fouling systems can gather 150 kg of fouling per square metre in less than six months of being at sea. On a very large crude carrier with 40,000 square metres this would translate to about 6,000 tonnes of fouling. "Anti-fouling Systems" (2002) Discussion Paper, 3, online: International Maritime Organization <http://www.imo.org/includes/blastDataOnly.asp/data_id%3D7986/FOULING2003.pdf> ["Anti-fouling Systems"].

From the early days of sail, chemicals such as lime and arsenic were used to protect the hull.[227] Even though new paint technologies were developed, it was necessary that the useful life of the paint be prolonged as much as possible to minimise the need for dry docking for hull cleaning thereby reducing costs.[228] A breakthrough came with the development of self-polishing anti-fouling paints.[229] More expensive than the traditional paints, these thick coatings smooth out over time providing up to five years of protection between dry dockings.[230] These benefits are attributable to the chemical properties of their major ingredient, the organotin compound tributyltin (TBT).[231] An extremely toxic chemical, TBT[232] persists in water, harming marine life and the marine environment,[233] and can even enter the food chain thereby endangering human health.[234]

The *International Convention on the Control of Harmful Anti-fouling Systems on Ships 2001* (AFC 2001) recognises the importance of "protecting the marine environment and human health from the adverse effects of anti-fouling systems".[235] To this end, parties are to prohibit or restrict the use of harmful anti-fouling systems on ships that fly their flag or those that operate under their authority, as well as on ships that enter its ports, shipyards or offshore terminals.[236]

---

[227] *Anti-fouling Systems*, online: International Maritime Organization <http://www.imo.org/>.

[228] D.J. Eyres, *Ship Construction*, 6th ed. (Boston: Butterworth-Heinemann, 2007) at 320 (ScienceDirect).

[229] *Ibid.*

[230] *Ibid.*

[231] *Ibid.*

[232] *Anti-fouling Systems*, *supra* note 227.

[233] TBT is known to cause shell deformations in oysters; sex changes (imposex) in whelks; and immune response, neurotoxic and genetic defects in other marine species. See "Anti-fouling Systems", *supra* note 226 at 1. See generally Berend P. Mensink *et al.*, "Tributyltin Exposure in Early Life Stages Induces the Development of Male Sexual Characteristics in the Common Whelk, *Buccinum undatum*" (1996) 42:I-4 Mar. Envtl. Research 151 (ScienceDirect) (pointing out that Common Whelks (*Buccinum undatum*) from the open North Sea show imposex, the development of male sexual characteristics in female prosobranch gastropods. Tributyltin used in anti-fouling paints is expected to be the cause).

[234] See generally S.M. Evans, T. Leksono & P.D. McKinnell, "Tributyltin Pollution: A Diminishing Problem Following Legislation Limiting the Use of TBT-Based Anti-fouling Paints (1995) 30:1 Mar. Pollution Bull. 14 (ScienceDirect) (the first evidence of the harmful effects of TBT on oysters came from Arcachon Bay on the west coast of France in the mid-1970s. TBT contamination from boats is believed to have caused high mortalities of oyster larvae and severe malformations of the shells of adults rendering them unmarketable).

[235] See *International Convention on the Control of Harmful Anti-fouling Systems on Ships*, 5 October 2001 (entered into force on 17 September 2008), preamble ¶ 6 [*AFC 2001*], online: Australian Government <http://www.environment.gov.au/coasts/pollution/antifouling/pubs/convention.pdf>. This convention has been ratified by 40 ratifications as on 31 August 2009. See also *Summary of Status of Convention: As at 31 August 2009*, online: International Maritime Organization <http://www.imo.org/conventions/mainframe.asp?topic_id=247>. See also *Report of the United Nations Conference on Environment and Development*, UN Doc. A/CONF.151/26 (Vol. II), (1992) ¶ 17.32 (calling on states to take measures to reduce water pollution caused by organotin compounds used in anti-fouling paints).

[236] *AFC 2001*, *ibid.*, art. 3(1).

Ships' anti-fouling systems on hulls or external parts are not to contain organotin compounds. In cases where ships do have coatings containing organotins, they are to have a barrier to prevent its leaching.[237]

In the most commonly used method of shipbreaking on the Indian subcontinent, *i.e.,* beaching, the anti-fouling paint scraps off from the ship's hull and gets mixed with the soil and water. Due to tidal action and ecosystem processes, the mixture gets dispersed, contaminating the coastal environment.[238] It has been reported that fish from areas that have a thriving ship scrapping business show a high presence of TBT.[239] Moreover, the cutting of the ship into manageable pieces requires the use of oxy-acetylene torches, releasing toxic fumes from the paint which is injurious to the workers' health.[240] Furthermore, the steel pieces obtained from the breaking of the hull are rarely cleaned before they are fed into the furnace for melting, releasing toxins into the atmosphere.

The provisions of the AFC 2001 mandating over-coating of existing TBT paint with sealer coat are problematic for shipbreakers. Sealer coats make it virtually impossible to remove the bottom layer of TBT paint[241] and because of this the risks associated with recycling the ship's hull and other parts encrusted with organotin are not reduced. Therefore, it should be a mandatory requirement that TBT sealer coats on the hull are sand blasted and removed before a ship is exported for recycling.[242]

D. *Aquatic Nuisance Species, the Ballast Water Convention and Ship Recycling*

Marine pollution generally brings to mind images of bleeding oil tankers with oil spurting out of its hull and fish and other marine life swimming helter-skelter to escape the deadly oil slick.[243] In recent times, however, the international community

---

[237] *Ibid.,* annex 1.

[238] Andrew P. Negre *et al.*, "Understanding Ship-grounding Impacts on a Coral Reef: Potential Effects of Anti-foulant Paint Contamination on Coral Recruitment" (2002) 44 Mar. Pollution Bull. 111 at 111 (ScienceDirect). The Malaysian cargo ship 'Bunga Teratai Satu' ran aground on Sudbury Reef, within the Great Barrier Reef World Heritage Area. Even though the ship was refloated, the reef sustained extensive structural damage and anti-fouling paint was scrapped during the initial grounding and subsequent refloating of the hull. *Ibid.*

[239] See generally Kurunthachalam Kannan *et al*, "Butyltins in Muscle and Liver of Fish Collected from Certain Asian and Oceanian Countries" (1995) 90:3 Envtl. Pollution 279 (ScienceDirect).

[240] *Recycling of Ships: Report of the Correspondence Group*, Submitted by the Co-ordinator of the Correspondence Group, IMO/MEPC 46/7, 18 January 2001, ¶ 7.1.6 (KR-CON) [*Report of the Correspondence Group*].

[241] Recycling of Ships, Comments on document MEPC 48/3 Submitted by India Marine Environment Protection Committee 48th Session Agenda Item 3 MEPC 48/3/2, 9 Aug. 2002, ¶ 2.4 (KR-CON).

[242] Comments on the Report, *supra* note 88 ¶ 1.4.

[243] For an overview of oil spills and their impact on the marine environment, see Christine Cartwright, "Natural Resource Damage Assessment: The Exxon Valdez Oil Spill and Its Implications,"

has focused on another kind of pollution that is as insidious as oil spills but which is more rampant and causing greater havoc to marine biodiversity. This is the introduction of invasive species or hitch-hiking harmful aquatic organisms through the discharge or exchange of ballast waters into new environments.[244] In fact, ballast water discharge has been identified as one of the four greatest threats to the world's oceans.[245]

Ballast is basically any material that is used to balance an object to maintain its buoyancy.[246] An essential element in ship safety, ballast controls the draft and trim of the ship while under operation.[247] A ship carrying little or no cargo rides high in the water rendering it vulnerable to being knocked over by waves and winds. Consequently, at the start of a voyage, a ship will take on huge quantities of water from the area where it is floating. At the end of the voyage, the ballast water is discharged.

As ships move from one port to another, loading and unloading goods and taking up and discharging water for ballast, they can inadvertently introduce invasive species in different locations.[248] Literally thousands of marine species are pumped along with the water into the ship's ballast tanks. Resembling

---

Note, (1991) 17 Rutgers Computer & Tech. L.J. 451 (WLeC); Bill Shaw, Brenda J. Winslett & Frank B. Cross, "The Global Environment: A Proposal to Eliminate Marine Oil Pollution" (1987) 27 Nat. Resources J. 157 (HeinOnline).

[244] In contrast to other forms of marine pollution, e.g., oil spills, where ameliorative action can be taken and from which the environment will eventually recuperate, the impacts of invasive marine species are most often irreversible. See generally *The Problem*, online: GloBallast Partnerships <http://globallast.imo.org/index.asp?page=problem.htm&menu=true>.

[245] See "Global Ballast Water Management Programme" *Ballast Water News* Issue 3 (October-December 2000) 1, online: Ballast Water News <http://globallast.imo.org/BallastWaterNews3.pdf>.

[246] This problem has been addressed at the international level by the *International Convention for the Control and Management of Ships' Ballast Water and Sediments, 2004* (not in force) [*BWC*]. For the text, see generally Adoption of the Final Act and Any Instruments, Recommendations and Resolutions Resulting from the Work of the Conference, International Convention for the Control and Management of Ships' Ballast Water and Sediments, 2004, IMO BWM/CONF/36, 16 February 2004. Ballast is defined under the BWC as "water with its suspended matter taken on board a ship to control trim, list, draught, stability or stresses of the ship," *Ibid*., art. 1(2). Article 1(3) defines ballast water management as the "mechanical, physical, chemical, and biological processes, either singularly or in combination, to remove, render harmless, or avoid the uptake or discharge of Harmful Aquatic Organisms and Pathogens within Ballast Water and Sediments". Harmful aquatic organisms and pathogens means "aquatic organisms or pathogens which, if introduced into the sea including estuaries, or into fresh water courses, may create hazards to the environment, human health, property or resources, impair biological diversity or interfere with other legitimate uses of such areas". *Ibid*., art. 1(8).

[247] Jeremy Firestone & James J. Corbett, "Coastal and Port Environments: International Legal and Policy Responses to Reduce Ballast Water Introductions of Potentially Invasive Species" (2006) 7:1 S.D. L. & Pol'y 45 at 45 (WLeC).

[248] Sarah McGee, "Proposals for Ballast Water Regulation: Biosecurity in an Insecure World" [2001] Colo. J. Int'l Envtl. L. & Pol'y 141 at 142 (WLeC); Lisa A. Brautigam, "Control of Aquatic Nuisance Species Introductions Via Ballast Water in the United States: Is the Exemption of Ballast Water Discharges from Clean Water Act Regulation a Valid Exercise of Authority by the Environmental Protection Agency?" (2001) 6 Ocean & Coast. L.J. 33 at 38 (WLeC).

"floating aquariums",[249] these species include bacteria and other microbes, small invertebrates and eggs, cysts and larvae of various species, and, sometimes, even fish.[250]

Ballasting and de-ballasting and the conditions inside the ballast tanks can be intimidating to organism survival and, consequently, the vast majority of aquatic species do not endure the journey.[251] But for those which make it through this ordeal, the next challenge is survival in the new environment.[252] It may happen that all factors are favourable, and an introduced species survives and establishes a reproductive population in the host environment.[253] It may become part of the background flora and fauna, or it may be invasive and out-compete native species, and threaten already endangered species and human health.[254] The need to regulate the discharge of ballast waters thus becomes critical.[255]

Managing the vector of ballast water has been on the international maritime agenda for quite some time. Sporadic attempts dealt with the problem through international agreements and declarations. Countries also have obligations under articles 196(1), 192, 194(1)(5), 195(2) of the LOS Convention and article 8(h) of the Convention on Biological Diversity, 1992, to control the spread of invasive species. Eventually, the IMO concluded an international instrument, the

---

[249] Sharonne O'Shea & Allegra Cangelosi, "Trojan Horses in Our Harbors: Biological Contamination from Ballast Water Discharge" (1996) 27 U. Tol. L. Rev. 381 at 381 (WLeC).

[250] Steve Raaymakers, "The Ballast Water Problem: Global Ecological, Economic and Human Health Impacts" (Paper Presented to the RECSO /IMO Joint Seminar on Tanker Ballast Water Management & Technologies, Dubai, UAE, 16–18 December 2002) 4, online: International Maritime Organization <http://www.imo.org/includes/blastDataOnly.asp/data_id%3D8595/RaaymakersGlobalImpactsPaper.pdf>.

[251] Moira L. McConnell, "Ballast and Biosecurity: The Legal, Economic and Safety Implications of the Developing International Regime to Prevent the Spread of Harmful Aquatic Organisms and Pathogens in Ships' Ballast Water" (2003) 17 Ocean Y.B. 213 at 218 (noting that there are estimates that such species can survive even for three months. The intake of organisms is increased if the ballast is conducted in shallow or turbulent waters close to the shore and at night, when species move to the ocean surface).

[252] Most of the organisms die in the ballast tanks as they are deprived of light, which they need for photosynthesis. But some other organisms or species, e.g., dinoflagellates, commonly form cysts when conditions are unfavourable. These cysts tend to accumulate in the sediments of the ballast tanks where they remain in a state of dormancy until they are deposited in a suitable environment. Sue Matthews & Kobie Brand, "Africa Invaded: The Growing Danger of Invasive Alien Species" *The Global Invasive Species Programme*, 2004, at 41, online: GiSP <http://www.gisp.org/publications/invaded/gispAfrica.pdf>.

[253] Raaymakers, *supra* note 250.

[254] In 1991, a cholera epidemic struck Peru, the first in Latin America for more than a century. Several million people were infected and more than 10,000 died. Over the next four years, Latin American governments poured more than USD200 billion into emergency repairs of sewage and drinking water systems. Peru lost USD1 billion in seafood exports and tourist income. The type of cholera was traced to ballast water carried from Bangladesh. Tim Low, "Ballast Invaders: the Problem and Response" *Invasive Species Council, Australia* (September 2003) 2, online: Invasive Species Council <http://www.invasives.org.au/downloads/tl_ballast.pdf>.

[255] Rebecca Adams, "The Ecological Decline of the Black Sea," Notes, [1997] Colo. J. Int'l Envtl. L. & Pol'y 209 at 213 (WLeC).

International Convention for the Control and Management of Ships' Ballast Water and Sediments (BWC).[256]

The BWC contains several measures for ballast water management. The main mode envisaged to control the introduction of invasive species is ballast water exchange (discharge of port/coastal water and take up of new water). Ballast water exchange is to be conducted at a distance of at least 200 nautical miles from the nearest land and in water which is at least 200 metres in depth. In cases where the ship is unable to do this, the exchange can be conducted in areas at least 50 nautical miles from the nearest land and where the depth of the water is at least 200 metres.[257]

Generally, even on their final voyage, ships carry residual cargo which is offloaded before it moves to the dismantling yard. In the process, the ship takes up ballast water that is discharged prior to beaching the ship to make it as light as possible. Unfortunately, this dimension of shipbreaking is yet to receive the attention it deserves since most of the major shipbreaking nations are yet to sign the BWC.[258] Interestingly, the IMOGSR requires that the last owner adopt measures to ensure that the ship's ballast water is managed in accordance with relevant international and national standards.[259] But since the IMOGSR is rarely followed by shipowners and recyclers in recycling operations, the introduction of invasive species through ballast waters in the coastal waters adjacent to the recycling yards continues to be an issue of concern.

E. *Ship Recycling and MARPOL 73/78*

The *International Convention for the Prevention of Pollution from Ships, 1973* and its *Protocol of 1978* (MARPOL 73/78), often hailed as one of the most effective international legal instruments to combat the menace of vessel-source marine pollution, deals with certain matters that have relevance in shipbreaking.[260] In fact, MARPOL 73/78 has been the catalyst to reforming ship recycling practices through its timetable to phase-out single-hull vessels. Under the revised Regulation

---

[256] *BWC, supra* note 246.

[257] *Ibid.*, reg. B-4(1).

[258] However the situation in India seems to be slightly different because the judiciary-sponsored Guidelines on shipbreaking requires that as part of the ship specific dismantling plan states have to address this issue. See *Research Foundation for Science Technology and Natural Resource Policy v. Union of India* (2007), [2007] A.I.R. S.C. 3118 (India S.C.) (affirming the recommendations of the High Power Committee appointed to study the hazards associated with shipbreaking, one of which related to ballast exchange in the high seas before the ship is brought for dismantling).

[259] IMOGSR, *supra* note 184 s. 8.3.3.1.5.

[260] For more details see *International Convention for the Prevention of Pollution from Ships*, 2 November 1973, 1340 U.N.T.S. 184, 12 I.L.M. 1319 as modified by the *Protocol of 1978 Relating to the International Convention for the Prevention of Pollution from Ships of 1973*, 17 February 1978, 1340 U.N.T.S. 61, 17 I.L.M. 546 [*MARPOL 73/78*].

13G of annex I to MARPOL 73/78, the final phase-out date for Category 1 tankers (pre-MARPOL tankers) was brought forward to 2005 from 2007. The final phase-out date for Category 2 and 3 tankers (MARPOL tankers and smaller tankers) has also been brought forward to 2010, from 2015.[261]

As noted earlier, the Basel Convention does not apply to wastes derived from the normal operations of a ship[262] as these are covered by MARPOL 73/78.[263] The IMOGSR, as will be seen below, requires the development of a green passport consisting of three parts. Similarly, the Ship Recycling Convention also calls upon parties to develop an inventory of hazardous materials (IHM) that consists of three parts. MARPOL 73/78 annexes are relevant[264] in developing parts two and three of the green passport and the IHM dealing with operationally-generated wastes and stores.

MARPOL 73/78 also exerts influence on shipbreaking by requiring parties to establish port reception facilities. Regulation 12(1) of annex I require governments to provide reception facilities at oil loading terminals, repair ports, and in other ports sufficient to meet the needs of ships that have oily residues to discharge. Regulation 12(2)(c) extends this obligation to all ports having ship repair yards or tank cleaning facilities. Regulation 17(1)(c) of annex VI requires parties to the 1997 Protocol to ensure provision of adequate port reception facilities for ozone-depleting substances and equipment in ship recycling yards. Even though this is the only provision in MARPOL 73/78 that explicitly mentions the need for reception facilities at recycling yards, there are practical difficulties in applying these provisions. It is doubtful whether most of the recycling sites, particularly those on the Indian subcontinent which are mainly beachheads, can be deemed as "ports" and thus be required to have reception facilities called for under MARPOL 73/78.[265]

F. *The IMO Guidelines on Ship Recycling, 2003*

Pursuant to a proposal by Norway, the Marine Environment Protection Committee (MEPC) of the IMO in its 43rd session[266] decided to include ship recycling on its agenda. Thereafter, the 44th session provided a platform to discuss, for the first time, the role of IMO in ship recycling and the terminology to describe this

---

261 *2003 Amendments to MARPOL*, 4 December 2003, entry into force April 2005, online: IMO <http://www.imo.org/Conventions/contents.asp?doc_id=678&topic_id=258#2001>.

262 *Basel Convention, supra* note 3 art. 1(4).

263 See *MARPOL 73/78, supra* note 260 annexes IV–V.

264 *Ibid.*

265 *Report of the Correspondence Group, supra* note 240 ¶ 4.3.1.3 (noting that once the ship is beached or moored it is doubtful whether MARPOL 73/78 applies).

266 *Marine Environment Protection Committee – 43rd session: 28 June – 2 July 1999*, online: International Maritime Organization <http://www.imo.org/newsroom/mainframe.asp?topic_id=109&doc_id=357>.

activity.[267] The 44th session also established a correspondence group, chaired by Bangladesh, to report on current ship recycling practices and to identify areas where the IMO could meaningfully contribute.[268]

At MEPC's 46th session, it was decided that the correspondence group should work on a new set of issues.[269] The 47th session reaffirmed the important role which IMO was to play in ship recycling. Recognising the need for co-ordination between the IMO, the ILO and the Basel Convention in recycling matters,[270] it was agreed that MEPC should develop recommendatory guidelines on ship recycling to be adopted by an Assembly resolution. Accordingly, a working group was established to develop a draft outline of the IMOGSR.[271] The 48th session of the MEPC discussed and readied the draft IMOGSR,[272] which was finalised by the 49th session[273] and adopted as a resolution by the 23rd Assembly.[274]

The IMOGSR is a voluntary non-binding code of "best practices" on ship recycling. Even though it recognises ship recycling as the best option available for the sustainable disposal of time-expired tonnage, it re-affirms that "the working practices and environmental standards in the recycling facilities often leave much to be desired".[275] The IMOGSR, while assigning responsibility to ensure environmental and worker protection to the recycling facility and the regulatory authorities of the country in which these are situated, notes that other stakeholders also have a responsibility to address these issues and can contribute meaningfully towards achieving results.[276] Some of the major features of the IMOGSR are discussed below.

The IMOGSR introduces the concept of a green passport[277] for ships. This is a document that is to accompany a ship throughout its operating life. It provides

---

[267] *Marine Environment Protection Committee (MEPC): 46th session: 23–27 April 2001*, online: International Maritime Organization <http://www.imo.org/newsroom/mainframe.asp?topic_id=109&doc_id=774#single> [*MEPC: 46th session*].

[268] *Marine Environment Protection Committee – 44th session: 6–13 March 2000*, online: International Maritime Organization <http://www.imo.org/newsroom/mainframe.asp?topic_id=109&doc_id=358>.

[269] *MEPC: 46th session, supra* note 267.

[270] *Marine Environment Protection Committee (MEPC), 47th session: 4–8 March 2002*, online: International Maritime Organization <http://www.imo.org/newsroom/mainframe.asp?topic_id=109&doc_id=1753#otherissues>.

[271] *Ibid.*

[272] *Marine Environment Protection Committee (MEPC), 48th session: 7–11 October 2002*, online: International Maritime Organization <http://www.imo.org/newsroom/mainframe.asp?topic_id=109&doc_id=2543>.

[273] *Marine Environment Protection Committee (MEPC), 49th session: 14–18 July 2003*, online: International Maritime Organization <http://www.imo.org/newsroom/mainframe.asp?topic_id=109&doc_id=2798>.

[274] For the text, see IMOGSR, *supra* note 184.

[275] *Ibid.*, s. 1.4.

[276] *Ibid.* See also *ibid.*, s. 1.7.

[277] Recycling of Ships: The Development of a Green Passport for an Existing Ship, Submitted by Canada Marine Environment Protection Committee, IMO/MEPC 53/INF.17, 13 May 2005, ¶ 13

details about the ship[278] and, more importantly, it serves as an inventory[279] of potentially hazardous materials that have been used in its construction, equipment and systems.[280] The passport is divided into three parts.[281] For new ships, the shipbuilder is to prepare Part 1 in consultation with equipment manufacturers and thereafter, it has to be passed on to the shipowner.[282] For existing ships, Part 1 is to be prepared by the shipowner drawing upon the blueprint of the ship in consultation with the shipbuilder, equipment manufacturers, *etc.*[283] Parts 2 and 3 are to be prepared by the shipowner prior to the final voyage of the ship to the recycling facility.[284] Successive owners of the ship are to maintain the accuracy of the green passport and incorporate all relevant design and equipment changes, with the final owner delivering the document, along with the ship, to the recycling facility.[285]

Most problems associated with ship recycling emanate from the usage of hazardous materials in ship construction. Accordingly, the IMOGSR identifies steps to reduce their presence in new ship construction. While designing and constructing a ship, ship designers and builders are to bear in mind the ship's ultimate disposal.[286] The IMOGSR reflects a "cradle to grave"[287] approach to ship recycling. It also calls for "product reformulation" where potentially hazardous materials that are as a matter of routine incorporated into the ship structure and equipment are replaced with less hazardous alternatives.[288] This obligation is taken a step further by requiring that competent authorities of ship-building states encourage research into the use of potentially less hazardous materials for ship construction. The IMOGSR envisages a scenario where shipowners and operators ask for designs that "minimise or prevent waste at source and at the end of

(KR-CON) (in its submission, Canada noted that the total cost for the development and implementation of the green passport was approximately USD16,155).

[278] IMOGSR, *supra* note 184 s. 5.2.1.

[279] *Ibid.*, app. 3 (the format of the inventory).

[280] *Ibid.*, s. 5.1.

[281] *Ibid.*, s. 5.2.2 (Part 1 provides details on the potentially hazardous materials in the ship's structure and equipment, Part 2 details the operationally-generated wastes, and Part 3 presents information on the stores).

[282] *Ibid.*, s. 5.5.1.

[283] *Ibid.*, s. 5.5.2.

[284] *Ibid.*, s. 5.6.

[285] *Ibid.*, s. 5.1.

[286] *Ibid.*, s. 6.1.4. See also s. 6.2.1 (stating that ship designers are to take into account aspects like structural designs that could facilitate ship recycling; equipment designs that facilitate removal from ships during recycling; the use of structural materials that can readily be recycled; providing to the new owner a brief technical document advising on the optimal approach for recycling the ship; the use of recycled materials in ship or equipment designs; limiting the use of materials that are difficult to separate into their specific individual substances or components; and taking steps to facilitate the removal of such materials).

[287] *Ibid.*, s. 1.1.

[288] *Ibid.*, s. 6.1.1. See also *ibid.,* s. 6.1.8 (prohibiting the use of substances that are prohibited or restricted by international conventions like the *Stockholm Convention on Persistent Organic*

the operating life of the ship", while ship and equipment designers recommend such designs to operators.[289]

The IMOGSR also seeks to minimise the generation of hazardous materials and its retention on a ship during its operational phase and at the end of its life.[290] This is to be achieved through the incorporation of processes like "cleaner production technologies", "process modification", "input substitution",[291] and "on-site, closed-loop recycling".[292] Shipowners are to minimise the amount of potentially hazardous materials on board the ship that exist in the form of stores.[293] A similar procedure is envisaged for existing ships as well as for new constructions.[294]

Selection of recycling facilities under the IMOGSR is the responsibility of the shipowner who must do so in consultation with the competent authorities of the recycling state. The shipowner should take into account the practices and facilities in the recycling yard[295] and the capability of the facility to manage the hazardous materials that arise from the recycling process.[296] Other considerations in the selection process include whether there is provision for training in the recycling yards, whether personal protection and safety equipment are provided to the workers, and the ability of the recycling facility to maintain a gas-free-for-hot-work condition throughout the recycling process.[297] After selection, the shipowner has to convey this decision to the maritime administration and also to the competent authority in the recycling state.[298] The IMOGSR also imposes other obligations on the shipowner when selecting the appropriate recycling yard. If it emerges that a particular recycling yard does not possess the capability to manage the hazardous wastes in a manner consistent with national and international standards and guidelines, the shipowner has to arrange for the removal of these materials at another facility.[299] The shipowner is also empowered to specify the

---

Pollutants, 2001, the *Montreal Protocol on Substances that Deplete the Ozone Layer, 1987* and AFC 2001 in the construction, refit and repair of ships).

[289] *Ibid.*, s. 6.1.5.

[290] *Ibid.*, ss. 6.1.2, 6.5.

[291] *Ibid.*, s. 6.1.3.4 (input substitution involves the use of consumables that are potentially less hazardous, resulting in the generation of less wastes).

[292] *Ibid.*, s. 6.1.3.5 (here, wastes are recycled on board the ship).

[293] *Ibid.*, s. 6.4.

[294] *Ibid.*, ss. 7.3, 7.2.1, 7.2.2 (referring to the procedures outlined in 6.1.3).

[295] *Ibid.*, s. 8.1.3 Some factors that need to be considered include the ability of the recycling facility to handle safely and dispose of properly potentially hazardous materials present in the ship; provisions for appropriate and sufficient personal protection and safety equipment; the ability of a recycling facility to maintain and monitor the ship in a gas free condition and approved fit for hot work during the whole process of ship recycling; and information relating to safety records, training programmes for workers and assessment of work quality.

[296] *Ibid.*, s. 8.1.6.

[297] *Ibid.*, s. 8.1.3.

[298] *Ibid.*, s. 8.1.8.

[299] *Ibid.*, ss. 8.1.5, 8.1.4.

recycling method in the contract,[300] reserve the right to monitor the recycling process,[301] and provide incentive payments for adhering to the prescribed mandates on recycling.[302]

The IMOGSR calls upon the recycling facility to develop a ship recycling plan in consultation with the shipowner taking into account factors like the potential hazards that can arise during the process of recycling, the infrastructural facilities available at the recycling yard, and relevant national and international requirements.[303] This plan helps to ensure that the ship is prepared to the maximum extent possible prior to recycling and that safety aspects have been duly considered.[304] The ship recycling plan helps to ensure that wastes that can potentially contribute to environmental pollution or jeopardise worker safety and health have been properly identified to facilitate their handling.[305] The IMOGSR further requires that an agreement to implement the ship recycling plan has to be incorporated into the contract between the shipowner and the facility.[306]

In preparing a ship for recycling, the IMOGOSR includes provisions regarding operational considerations during its final journey[307] and various occupational, safety, health, and environmental[308] aspects that also need to be addressed. It requires that the last owner of a ship, consistent with its safe operation, minimise the amount of fuel, diesel and other oils and chemicals on board,[309] and ensure that the green passport is complete in all respects.[310] Steps to manage ballast water[311] and measures to facilitate controlled drainage of potentially harmful liquids from the ship are to be taken.[312] If the recycling facility lacks necessary reception facilities, these liquids are to be emptied elsewhere before the ship is delivered for recycling.[313] Prior to recycling, the shipowner is to arrange for gas-free-for-hot-work-safe certification[314] and ensure that critical support structures are dismantled in a way as to avoid accidental structural collapse.[315] The

---

300 *Ibid.*, s. 8.1.7.2.
301 *Ibid.*, s. 8.1.7.3.
302 *Ibid.*, s. 8.1.7.4.
303 *Ibid.*, s. 8.3.2.2. In addition to these requirements, the green passport, technical advice from the shipbuilder, potential hazards to worker safety, etc. are also to be considered. *Ibid.*, s. 8.3.2.4. See also Recycling of Ships: Proposed Technical Considerations for Preparing a Ship Recycling Plan, Submitted by Canada, IMO/MEPC 51/3/2, 22 January 2004 (KR-CON).
304 IMOGSR, *supra* note 184 s. 8.3.2.2.
305 *Ibid.*, s. 8.3.2.3.
306 *Ibid.*, s. 8.3.2.5.
307 *Ibid.*, s. 8.3.1.1.
308 *Ibid.*, s. 8.3.3.2.
309 *Ibid.*, s. 8.3.3.1.1.
310 *Ibid.*, s. 8.3.3.1.3.
311 *Ibid.*, s. 8.3.3.1.5.
312 *Ibid.*, s. 8.3.3.1.4.
313 *Ibid.*, s. 8.3.3.1.2.
314 *Ibid.*, s. 8.3.4.1.1.
315 *Ibid.*, s. 8.3.4.1.4.

shipowner must ensure that oxygen deficient compartments are clearly marked,[316] and that areas with structural integrity problems are identified and their locations indicated.[317] The shipowner should also provide technical advice available from the shipbuilder and equipment suppliers that may be of assistance in dismantling.[318] Arrangements should also be made to ensure that the final journey is covered by insurance and that contingency plans are in place to meet situations where the ship is unable to complete its voyage or where the recycling facility is unable to accept delivery.[319]

Yet another important feature of the IMOGSR is that it seeks to foster a cooperative ethic among the different stakeholders in ship recycling.[320] In this regard, it identifies a series of stakeholders and specifies their roles. These stakeholders include the flag, port and recycling states as well as both the shipping and ship recycling industries.

Three major duties are identified for flag states: First, they are to ensure that the ship, as long as it is operational, complies with applicable IMO conventions and other relevant requirements. Second, flag states are to promote the use of a ship recycling sale and purchase contract, such as DEMOLISHCON. Third, they are to cooperate with recycling states to facilitate the implementation of the IMOGSR.

Like any other ship, ships that are destined for recycling are also subject to port state controls under applicable international regulations; an obligation re-affirmed by the IMOGSR.[321] Furthermore, port states are also required to promote the widespread use of the IMOGSR and are to cooperate with the flag and recycling states to facilitate their implementation.[322]

Recycling states are encouraged by the IMOGSR to introduce and implement legislation on ship recycling. Such legislation is to include provisions for licensing of recycling facilities,[323] identifying conditions under which ships can be accepted, defining worker health and safety requirements,[324] and requiring gas-free-and-hot-work-safe certification.[325] Since acceptance of ownership of a vessel by a recycling facility assumes that the responsibility for the proper handling of onboard wastes is on the recycling facility, the IMOGSR requires that the recycling state ensure that the recycling facility ascertains that the actual

---

316 *Ibid.*, s. 8.3.4.1.2.
317 *Ibid.*, s. 8.3.4.1.3.
318 *Ibid.*, s. 8.2.1.
319 *Ibid.*, s. 8.2.2.
320 *Ibid.*, s. 1.6.
321 *Ibid.*, s. 9.3.1.1.
322 *Ibid.*, s. 9.3.2.
323 *Ibid.*, s. 9.4.4.1.
324 *Ibid.*, ss. 9.4.3.1, 9.4.1.2.
325 *Ibid.*, s. 9.4.3.2.

condition of the ship corresponds to the purchase contract and that national requirements are fulfilled before the ship is accepted.[326] The recycling state is also to provide support to recycling facilities in their decision to accept or reject a ship for recycling.[327] In addition, the IMOGSR recommends that competent authorities in the recycling state assess the capabilities of the recycling facilities and make available the results of these assessments.[328] The recycling state is also to ensure that recycling facilities utilise standard ship recycling contracts like DEMOLISHCON in the conduct of their business.[329] They are also to ensure the availability of adequate reception facilities as mandated by MARPOL 73/78[330] and provide for the proper handling and disposal of asbestos, oils and other hazardous substances.[331]

With regard to the shipping industry, the IMOGSR requires that the industry cooperate with other stakeholders to ensure safe and environmentally sound decommissioning of ships.[332] Further, they are to develop an Industry Code of Practice on Ship Recycling.[333] The IMOGSR also directs Baltic and International Maritime Council to revise the DEMOLISHCON to incorporate its standards.[334]

The IMOGSR requires the ship recycling industry to establish "control standards" in their facilities to contribute to the safe and environmentally sound disposal of "time-expired ships".[335] The ship recycling industry is to develop a code of practice appropriate to its functioning and encourage its endorsement by international bodies.[336] The recycling industry is also directed to improve quality management systems in the recycling yards,[337] and encourage recycling facilities to provide details regarding procedures for the safe handling of hazardous materials and working practices.[338] The IMOGSSR also recognises the roles of the Basel Convention,[339] the ILO,[340] and the London Convention and its 1996 Protocol[341] in ship recycling.

Despite the recognition of the contributions of the IMOGSR to the industry, there has been criticism of its provisions. Greenpeace has suggested that these

---

326 *Ibid.*, s. 9.4.3.4.
327 *Ibid.*, s. 9.4.4.3.
328 *Ibid.*, s. 9.4.1.5.
329 *Ibid.*, s. 9.4.3.3.
330 *Ibid.*, s. 9.4.2.
331 *Ibid.*, s. 9.4.4.2.
332 *Ibid.*, s. 9.8.
333 *Ibid.*, s. 9.8.1.3.
334 *Ibid.*, s. 9.8.2.
335 *Ibid.*, s. 9.9.2.1.
336 *Ibid.*, ss. 9.9.2.2, 9.9.2.3.
337 *Ibid.*, s. 9.9.2.5.
338 *Ibid.*, s. 9.9.2.4.
339 *Ibid.*, s. 9.5.
340 *Ibid.*, s. 9.6.
341 *Ibid.*, s. 9.7.

Guidelines run counter to the Basel Convention.[342] The Greenpeace objections center on three issues: deflection of legal responsibility from those responsible for waste generation; insufficient provisions for prior decontamination; and consideration of ships as waste.

One of the major criticisms levelled against the IMOGSR is that under this instrument the obligation to ensure environmental and worker safety in ship recycling facilities rests with the recycling facility and with the regulatory authorities of the countries in which these facilities operate.[343] Greenpeace considers this to be a deflection of legal responsibility from shipowners, which is, in essence, a violation of the polluter pays principle.[344] Under the Basel Convention, the duty to ensure ESM of hazardous and other wastes cannot be transferred to the importing or transit state.[345] Clearly, the IMOGSR reduces the rigour of this requirement. However, a closer examination reveals that the responsibility is not transferred all of a sudden to the recycling state or that the shipowner is exonerated of all responsibilities. The IMOGSR requires that the recycling facility examine the ship to ensure that its "actual condition" is consistent with relevant international guidelines and that the terms as mentioned in the purchase contract and national requirements are fulfilled.[346] It is only after completion of these steps that the recycling facility accepts ownership of the ship and the responsibility for the proper handling of wastes is transferred.[347] As well, the IMOGSR requires that the shipowner retain some control over the recycling process *via* the recycling contract. If the recycling yard is unable to process the waste, the shipowner has to arrange for its removal at another facility with reprocessing capabilities. Thus the IMOGSR sets out a gradual procedure preceding transfer of responsibility to the recycling yard.

Another criticism levelled on the IMOGSR is that it ignores the principle of prior decontamination of vessels.[348] Although it does not insist on prior decontamination in all cases, the Guidelines do call for partial prior decontamination in certain situations. The IMOGSR requires that the shipowner remove hazardous materials elsewhere or minimise the amount and its location if the facility is not equipped to handle the wastes.[349] It seems that the drafters adopted a cautious approach, recognising that insisting on a complete prior decontamination for all

---

[342] Recycling of Ships: Proposals for Review and Further Development of the IMO Guidelines, Submitted by Greenpeace International, IMO/MEPC 52/3/3, 2 August 2004, ¶ 2 (KR-CON) [Review & Development of IMO Guidelines] (holding that though not adequate to protect the rights of the workers and the environment, the guidance offered by the Guidelines definitely have "value").

[343] Analysis of Inconsistencies, *supra* note 94 ¶ 18.

[344] *Ibid.*, ¶ 19.

[345] *Basel Convention*, *supra* note 3 art. 4(10).

[346] IMOGSR, *supra* note 184 s. 9.4.3.4.

[347] *Ibid.*

[348] Analysis of Inconsistencies, *supra* note 94 ¶ 11–14.

[349] IMOGSR, *supra* note 184 ss. 8.3.1.2, 8.3.3.1.2.

ships could lead to the economic collapse of the ship recycling industry on the Indian subcontinent. Moreover, ships must retain their ability to operate in the marine environment.[350] A complete prior decontamination would mean that ships would have to be towed from all corners of the world through some of the busiest shipping lanes to the subcontinent, raising issues of maritime safety.

Greenpeace claims that the "prevailing sentiment" in the IMO that ships are not waste is fortified by the IMOGSR. As such, the Guidelines are "fundamentally flawed and seemingly ignorant".[351] For example, the IMOGSR defines the "operating life of a ship" as "the time when it is capable of performing its current functions."[352] Greenpeace contends that this stands in opposition to the Basel Convention. However, Decision VII/26 of the Joint Working Group, which affirmed that a ship can become waste as defined under the Basel Convention and, at the same time, continue to remain a ship under other international rules, would seem to refute the argument.[353] Moreover, the fact that obsolete ships destined to be scrapped are still characterised as ships brings them under the regulatory ambit of several IMO conventions.

The term "green passport" has also come in for some trenchant criticism by Greenpeace:

> The title "Green Passport" is misleading and cynical. No hazardous waste is eliminated from the ship through the use of the passport, thus, the term "green" should not be used. The term passport implies that it is cleared for sailing without problems. Again this is highly misleading when all it is, is an inventory of the problems that remain on board. This should simply be called Hazard Inventory or Toxic Inventory and in fact NO contaminants should be on board when the ship is delivered to the recycling yards. This decontamination prior to export principle would equate to a genuine Green Passport. The way it is used here is simply a "green wash".[354]

Even though there is merit in the Greenpeace criticism on the use of the term "green passport", this should not blind one to the utility of this document in enabling the recycling facility to prepare itself for the delivery of the ship. The green passport acts as a road map, enabling the recycling facility to shred the ship in a manner that causes least damage to the environment. Also, by requiring shipowners to maintain its accuracy,[355] the IMOGSR in effect insists on substantial rather than formal compliance with the rules.

---

[350] TGSD, *supra* note 79 at 31.

[351] Analysis of Inconsistencies, *supra* note 94 ¶ 15.

[352] IMOGSR, *supra* note 184 s. 3.

[353] *VII/26*, *supra* note 69.

[354] Greenpeace International/Basel Action Network (BAN), "The IMO Guidelines on Ship Recycling Annotated", 21 November 2003, online: Basel Action Network <http://www.ban.org/Library/greenpeace-ban-iii-3e.pdf>.

[355] IMOGSR, *supra* note 184 s. 5.1.

In spite of these noteworthy elements, the primary loophole in the IMOGSR emanates from the definition of the term shipowner, which includes "those who have ownership of the ship for a limited period pending its sale to a recycling facility".[356] By broadening the definition, several of the obligations which the IMOGSR seeks to impose on the shipowner may fall on downstream entities like cash buyers. This may enable the real owners to avoid responsibility since the flag state may not be able to exercise control over these entities.

On the whole, the IMOGSR constitutes a positive step in establishing a sustainable framework for the recycling of ships. Although non-binding, the IMOGSR is a comprehensive and practical instrument that provides guidance to the different stakeholders on how best to manage issues related to the export of ships for scrapping. It contains principles grounded in the precautionary approach which, if duly observed, can help to ensure the development of a sustainable ship recycling industry. Indeed, sustainability is a key message throughout the Guidelines. Measures such as the introduction of green passports, identification of stakeholders, responsibilities of the shipowner, the recycling state and recycling facilities, the duty to minimise the use of hazardous substances in the construction of new ships and their equipment, minimisation of waste generation, developing the ship recycling plan, and measures to prevent pollution and protect occupational safety and health and technical cooperation[357] enhance the sustainability of the ship recycling industry. Much could have been achieved had the initiatives that sought to transform this voluntary code into a mandatory binding instrument come to fruition.[358]

VI. Conclusion

Contrary to the popular belief that current international law does not address shipbreaking issues, there is indeed a voluminous body of international law that

[356] *Ibid.*, s. 3.

[357] IMOGSR, *supra* note 184 s. 10.

[358] Recycling of Ships: Mandatory Requirements for the IMO Guidelines, Submitted by India, IMO/MEPC 53/3/6, 13 May 2005, ¶ 2 (KR-CON) (pointing out that developed countries opposed demands for at least a partial application of the IMOGSR as they would have had to take responsibility as shipowners for the consequences of not adhering to the IMOGSR if these were made mandatory). See also Recycling of Ships, Comments on Future Work Items on Ship Recycling: Mechanism to Promote the Implementation of the IMO Guidelines on Ship Recycling, Submitted by Japan, IMO/MEPC 51/3/4, 23 January 2004, ¶ 1.2 (KR-CON) (noting that the global implementation of the Guidelines is the most appropriate measure to reduce the environmental and occupational safety and health risks related to ship recycling); Review & Development of IMO Guidelines, *supra* note 342 ¶ 2 (pointing out the inadequacy of the voluntary Guidelines to prevent accidents by referring to the explosion which occurred while recycling the LPG carrier 'Gaz Med.'); Proposals for Mandatory Guidelines, *supra* note 95 ¶ 14 (arguing that the MEPC adopt a policy to make the IMOGSR as a whole, or at least certain parts mandatory).

has relevance to shipbreaking. Beginning with the Basel Convention and the LOS Convention, extending to the London Convention and its 1996 Protocol, the WRC, the AFC 2001, the BWC, the MARPOL 73/78 to the more recent non-binding IMOGSR, as well as ILO conventions, recommendations and guidelines, a plethora of international instruments impose obligations on states as well as on other stakeholders engaged in shipbreaking operations. However, this legal regime is muddled and lacks coherence between the instruments and the obligations they prescribe. Some of the international instruments are non-binding, while others have only remote application to shipbreaking. It must be reiterated that shipbreaking involves a completely different set of factual situations and many of these instruments, due to their indirect application to the issues posed by shipbreaking, may not always provide appropriate regulatory tools. As well, many of the instruments contemplate different sets of authorities to further complicate the legal framework. At best, the current international legal regime reflects a cumulative approach, not an integrated one. However, these instruments will fortify an international legal framework that is devoted exclusively to shipbreaking. Recognising the need for an integrated, stand-alone internationally-binding legal instrument directed to ensuring sustainability in ship recycling, the international community responded with the recently adopted Ship Recycling Convention, which is the focus of the next chapter.

Chapter 5

Deciphering the Ship Recycling Convention

I. Introduction

Shipbreaking caught the attention of the international community following the investigative reporting series by Gary Cohn and Will Englund in "The Baltimore Sun", for which they won the Pulitzer Prize in 1998.[1] Since then, much print has been devoted to highlighting the working conditions in these demolition yards. As well, environmental non-governmental organisations (ENGOs) have played an important role in placing the global spotlight on shipbreaking by vigorously campaigning for change. Despite all these efforts, and growing revulsion to one of the most sinister forms of environmental racism, there have only been superficial changes in these yards. We are yet to witness an international legal regime devoted exclusively to the issue of breaking ships. As outlined in the preceding chapter, controls on the transboundary movement of dilapidated ships can emanate from various international instruments,[2] the most important of these being the Basel Convention. However, as seen, this control regime is fragmented and incomplete. It was also recognised that the unique characteristics of this trade required a binding, stand-alone international legal instrument setting out clear minimum standards to regulate all facets of shipbreaking. Following acceptance of the IMO Guidelines on Ship Recycling (IMOGSR)[3], the IMO turned to developing a binding legal regime, the recently adopted *Hong Kong International Convention for the*

---

[1] *The 1998 Pulitzer Prize Winners Investigative Reporting*, online: The Pulitzer Prizes <http://www.pulitzer.org/works/1998,Investigative+Reporting>.

[2] For an analysis on the international framework, see Chapter Four entitled, "Contemporary International Law and Ship Recycling".

[3] See generally IMO Guidelines on Ship Recycling: Res. A.962(23) Adopted on 5 December 2003 (Agenda Item 19), A 23/Res.962, 4 March 2004 (KR-CON) [IMOGSR]. See also Recycling of Ships: Comments on the Report of the Correspondence Group, Submitted by Greenpeace International, IMO/MEPC 47/3/5, 11 January 2002 (KR-CON); Recycling of Ships: Comments on the Report of the Correspondence Group: Selected Cases of Decommissioning of Vessels Indicating the Need for Mandatory Requirements, Submitted by Greenpeace International, IMO/MEPC 49/3/2, 9 May 2003 (KR-CON) (highlighting the inconsistent approach of different stakeholders in relation to the four cases involving the 'Sandrien', the 'Sea Beirut', the 'Forthbank' (*alias* 'Pacific Emerald'), and the 'Silver Ray' (renamed *Naxos* 1) that re-emphasise the need for mandatory requirements for ship recycling) [Selected Cases of Decommissioning of Vessels].

Safe and Environmentally Sound Recycling of Ships, 2009 (Ship Recycling Convention).[4]

Under article 11 of the Basel Convention, any new treaty that seeks to supplant existing controls over the transboundary movement of wastes, which includes decaying ships intended for breaking, will have to establish an "equivalent level of control". In effectuating this principle of equivalency, the new agreement must stipulate conditions that are no less environmentally sound than those provided for by the Basel Convention, taking into account the interests of developing countries.[5] Accordingly, it is expected that the recent IMO-sponsored regime will encompass the obligations and controls set out in the Basel Convention, as well as filling in the gaps and loopholes created when the more general provisions of the Basel Convention are applied to regulating the specific waste stream of end-of-life ships.[6] It is therefore useful to consider whether the Ship Recycling Convention establishes a comparable set of controls to the existing regime, or whether it merely is an

---

[4] For the text of the convention, see International Conference on the Safe and Environmentally Sound Recycling of Ships: Agenda Item 8, Adoption of the Final Act and Any Instruments, Recommendations and Resolutions Resulting from the Work of the Conference: Hong Kong International Convention for the Safe and Environmentally Sound Recycling of Ships, 2009, Text Adopted by the Conference, IMO/SR/CONF/45, 19 May 2009, online: SJÖFARTSVERKET <http://www.sjofartsverket.se/pages/19514/sr-conf-45.pdf> (opened for signature 1 September 2009) [*Ship Recycling Convention*]. For the first draft of the Ship Recycling Convention developed by Norway see Recycling of Ships: Proposal for a New Legally-binding Instrument on Recycling of Ships, Submitted by Norway, IMO/MEPC 54/3, 5 December 2005 (KR-CON). See also Sveinung Oftedal, *Development of the Draft International Convention for the Safe and Environmentally Sound Recycling of Ships*, online: European Commission Environment <http://ec.europa.eu/environment/greenweek2007/sources/files/day2_session8_2/Oftedal.pdf>. Nikos Mikelis, "Developments and Issues on Recycling of Ships" (Paper presented to The East Asian Seas Congress, Haikou City, China, 12 December 2006) at 4, online: International Maritime Organization <http://www.imo.org/includes/blastDataOnly.asp/data_id%3D17980/Developments.pdf> [Mikelis, "Developments on Recycling"]. See also Nikos Mikelis, "Development of the International Convention for the Safe and Environmentally Sound Recycling of Ships" (PowerPoint presented to the Preparatory Discussions on Promoting Sustainable Ship Recycling through the Global Programme, Dhaka, Bangladesh, 13 January 2008), online: Basel Convention <http://www.basel.int/ships/gpssr/index.html>.

[5] *Basel Convention on the Control of Transboundary Movements of Hazardous Wastes and their Disposal*, 22 March 1989, 1673 U.N.T.S 126, 28 I.L.M. 657 (entered into force 5 May 1992), art. 11(1) [*Basel Convention*]. See also Recycling of Ships: Equivalent Levels of Control Established under the Basel Convention and the Draft IMO Ship Recycling Convention, Submitted by the Secretariat of the Basel Convention, IMO/MEPC 57/3/3, 25 January 2008, ¶ 7 (KR-CON) (noting that the conference of the parties to the Basel Convention in its tenth meeting to be held in 2010 will make a final determination as to whether the Draft Convention provides an equivalent level of control to that established under the Basel Convention).

[6] France, Interdepartmental Committee on the Dismantling of Civilian and Military End-of-Life Ships, *Le rapport de la Mission Interministérielle portant sur le Démantèlement des Navires civils et militaires en fin de vie*, at 16 (27 March 2007, Chair: Xavier de la Gorce), online: SGMer <http://www.sgmer.gouv.fr/article.php3?id_article=52> [*MIDN*]. See generally *Obligations and Opportunities for a Mandatory Alternate or Additional Instrument to the Basel Convention for End-of-Life Ships*, Submitted by Greenpeace & the Basel Action Network, 26 May 2005, online: basel action network <http://www.ban.org/Library/mepc_submission_long.pdf> [*Alternate or Additional Instrument to the Basel Convention*].

attempt by certain vested shipping interests to wrest control over matters involving ships as waste which currently are governed by the stringent Basel Convention.

A brief review of the provisions of the Ship Recycling Convention and a comparison of the protections that it guarantees *vis-à-vis* the Basel Convention provide the basis for considering these questions and evaluating this instrument's effectiveness in regulating the shipbreaking industry. It is also useful to consider whether the Ship Recycling Convention in any measure adds value to the IMOGSR or does it derogate from its stipulations.

II. Salient Features of the Ship Recycling Convention

Pursuant to the recommendation of the 55th session of the IMO Marine Environment Protection Committee in October 2006, the concurrent decision of the IMO Council at its 97th session in November 2006, and the endorsement of that decision by the Assembly at its 25th regular session (Resolution A. 991(25), November 2007), the IMO convened an International Conference on the Safe and Environmentally Sound Recycling of Ships, which took place at Hong Kong, from 11 to 15 May 2009.[7] Representatives of 63 states, two associate members of the IMO namely, Hong Kong and Macau, the United Nations Environment Programme, the International Labour Organization, the European Commission, and eight non-governmental organisations with consultative status participated in the conference.[8] As a result of its deliberations, the conference adopted the *Hong Kong International Convention for the Safe and Environmentally Sound Recycling of Ships, 2009* and certain resolutions.[9]

---

[7] See generally *International Maritime Organization Diplomatic Conference on Ship Recycling*, online: IMO CONF2009 <http://www.imoconf2009.hk/>.

[8] The non-governmental organisations that participated are: International Chamber of Shipping, BIMCO, International Association of Classification Societies, Oil Companies International Marine Forum, Friends of the Earth International, International Association of Independent Tanker Owners, Greenpeace International, and the International Parcel Tankers Association.

[9] Resolution 1: Expression of appreciation to the host Government; Resolution 2: Contribution of the Parties to the Basel Convention and the International Labour Organization in the development of the Hong Kong International Convention for the Safe and Environmentally Sound Recycling of Ships, 2009; Resolution 3: Promotion of technical co-operation and assistance; Resolution 4: Future work by the Organization pertaining to the Hong Kong International Convention for the Safe and Environmentally Sound Recycling of Ships, 2009; Resolution 5: Early implementation of the technical standards of the Hong Kong International Convention for the Safe and Environmentally Sound Recycling of Ships, 2009; Resolution 6: Exploration and monitoring of the best practices for fulfilling the requirements of the Hong Kong International Convention for the Safe and Environmentally Sound Recycling of Ships, 2009. See International Conference on the Safe and Environmentally Sound Recycling of Ships: Agenda Item 8, Adoption of the Final Act and Any Instruments, Recommendations and Resolutions Resulting from the Work of the Conference: Final Act of the International Conference on the Safe and Environmentally Sound Recycling of Ships, Text Adopted by the Conference, SR/CONF/46, 19 May 2009, online: SJÖFARTSVERKET <http://www.sjofartsverket.se/pages/19514/sr-conf-46.pdf> [Work of the Conference].

Fifty-nine states signed the Final Act of the conference, which will be open for signature at the IMO headquarters from 1 September 2009 and until 31 August 2010. Thereafter, it will be opened for accession.[10] In accordance with article 17, the Ship Recycling Convention enters into force 24 months after the date on which the following conditions are met: not less than 15 states have either signed it without reservation, or have deposited the requisite instrument of ratification, acceptance, approval or accession in accordance with article 16; the combined merchant fleets of these states constitute not less than 40 percent of the gross tonnage of the world's merchant shipping; and the combined maximum annual ship recycling volume of these states during the preceding ten years constitutes not less than three percent of the gross tonnage of the combined merchant shipping of the same states. Thus, apart from the traditional entry into force factors like the number of states and the combined merchant fleet tonnage, the convention also includes a ship recycling tonnage criterion.[11]

The Ship Recycling Convention's preamble demonstrates the resolve of the IMO to

> effectively address, in a legally-binding instrument, the environmental, occupational health and safety risks related to ship recycling, taking into account the particular characteristics of maritime transport and the need to secure the smooth withdrawal of ships that have reached the end of their operating lives.[12]

To achieve this objective, it calls upon parties "to give full and complete effect" to its terms "in order to prevent, reduce, minimise and, to the extent practicable, eliminate accidents, injuries and other adverse effects on human health and the environment" consequent to ship recycling "and enhance ship safety, protection

---

[10] *Report on the 2009 International Conference on the Safe and Environmentally Sound Recycling of Ships*, Note by the Secretary-General, C 102/13, 20 May 2009. See also Recycling of Ships: Outcome of the 2009 International Conference on the Safe and Environmentally Sound Recycling of Ships, Note by the Secretariat, IMO/MEPC 59/3/3, 20 May 2009 (KR-CON).

[11] See International Conference on the Safe and Environmentally Sound Recycling of Ships: Agenda Item 6, Consideration of the Draft International Convention for the Safe and Environmentally Sound Recycling of Ships: Proposal on the Conditions on Entry into Force, Submitted by Japan, SR/CONF/41, 8 May 2009, online: SJÖFARTSVERKET <http://www.sjofartsverket.se/pages/19514/sr-conf-41.pdf>. But see International Conference on the Safe and Environmentally Sound Recycling of Ships: Agenda Item 6, Consideration of the Draft International Convention for the Safe and Environmentally Sound Recycling of Ships: Ensuring Sustainable Green and Safe Ship Dismantling – Concerning the Entry into Force Provisions, Submitted by Greenpeace International and FOEI, SR/CONF/13, 9 February 2009, online: SJÖFARTSVERKET <http://www.sjofartsverket.se/pages/19514/sr-conf-13.pdf> (noting that acknowledging the practice of unsound dismantling of end-of-life ships on tidal beaches as "capacity" for the purpose of entry into force of the convention will send a wrong signal that environmental degradation and human suffering caused by this worst form of ship dismantling is acceptable with regard to the requirements of the convention).

[12] See *Ship Recycling Convention*, *supra* note 4 preamble.

of human health and the environment throughout a ship's operating life".[13] Accordingly, parties are to adopt effective measures to ensure that ships entitled to fly their flags or operating under their authority, or ship recycling facilities under their jurisdiction, comply with the requirements as set forth in the convention.[14]

The preamble and the 21 articles enumerate the basic principles for safe and environmentally sound ship recycling. They also set out the general rights and obligations of parties and administrative provisions.[15] The annex entitled "Regulations for Safe and Environmentally Sound Recycling of Ships" is the key element in this convention,[16] and specifies "the main technical requirements" in four chapters.[17] The design, construction, survey, certification, operation and recycling of ships have to be performed in accordance with this annex.[18] The annex is further supported by appendices setting out further technical details.[19] The convention also refers to guidelines which are to be developed in due course by the IMO.[20] The Ship Recycling Convention incorporates control and enforcement measures from two perspectives: the first set of controls apply to ships during their life cycle, and the second details standards in relation to the operation of ship recycling facilities.[21]

---

[13] *Ibid.*, art. 1(1).

[14] *Ibid.*, art. 4.

[15] *Ibid.*, arts. 18 (deals with amendment), 17 (deals with entry into force).

[16] See also *ibid.*, art. 1(5) (affirming that the annex is an integral part of the convention).

[17] Chapter One contains general provisions. Chapter Two specifies the requirements for ships, and is organised into Part A (design, construction, operation and maintenance of ships); Part B (preparation for ship recycling); and Part C (surveys and certification). Chapter Three enumerates requirements for ship recycling facilities, and Chapter Four specifies the reporting requirements.

[18] *Ibid.*, reg. 2.

[19] The first enumerates the control measures on the hazardous materials that are regulated by the Ship Recycling Convention; the second lays down the minimum list of items for the inventory of hazardous materials; and appendices three to seven provide different formats for relevant certificates and other documents.

[20] As a matter of urgency, the Hong Kong Ship Recycling Conference called upon the IMO to develop nearly 6 sets of Guidelines namely, Guidelines for the Development of the Inventory of Hazardous Materials; Guidelines for Survey and Certification; Guidelines for Inspection of Ships; Guidelines for the Authorisation of Ship Recycling Facilities; Guidelines for Safe and Environmentally Sound Ship Recycling; and Guidelines for the Development of the Ship Recycling Plan. In addition, the IMO is also to develop other guidelines or circulars as may be identified by the Marine Environment Protection Committee. Work of the Conference, *supra* note 9 res. 4. See also Recycling of Ships: Proposal on the Drafting of a Code for Safe and Environmentally Sound Management of Ship-scrapping Facilities, Submitted by France, IMO/MEPC 55/3/4, 28 July 2006 (KR-CON).

[21] EU Comments Received Pursuant to OEWG6 Decisions, "Assessment on Ship Dismantling with Particular Reference to the Levels of Control and Enforcement Established by the Basel Convention and the Expected Level of Control and Enforcement to be provided by the Draft Ship Recycling Convention in their Entirety" ¶ 49, online: Basel Convention <http://www.basel.int/ships/commentsOEWG6/oewg6.html> [EU Comments].

A. *Applicability*

The Ship Recycling Convention applies to "ships"[22] and to "ship recycling facilities"[23] of parties. However, ships less than 500 gross tonnage (GT),[24] warships, naval auxiliary or other ships owned or operated by a party and used only for government non-commercial service, and ships that are "operating throughout their life solely in waters subject to the sovereignty or jurisdiction of the state whose flag the ship is entitled to fly", if recycled in the same state where they have operated, are exempt from its terms.[25] Even so, the parties are to adopt "appropriate measures" to ensure that recycling of these ships is performed in a manner consistent, as far as "reasonable and practicable", with the Ship Recycling Convention.[26] With regard to non-party ships, parties are to apply the Ship Recycling Convention in a manner so as not to afford any favourable treatment.[27]

Generally, ship recycling is understood to mean not only processes like anchoring, scrapping and recovery of materials, but also downstream activities like reprocessing. However, the Ship Recycling Convention defines "ship recycling" as

> the activity of complete or partial dismantling of a ship at a Ship Recycling Facility in order to recover components and materials for reprocessing and re-use, whilst taking care of hazardous and other materials, and includes associated operations such as storage and treatment of components and materials on site, but not their further processing or disposal in separate facilities.[28]

Thus, the scope of the Ship Recycling Convention is confined only to activities like scrapping and storage in the recycling yard and does not extend to the remaining steps in the recycling chain.[29]

---

[22] *Ship Recycling Convention*, *supra* note 4 art. 2(7). See also *ibid.*, art. 3(1)(1) (the convention applies to ships that are entitled to fly the flag of a party or those that operate under its authority).

[23] See *ibid.* art. 2(11) (defining "Ship Recycling Facility" to mean a defined area that is a site, yard or facility used for the recycling of ships.). See also *ibid.*, art. 3(1)(2).

[24] The "World Fleet Statistics 2005" by Lloyd's Register reports that there are 47,258 ships above 500 GT. Mikelis, "Developments on Recycling", *supra* note 4 at 8.

[25] *Ship Recycling Convention*, *supra* note 4 art. 3(2)–(3). See also Recycling of Ships: Comments on the Report of the Second Intersessional Meeting of the Working Group on Ship Recycling, Submitted by Belgium and the Netherlands, IMO/MEPC 56/3/16, 17 May 2007 (KR-CON) (pointing out to the need to exclude vessels built for use in internal waterways (barges)). Barges differ from sea-going ships not only in terms of concept, but more importantly, in respect of applicable regulations (they do not sail under the IMO conventions), tonnage measurement, and survey and certification. Barges cannot sail overseas to a different recycling yard; in fact, the sound recycling of barges is an environmental protection matter of local countries). *Ibid.*

[26] *Ship Recycling Convention*, *supra* note 4 art. 3(3).

[27] *Ibid.*, art. 3(4).

[28] *Ibid.*, art. 2(10).

[29] Recycling of Ships: Proposal to Clarify the Scope of the Convention with Respect to Land-based Facilities, Submitted by France, IMO/MEPC 55/3/14, 17 August 2006, ¶ 7 (KR-CON). See also International Conference on the Safe and Environmentally Sound Recycling of Ships: Agenda Item 6, Consideration of the Draft International Convention for the Safe and Environmentally

B. *Ban on Trade between Parties and Non-Parties*

Ships can only be recycled at ship recycling facilities authorised by parties in accordance with the convention.[30] Authorised ship recycling facilities are only to accept ships (party ships or non-party ships treated similarly pursuant to article 3(4)) that comply with the requirements mandated by the Ship Recycling Convention. Moreover, ship recycling facilities can only accept ships that they are authorised to recycle. A combined reading of these stipulations reveals that the Ship Recycling Convention contemplates mainly party-party situations. However, in certain cases non-party ships can be recycled at authorised party facilities. These provisions can thus be applied in a manner as to practically ban the trade in derelict ships between parties and non-parties even though technically, non-party ships can continue to operate solely between ports and recycling facilities located in non-party states.

C. *Identification of Major Actors*

The Ship Recycling Convention identifies two major actors in ship recycling: the "Administration", which is the government of the state whose flag the ship is entitled to fly or under whose authority the ship operates,[31] and the state where the recycling facilities are situated. In addition, port states,[32] the recycling company,[33] the shipowner,[34] and workers[35] also have obligations under the Ship Recycling Convention.

Sound Recycling of Ships: Proposal to Amend Article 2, Submitted by the United States, SR/CONF/43, 10 May 2009, online: SJÖFARTSVERKET <http://www.sjofartsverket.se/pages/19514/sr-conf-43.pdf>.

[30] *Ship Recycling Convention*, *supra* note 4 reg. 8(1)(1).

[31] *Ibid.*, art. 2(2).

[32] *Ibid.*, art. 8.

[33] "Recycling Company" means the owner of the Ship Recycling Facility or any other organization or person who has assumed the responsibility for operation of the Ship Recycling activity from the owner of the Ship Recycling Facility and who on assuming such responsibility has agreed to take over all duties and responsibilities imposed by this convention. *Ibid.*, art. 2(12).

[34] The definition of "Shipowner" reads:

> "Shipowner" means the person or persons or company registered as the owner of the ship or, in the absence of registration, the person or persons or company owning the ship or any other organization or person such as the manager, or the bareboat charterer, who has assumed the responsibility for operation of the ship from the owner of the ship. However, in the case of a ship owned by a State and operated by a company which in that State is registered as the ship's operator, "owner" shall mean such company. This term also includes those who have ownership of the ship for a limited period pending its sale or handing over to a Ship Recycling Facility. *Ibid.*, reg. 1(8).

[35] "Worker" means any person who performs work, either regularly or temporarily, in the context of an employment relationship including contractor personnel. *Ibid.*, reg. 1(12).

1. *Responsibilities of the Flag State*

A flag state administration has the duty to adopt effective measures to ensure that ships that fly its flag comply with the requirements set out in the Ship Recycling Convention.[36] The Ship Recycling Convention identifies two primary responsibilities for flag states: preparing the inventory of hazardous materials (IHM), and surveys and certification.

The key to an efficient strategy for environmentally-sound ship disposal lies in prior knowledge of the hazardous and polluting substances on board. This enables the recycling yard to customise its processes to suit the destructive requirements of a particular ship in order that the dismantling does not contaminate the environment. The Ship Recycling Convention requires a new ship[37] to carry on board the IHM specific to it, verified by its flag authority or any person authorised by it in accordance with the IMO guidelines (to be developed, which are to also specify the threshold values and exemptions).[38]

The IHM is comprised of three parts. Part I requires listing of the hazardous materials[39] in the ship's structure or equipment in accordance with appendices 1 and 2 of the Ship Recycling Convention, their location, and the approximate quantity.[40] This listing is to be maintained and updated throughout the operational life of the ship and must reflect relevant changes to the ship's structure and equipment including new installations that contain hazardous materials listed in appendix 2.[41] Moreover, the IHM is also to clarify that the ship complies with regulation 4, which outlines controls on a ship's hazardous materials.[42] Existing ships are to develop such a listing "as far as practicable" within five years from the date on which the Ship Recycling Convention enters into

---

[36] *Ibid.*, art. 4(1).

[37] *Ibid.*, reg. 1(4).

[38] *Ibid.*, reg. 5(1). See also International Conference on the Safe and Environmentally Sound Recycling of Ships: Agenda Item 6, Consideration of the Draft International Convention for the Safe and Environmentally Sound recycling of Ships: Proposed Amendment to Draft Regulation 5 – Inventory of Hazardous Materials, Submitted by the Marshall Islands, Panama, Ukraine, ICS, BIMCO, IACS, OCIMF, INTERTANKO, INTERCARGO and IPTA, SR/CONF/23, 1 April 2009, online: SJÖFARTSVERKET <http://www.sjofartsverket.se/pages/19514/sr-conf-23.pdf> (arguing for threshold values to be incorporated into the text of the Ship Recycling Convention).

[39] *Ship Recycling Convention*, *supra* note 4 art. 2(9) (which defines hazardous material to mean any material or substance that is liable to create hazards to human health and/or the environment). Appendix 1 entitled "Controls of Hazardous Materials" lists the following hazardous materials, namely, asbestos, ozone-depleting substances, polychlorinated biphenyls, and anti-fouling compounds and systems.

[40] *Ibid.*, reg. 5(1)(1). See also *ibid.*, regs. 6–7 (which envisages a special procedure to amend appendices 1 and 2).

[41] *Ibid.*, reg. 5(3). See also *ibid.*, app. 2 (providing the "Minimum List of Items for the Inventory of Hazardous Materials", which includes the hazardous materials in appendix 1, cadmium and cadmium compounds, radioactive substances, lead and lead compounds, polybrominated biphenyls, hexavalent chromium and hexavalent chromium compounds).

[42] *Ibid.*, reg. 5(1)(2).

force, or before a ship proceeds for recycling prior to the end of this five year period taking into account the IMO guidelines to be developed in this regard and its harmonised system of survey and certification.[43] Since difficulties could be encountered while developing Part I of the IHM for existing ships (which includes the listing of hazardous materials listed out in appendices I and II), the Ship Recycling Convention provides that, for such ships, at a minimum, the hazardous materials listed in appendix I are to be identified.[44] Furthermore, it has been provided that existing ships also prepare a plan to describe the "visual/sampling check" that was undertaken during the development of the IHM.[45] Both new and existing ships are expected to properly maintain and update Part I of the IHM.[46]

Since Part II (operationally generated wastes) and Part III (stores) of the IHM are only relevant if they are still present on the ship at the time of its recycling, these listings need only be developed prior to the recycling.[47] Both parts have to be verified either by the administration or by any person/organisation authorised by it taking into account relevant IMO guidelines.[48]

Under the Ship Recycling Convention, ships are subject to a series of surveys and certifications,[49] for the purpose of enforcement of the provisions of the convention. Surveys are to be carried out by officers of the administration, or by surveyors or recognised organisations nominated by the administration in accordance with the IMO guidelines (to be developed).[50] However, in all cases, it is the responsibility of the administration to ensure the "completeness and efficiency of the survey", and to this end, it is empowered to adopt "necessary arrangements" to satisfy this obligation.[51]

An initial survey is to be undertaken before a ship is placed in service, or before the IHM is issued.[52] Its purpose is to verify whether Part I of the IHM is in accordance with the stipulations mandated by the Ship Recycling Convention.[53] Thereafter, renewal surveys at intervals (to be specified by the administration, but not exceeding five years) to verify whether Part I of the IHM still complies with the requirements of the Ship Recycling Convention are to be conducted.[54] Upon the successful completion of the initial or renewal survey, the administration or

[43] *Ibid.*, reg. 5(2).
[44] *Ibid.*
[45] *Ibid.*
[46] *Ibid.*, reg. 5(4).
[47] *Ibid.*
[48] *Ibid.*
[49] *Ibid.*, art. 5.
[50] *Ibid.*, reg. 10(2).
[51] *Ibid.*, reg. 10(4).
[52] *Ibid.*, reg. 10(1)(1).
[53] *Ibid.*
[54] *Ibid.*, reg. 10(1)(2).

any person or organisation authorised by it, issues the International Certificate on Inventory of Hazardous Materials.[55]

Depending upon the circumstance, additional surveys (general or partial) can be carried out at the request of the shipowner whenever there is a "change, replacement, or significant repair" involving the ship's structure, or its equipment, systems, fittings, etc.[56] This survey also seeks to ensure that the ship continues to comply with the terms of the Ship Recycling Convention and that Part I of the IHM has been suitably amended so as to reflect the necessary changes.[57] Once an additional survey is successfully completed, the administration or any person or organisation authorised by it, will have to endorse the International Certificate on Inventory of Hazardous Materials.[58]

Prior to a ship being taken out of service and before its recycling, there must be a "final survey" to verify whether the IHM is in accordance with the requirements of the convention and that the ship recycling plan (SRP) contains information relating to the IHM and for safe-for-entry and safe-for-hot-work.[59] The final survey also examines whether the ship recycling facility where the ship is to be recycled holds a valid authorisation in accordance with the Ship Recycling Convention.[60] Upon its successful completion, the administration, or any person or organisation authorised by it, issues the International Ready for Recycling Certificate (IRRC).[61] The IRRC certifies that the ship has been duly surveyed, possesses a valid IHM, the SRP adequately reflects the IHM and contains information relating to safe-for-entry and safe-for-hot-work conditions, and that the ship recycling facility is authorised.[62] It is expected that the entire series of surveys, which culminates in the issuance of the IRRC, will form an important enforcement mechanism that will lead to improved health, safety, and better environmental protection in the shipbreaking yards.[63]

---

[55] *Ibid.,* reg. 11(1). See also *ibid.*, app. 3 (entitled "Form of the International Certificate on Inventory of Hazardous Materials"). In the international certificate on inventory of hazardous materials, Part I of the inventory of hazardous materials assumes absolute primacy. Being an essential part of this certificate on, Part I always has to accompany it. In fact, the international certificate on inventory of hazardous materials certifies that a ship has been surveyed in accordance with regulation 10 and that the survey shows that Part I of the inventory of hazardous materials, fully complies with the applicable requirements of the convention. *Ibid.*

[56] *Ibid.*, reg. 10(1)(3).

[57] *Ibid.*

[58] *Ibid.*, reg. 11(2). See also *ibid.*, app. 3 (entitled "Form of the International Certificate on Inventory of Hazardous Materials" includes the section "Endorsement for Additional Survey").

[59] *Ibid.*, reg. 10(1)(4)(1), (2).

[60] *Ibid.*, reg. 10(1)(4)(3).

[61] *Ibid.*, reg. 11(11). See also *ibid.*, reg. 8(6) (requiring that ships destined for recycling shall be certified as ready for recycling by the administration, or organisation recognised by it, prior to the recycling activity). See also *ibid.*, app. 4 (which provides the form for the International Ready for Recycling Certificate).

[62] *Ibid.*, app. 4.

[63] Recycling of Ships: Ready for Recycling Criteria, Submitted by ICS, BIMCO, INTERTANKO, INTERCARGO & IPTA, IMO/MEPC 56/3/9, 4 May 2007 (KR-CON).

2. *Recycling State*

The Ship Recycling Convention also establishes a series of obligations for the second major entity, namely, the recycling state. Each party under whose jurisdiction a "ship recycling facility"[64] operates has to ensure that such facilities are authorised.[65] Parties are to establish mechanisms to authorise such facilities subject to "appropriate conditions" that would enable such facilities to meet the requirements mandated by the Ship Recycling Convention.[66] Parties are also to provide for a mechanism to ensure that ship recycling facilities comply with the terms of chapter 3 of the Ship Recycling Convention, entitled Regulations for Safe and Environmentally Sound Recycling of Ships, which spells out the requirements for ship recycling facilities. Such a mechanism can include the establishment and effective use of provisions relating to inspection, monitoring and enforcement and even an audit scheme, the results of which has to be communicated to the IMO.[67] Even though the authorisation is to be carried out by competent authority(ies),[68] which includes "verification of documentation" and "site inspection", the competent authority(ies) are empowered to entrust this task to organisations recognised by it.[69] However, in all cases, the ultimate responsibility for the authorisation rests with the competent authority(ies).[70] This authorisation is subject to terms (relating to its issuance, renewal, amendment, suspension, or withdrawal) to be identified by the party and is valid for a period not exceeding five years.[71] Generally, for an authorisation, the ship recycling facility will have to meet the requirement that it be designed, constructed, and operated in a safe and environmentally sound manner. This includes complying with the relevant requirements of regulations 16–25 of the annex; having the ability to establish, maintain and monitor safe-for-hot-work and safe-for-entry conditions; and the appropriate measures in place to manage[72] hazardous materials.[73]

---

[64] *Ship Recycling Convention*, *supra* note 4 art. 2(11) ("Ship Recycling Facility" means a defined area that is a site, yard or facility used for the recycling of ships).

[65] *Ibid.*, art. 6. See also *ibid.*, regs. 8(1)(1), 16(1).

[66] *Ibid.*, reg. 15(2)–(3).

[67] *Ibid.*, reg. 15(3). See also Work of the Conference, *supra* note 9 res. 6 (inviting MEPC to explore possible ways to ensure best practices for complying with this regulation).

[68] *Ship Recycling Convention*, *supra* note 4 art. 2(3) defines "Competent Authority(ies)" to mean a governmental authority or authorities designated by a party as responsible, within specified geographical area(s) or area(s) of expertise, for duties related to ship recycling facilities operating within the jurisdiction of that party as specified in this convention.

[69] *Ibid.*, reg. 16(2). The authorisation is to be in accordance with the form set out in appendix 5 entitled "Form of the Authorization of Ship Recycling Facilities". *Ibid.*, reg. 16(4).

[70] *Ibid.*, reg. 16(3).

[71] *Ibid.*, reg. 16(5).

[72] The Ship Recycling Convention in the "Supplement to: Document of Authorization to Undertake Ship Recycling (DASR) in Accordance with the Hong Kong International Convention for the Safe and Environmentally Sound Recycling of Ships, 2009" identifies three methods in managing hazardous wastes namely, removal, storage and processing (which include incineration and reclamation of hazardous materials and treatment of oily residues). See *ibid.*, app. 5.

[73] *Ibid.*

Parties are also to develop a domestic legal framework to ensure that ship recycling facilities are "designed, constructed, and operated in a safe and environmentally sound manner ..."[74] They are to have "management systems, procedures and techniques" that do not pose health risks to the workers or to the populations living in the vicinity of the recycling yard.[75] To the extent practicable, these measures should also seek to reduce, minimise and eliminate the adverse effects of ship recycling on the environment.[76]

The Ship Recycling Convention identifies two specific instances where the authorisation can be suspended or withdrawn. The first relates to refusal to have the ship recycling facility inspected by the competent authority(ies) or by the recognised organisation.[77] Secondly, if "incidents or actions" at the ship recycling facility run counter to the conditions of authorisation, the facility is to inform the competent authority(ies) which can withdraw the authorisation, or require corrective measures.[78] In cases where a party requests an investigation and provides sufficient evidence to prove that a ship recycling facility is operating, has operated or is about to operate in violation of the Ship Recycling Convention, then the party within whose jurisdiction the ship recycling facility is operating has to conduct an investigation and make a report, a copy of which (it should include information on action taken or to be taken) has to be send to the party requesting it and to the IMO for appropriate action.[79]

3. *Authorised Ship Recycling Facilities*

Apart from the above identified major actors, the next important player is the authorised ship recycling facility. The Ship Recycling Convention also elaborates upon a set of obligations that such facilities are supposed to carry out. Firstly, authorised ship recycling facilities can only accept ships that comply with the terms of the Ship Recycling Convention. Secondly, they can only accept ships which they are authorised to recycle. Finally, they are to make available documentation relating to its authorisation to a shipowner who is considering recycling his/her ship at this recycling facility.[80] Additionally, other obligations include:

i. *Ship Recycling Facility Plan*

Authorised ship recycling facilities are to prepare a "Ship Recycling Facility Plan" (SRFP), which is to be adopted by the board or the appropriate governing

[74] *Ibid.*, reg. 15(1).
[75] *Ibid.*, reg. 17(1).
[76] *Ibid.*
[77] *Ibid.*, reg. 16(5).
[78] *Ibid.*, reg. 16(6).
[79] *Ibid.*, art. 9(4).
[80] *Ibid.*, reg. 17(2).

body of the recycling company.[81] The SRFP is to include, *inter alia*, policies for worker safety, measures to protect human health and the environment; systems to ensure the implementation of the requirements set out in the Ship Recycling Convention and the policy goals of the recycling company;[82] roles and responsibilities for employers and workers while conducting recycling operations;[83] a programme for providing appropriate information and training to workers;[84] an emergency preparedness and response plan;[85] a system to monitor the performance of the recycling operation;[86] a record-keeping system showing how the recycling was carried out;[87] and, provide for reporting systems in relation to discharges, emissions, incidents and accidents, and on occupational diseases, injuries, etc.[88]

ii. *Emergency Preparedness and Response Plan*

Authorised ship recycling facilities should have an emergency preparedness and response plan[89] that balances the location and environment of the ship recycling facility with the size and nature of its activities. This plan is to ensure, *inter alia*, that necessary equipment and procedures needed in an emergency are in place; drills are conducted regularly; there are communication channels with competent authority(ies), the neighbourhood and with emergency response services; relevant information and training is provided to all workers depending on their level of competence; and that there is provision for first-aid and medical assistance, pollution prevention, and fire-fighting.[90]

iii. *Worker Safety and Training*

As part of its obligation to provide for worker safety measures,[91] authorised ship recycling facilities are to ensure the availability, maintenance and use of personal protective equipment[92] and clothing needed for the recycling operations and to impart appropriate training so as to enable the workers to safely undertake all ship recycling operations.[93] The training programmes are to take into account the

---

[81] *Ibid.*, reg. 18. See also Recycling of Ships: Guidelines for the Development of the Ship Recycling Facility Management Plan, Submitted by the United States, IMO/MEPC 56/3/6, 6 April 2007 (KR-CON).

[82] *Ship Recycling Convention*, *supra* note 4 reg. 18(1)–(2).

[83] *Ibid.*, reg. 18(3).

[84] *Ibid.*, reg. 18(4).

[85] *Ibid.*, reg. 18(5).

[86] *Ibid.*, reg. 18(6).

[87] *Ibid.*, reg. 18(7).

[88] *Ibid.*, reg. 18(8)–(9).

[89] *Ibid.*, reg. 21.

[90] *Ibid.*

[91] *Ibid.*, reg. 22.

[92] The personal protective equipment includes head protection, face and eye protection, hand and foot protection, respiratory protective equipment, hearing protection, protectors against radioactive contamination, protection from falls, and appropriate clothing. *Ibid.*, reg. 22(2).

[93] *Ibid.*, reg. 22(1).

IMO guidelines (to be developed) and should be conducted by competent persons at regular intervals.[94] Since most ship recycling workers are illiterate, provision should be made for refresher training at regular intervals, and for evaluating their comprehension and ability to retain the lessons learnt.[95] The training is to be documented, periodically reviewed, and modified if necessary.[96]

iv. *Ship Recycling Plan*

Prior to the recycling of a ship, the ship recycling facility is to prepare a "ship-specific ship recycling plan" taking into account the information provided by the shipowner to ensure that a ship has been prepared to the maximum extent possible.[97] The plan must include information on safe-for-entry and safe-for-hot-work conditions and how the hazardous materials that are identified in the IHM will be managed.[98] In cases where more than one ship recycling facility is being used, the SRP is to identify all these ship recycling facilities and specify the recycling activities that each of them will carry out in the order in which they will occur.[99]

The SRP has to be approved by the competent authority that has authorised the ship recycling facility.[100] The mode of approval is subject to two possible courses of action, namely, explicit or tacit approval. A recycling state has to, at the time it expresses its intent to be bound by the Ship Recycling Convention, specify whether it requires explicit or tacit approval of the SRP.[101] This declaration can subsequently be revised by a notification to the Secretary-General of the IMO.[102]

On receipt of the SRP, the competent authority has to acknowledge it in writing, addressing the acknowledgement to the ship recycling facility, the shipowner, and the administration within three working days.[103] If the recycling state party requires explicit approval, the competent authority has to send a written notification of its decision to either approve or reject the SRP to the ship recycling facility, the shipowner and the administration.[104] In states where the mode of approval is tacit, the written acknowledgement by the competent authority will have to specify an end date for a 14 day review period.[105] If, within this timeframe,

---

[94] *Ibid.*, reg. 22(3).
[95] *Ibid.*, reg. 22(3)(3), (4).
[96] *Ibid.*, reg. 22(3)(5), (6).
[97] *Ibid.*, reg. 9(1). See also Reg. 8(4). See also International Conference on the Safe and Environmentally Sound Recycling of Ships: Agenda Item 6, Consideration of the Draft International Convention for the Safe and Environmentally Sound Recycling of Ships: The Role of the Shipowner in the Development of the Ship Recycling Plan, Submitted by India, SR/CONF 27, 2 April 2009, online: SJÖFARTSVERKET <http://www.sjofartsverket.se/pages/19514/SR-CONF-27.pdf>.
[98] *Ship Recycling Convention*, *supra* note 4 reg. 9(3).
[99] *Ibid.*, reg. 9(6).
[100] *Ibid.*, reg. 9(4).
[101] *Ibid.*, art. 16(6).
[102] *Ibid.*
[103] *Ibid.*, reg. 9(4).
[104] *Ibid.*, reg. 9(4)(1).
[105] *Ibid.*, reg. 9(4)(2).

no written objection is received by the ship recycling facility, the shipowner, and the administration, the SRP is deemed to have been approved.[106] Once approved, the plan is to be made available for inspection by the administration, or by the nominated surveyors or by organisations recognised by the administration.[107]

v. *Ensuring Safe and Environmentally Sound Management of Hazardous Wastes*
Authorised ship recycling facilities are to ensure that there is safe and environmentally sound removal of hazardous materials from certified ships.[108] Those in-charge of the recycling operations as well as the shipbreakers are required to be familiar with the mandates of the Ship Recycling Convention that are relevant to their tasks.[109] At all stages of hazardous wastes removal, they are to actively rely on the IHM and the SRP.[110] Prior to cutting the ship, the recycling facility is to identify, label, package, and remove all the hazardous materials[111] detailed in the IHM to the "maximum extent possible" by trained and equipped workers.[112]

Furthermore, the wastes generated are to be separated from the recyclable materials and are to be labelled and stored in appropriate conditions in order to avoid risks to the workers, human health and to the environment.[113] Waste management and disposal sites are to be identified[114] and the wastes should be transferred to an authorised waste management facility for their proper treatment and disposal.[115]

---

[106] *Ibid.* Several countries expressed strong objections to the tacit approval practice since in their opinion it could lead to uncertainty. The United States refuted these objections on the basis of the OECD tacit approval procedure for hazardous wastes shipments. See International Conference on the Safe and Environmentally Sound Recycling of Ships: Agenda Item 6, Consideration of the Draft International Convention for the Safe and Environmentally Sound Recycling of Ships: Proposal to Amend Regulation 9, Submitted by the United States, SR/CONF/44, 10 May 2009, online: SJÖFARTSVERKET <http://www.sjofartsverket.se/pages/19514/sr-conf-44.pdf>. See also International Conference on the Safe and Environmentally Sound Recycling of Ships: Agenda Item 6, Consideration of the Draft International Convention for the Safe and Environmentally Sound Recycling of Ships: Proposed Amendments to the Draft Convention, Improvement in the Communication of Information (Articles 9 and 12) and Clarification on the Ship Recycling Plan Approval Process (regulation 9), Submitted by France, SR/CONF/28, 6 April 2009, online: SJÖFARTSVERKET <http://www.sjofartsverket.se/pages/19514/sr-conf-28.pdf>.

[107] *Ship Recycling Convention*, *supra* note 4 reg. 9(5).

[108] *Ibid.*, reg. 20(1). The certification has to be in accordance with regulation 11 or 12.

[109] *Ibid.*

[110] *Ibid.*

[111] *Ibid.*, reg. 20(2). Some of the hazardous materials identified include hazardous liquids, residues and sediments; substances or objects containing heavy metals such as lead, mercury, cadmium and hexavalent chromium; paints and coatings that are highly flammable and/or lead to toxic releases; asbestos and materials containing asbestos; PCBs and materials containing PCBs; CFCs and halons; and other hazardous materials that are not listed and are not a part of the ship structure. *Ibid.*, reg. 20(2)(1)–(7).

[112] *Ibid.*, reg. 20(2).

[113] *Ibid.*, reg. 20(4).

[114] *Ibid.*, reg. 20(3).

[115] *Ibid.*, reg. 20(4).

vi. *Preventing Adverse Effects to Human Health and the Environment*
To prevent explosions, fires, and harm from dangerous atmospheres and other unsafe conditions, authorised ship recycling facilities are to establish and utilise procedures to facilitate the establishment and maintenance of safe-for-hot-work[116] and safe-for-entry[117] conditions throughout the recycling process.[118] As well, procedures to prevent spillage or emissions and to avert accidents, occupational diseases, and injuries, are also to be put in place to safeguard human health and the environment.[119]

4. *Shipowner*

The Ship Recycling Convention also imposes certain duties and rights on the shipowner. The shipowner initiates the entire process of recycling by notifying the administration of his/her intention to recycle a ship in order to enable the administration to meet the survey and certification requirements.[120] While the initial, renewal and final surveys are carried out as a matter of routine according to the rules mentioned therein, additional surveys can be carried out only at the request of the shipowner.[121] The shipowner also has an important role to play in the preparation of the ship-specific SRP developed by the ship recycling facility by providing the necessary information to the ship recycling facility.[122]

Regulation 8 which outlines the general requirements in preparing a ship for recycling states that "ships destined to be recycled shall" conduct operations to minimise cargo residues, fuel oil and wastes prior to the recycling,[123] and, in the case of tankers, ensure that the cargo tanks and pump rooms are ready for certification as safe-for-entry and safe-for-hot-work.[124] Since these obligations are triggered before the ship enters the recycling yard, it must be assumed that these responsibilities fall on the shipowner.

The Ship Recycling Convention also confers certain rights on the shipowner. To effectuate the selection of the most appropriate ship recycling yard, the convention provides that an authorised ship recycling facility provide documentation relating to its authorisation to a shipowner who is considering the recycling of his/her ship at that facility upon request.[125] The shipowner is also entitled to receive a written acknowledgement, and subsequently a written notification in the case of explicit

[116] *Ibid.*, reg. 1(7).
[117] *Ibid.*, reg. 1(6).
[118] *Ibid.*, reg. 19(1), (2).
[119] *Ibid.*, reg. 20(2)–(3).
[120] *Ibid.*, reg. 24(1).
[121] *Ibid.*, reg., 10(1)(3).
[122] *Ibid.*, reg. 9(1).
[123] *Ibid.*, reg. 8(2).
[124] *Ibid.*, reg. 8(3).
[125] *Ibid.*, reg. 17(2)(3).

approval from the competent authority of the recycling state, regarding grant of approval or non-approval to the ship-specific SRP.[126] In the case of tacit approval, the shipowner has to be notified by the competent authority of the recycling state in writing of objections to the ship-specific SRP within a 14 day review period.[127]

5. *Port State Control*

Generally, ships are subject to port state control measures in accordance with applicable international conventions and regulations. A supplement to flag state control, port state control extends to verifying whether the condition of foreign ships and their equipment, when in their ports, are in compliance with international maritime conventions and that the ship is manned in accordance with these rules.[128] Even a ship destined for recycling, while on its final journey, is subject to all these rules and regulations, and the port state has the power to examine whether the ship complies with these stipulations.

Apart from the general port state control powers, the Ship Recycling Convention provides for additional port state control obligations (the port state has to be a party to the Ship Recycling Convention) at three levels. First, a party is to prohibit or restrict the installation or use of hazardous materials on ships while in its ports, shipyards, ship-repair yards, or offshore terminals.[129] Second, a ship can be inspected at any port or offshore terminal of another party for a valid IHM or the IRRC.[130] In the absence of such certificates, or if there is disconnect between the physical conditions of the ship with the particulars mentioned in the certificates, a detailed inspection can be carried out.[131] Finally, as part of the scheme to detect violations, a party holding sufficient evidence can request another party to investigate the violation when the ship enters its port or offshore terminal.[132] In cases where the ship is determined to be in violation of the Ship Recycling Convention, the inspecting party can warn, detain, dismiss, or exclude the ship from entering its ports.[133]

More importantly, the Ship Recycling Convention bases the concept of port state control on the no-more-favourable treatment principle in line with earlier IMO conventions like MARPOL 73/78, the *International Convention on the Control of Harmful Anti-fouling Systems on Ships, 2001* (AFC 2001), and the *International Convention for the Control and Management of Ships' Ballast Water and Sediments, 2004.*[134] By doing so, the Ship Recycling Convention

---

[126] *Ibid.*, reg. 9(4)(1).
[127] *Ibid.*, reg. 9(4)(2).
[128] See IMOGSR, *supra* note 3 s. 9.3.
[129] *Ship Recycling Convention, supra* note 4 reg. 4(2).
[130] *Ibid.*, art. 8(1).
[131] *Ibid.*, art. 8(2).
[132] *Ibid.*, art. 9(1)–(2).
[133] *Ibid.*, art. 9(3).
[134] *Ibid.*, art. 3(4).

greatly enhances the port state control legal regime, particularly over non-party ships, thereby fostering the global applicability of IMO conventions to all ships engaged in international trade.[135]

6. *Reporting Requirements*

The Ship Recycling Convention also includes detailed reporting requirements. The process is initiated when the shipowner notifies the administration in writing of his intention to recycle a ship in "due time" in order to enable the administration to prepare itself for the survey and certification requirements as contemplated under the Ship Recycling Convention.[136] The ship recycling facility is also to notify in writing its competent authority(ies) of its intention to receive the ship.[137] Once the ship acquires the IRRC from its administration, the ship recycling facility has to report on the planned start of recycling to its competent authority(ies). It is only after the submission of this report that the ship can be recycled.[138]

Once the recycling (whether partial or complete) is completed in accordance with the convention, the recycling facility is to issue a "statement of completion" which has to be reported to the competent authority(ies).[139] This statement, which is to be issued within 14 days from the date of recycling, should report on any incidents or accidents during the breaking that could have damaged human health and/or the environment.[140] It is also to be copied to the administration which issued the IRRC.[141]

---

[135] Intersessional Meeting of the Working Group on Ship Recycling: Further Development of the Draft Convention: Consideration of the Use of Ship Recycling Facilities under the Jurisdiction of Non-parties, Submitted by Norway, MEPC-ISRWG 3/2/9, 14 December 2007, online: SJÖFARTSVERKET <http://www.sjofartsverket.se/pages/14272/3-2-9.pdf> [Use of Ship Recycling Facilities under the Jurisdiction of Non-parties].

[136] *Ship Recycling Convention*, *supra* note 4 reg. 24(1).

[137] *Ibid*., reg. 24(2). The notification is to include as a minimum the following ship details, namely, the name of the flag state; the date of registration with that state; the ship's IMO number; hull number on new-building delivery; the name and type of the ship; the port at which the ship is registered; the name and address of the shipowner along with the IMO registered owner identification number; the name and address of the company along with the IMO company identification number; the names of all classification society(ies) with which the ship is classed; the inventory of hazardous materials and the draft ship recycling plan; and the ship's main particulars like lightweight, cargo carrying capacity, gross and net tonnage, engine type, etc. *Ibid*., reg. 24(2)(1)–(12).

[138] *Ibid*., reg. 24(3). See also *ibid*., app. 6 (providing the "Form of Report of Planned Start of Ship Recycling").

[139] *Ibid*., reg. 25. See also *ibid*., app. 7 (providing the "Form of the Statement of Completion of Ship Recycling").

[140] *Ibid*., reg. 25. See also *ibid*., reg. 23. Authorised ship recycling facilities are also to report to its competent authority(ies) of incidents, accidents, occupational diseases, or chronic effects that endanger or have the potential to endanger worker safety, human health and the environment, the reasons for the same, and the consequences. The report is also to contain information on the response action taken, the consequences and what other corrective actions are needed. *Ibid*.

[141] *Ibid*., reg. 25.

7. *Hazardous Materials in the Design, Construction, Operation and Maintenance of Ships*

The Ship Recycling Convention seeks to ensure that hazardous materials do not make their way into the ship structure. As such, flag states are to prohibit or restrict the installation or use of hazardous materials enumerated in appendix 1 on ships entitled to fly its flag or operating under its authority.[142] Appendix 1 imposes certain controls on the use of hazardous materials on all new ships. New installation of materials containing asbestos or polychlorinated biphenyls is prohibited. Again, new installations containing ozone depleting substances are also to be prohibited, except those containing hydrochlorofluorocarbons, which are permitted till 1 January 2020. As far as anti-fouling compounds and systems that are regulated under annex I to the AFC 2001 are concerned, the regulation is twofold. Firstly, no ship may apply anti-fouling systems containing organotin compounds as a biocide or any other anti-fouling system whose application or use is prohibited by the AFC 2001. Secondly, no new ships or new installations on ships shall apply or employ anti-fouling compounds or systems in a manner inconsistent with the AFC 2001.[143] As noted above, a port state can also prohibit or restrict the use of such materials on ships while in its ports, shipyards, ship-repair yards, or offshore terminals.[144]

8. *Cooperation and Technology Transfer*

The Ship Recycling Convention calls for cooperation between parties "… for the purpose of effective implementation of, compliance with and enforcement of this Convention."[145] It also requires parties "to encourage the continued development of technologies and practices which contribute to safe and environmentally sound ship recycling".[146] Furthermore, it envisages the transfer of management systems and technology for the safe and environmentally sound recycling of ships, subject to national laws, regulations and policies.[147] Cooperation and technology transfer is further fortified by the provision that empowers parties to directly request technical assistance from other parties or through the IMO and other international bodies. Such assistance can take the form of personnel training, technology facilities and equipment transfer, or programmes for joint research and development.[148]

---

[142] *Ibid.*, reg. 4(1).
[143] See *ibid.*, app. 1 (entitled "Controls of Hazardous Materials).
[144] *Ibid.*, reg. 4(2).
[145] *Ibid.*, art. 1(3).
[146] *Ibid.*, art. 1(4).
[147] *Ibid.*, art. 13(2).
[148] *Ibid.*, art. 13(1).

9. *Enforcement and Compliance*

Violations of the Ship Recycling Convention by ships and ship recycling facilities are to be prohibited under national law.[149] Parties are to establish sanctions "adequate in severity" to discourage the same.[150]

As far as ships are concerned, sanctions can be imposed by the flag state irrespective of where the violation occurs.[151] When the administration is informed of a violation by a party of one of its ships, it must investigate the matter and can request the reporting party to provide it with additional evidence regarding the alleged violation.[152] If it is satisfied that there is sufficient evidence, it can initiate proceedings against the ship and promptly inform the party that reported the alleged violation, as well as the IMO, of any action taken.[153]

To discourage violations in ship recycling facilities, a party having jurisdiction over such facilities is to establish sanctions under its national law.[154] If informed by another party of an alleged violation by a recycling facility, the recycling state can investigate the matter and request the reporting party to furnish additional evidence.[155] If the recycling state is satisfied that there is sufficient evidence, proceedings can be initiated.[156] The recycling state is to promptly inform the reporting party, as well the IMO, of any action taken.[157]

In both the situations mentioned above, if the administration as well as the party does not take any action within a year of receiving the information, it has to inform the party that reported the alleged violation as well as the IMO of the reasons for the inaction.[158]

10. *Relation to Other International Instruments*

The Ship Recycling Convention does not prejudice the rights and obligations of parties under other relevant and applicable international agreements.[159] No provision in the Ship Recycling Convention is to be interpreted so as to prevent a party from adopting individually or jointly, more stringent measures consistent with international law, for the safe and environmentally sound recycling of ships.[160] Thus, the principles embodied in this convention represent only an irreducible

---

149 *Ibid.*, art. 10.
150 *Ibid.*, art. 10(3).
151 *Ibid.*, art. 10(1).
152 *Ibid.*
153 *Ibid.*
154 *Ibid.*, art. 10(1)(2).
155 *Ibid.*
156 *Ibid.*
157 *Ibid.*
158 *Ibid*, art. 10(1)(1), (2).
159 *Ibid.*, art. 15(2).
160 *Ibid.*, art. 1(2).

minimum for achieving the objective of safe and environmentally sound ship recycling.

The Ship Recycling Convention also articulates its relationship *vis-à-vis* the *United Nations Convention on the Law of the Sea, 1982* (LOS Convention), the Basel Convention and ILO instruments. It expressly recognises the paramount importance of the LOS Convention by providing that the rights and responsibilities under this convention and the customary international law of the sea continue to operate.[161] It has also been provided that in effectuating the "Regulations for Safe and Environmentally Sound Recycling of Ships", parties are to take into account "relevant and applicable" ILO instruments and technical standards, and recommendations and guidance developed under the Basel Convention.[162] Similarly, Resolution 2 adopted by the Hong Kong Ship Recycling Conference, which recognises the "Contribution of the Parties to the Basel Convention and the International Labour Organization in the Development of the Hong Kong International Convention for the Safe and Environmentally Sound Recycling of Ships, 2009", "invites the International Maritime Organization to continue co-operation on ship recycling with the Basel Convention and the International Labour Organization following the adoption of the Convention."[163]

11. *Communication and Exchange of Information*

To ensure effective implementation and greater transparency, parties are obliged to report to the IMO information on certain matters for dissemination.[164] Parties are also to provide to the IMO or to other parties "relevant information" on which they based their decision to authorise a particular ship recycling facility in a swift and timely manner.[165]

12. *Dispute Settlement*

Article 14 provides the dispute settlement mechanism for the convention. Parties are called upon to settle any dispute relating to "the interpretation or application of this Convention by negotiation or any other peaceful means agreed upon

---

[161] *Ibid.*, art. 15(1).

[162] *Ibid.*, reg. 3.

[163] Work of the Conference, *supra* note 9 res. 2.

[164] *Ship Recycling Convention*, *supra* note 4 art. 12. The information relates to the list of authorised ship recycling facilities and contact details of competent authority(ies) including a single contact point; the list of recognised organisations and nominated surveyors and their specific responsibilities, aspects relating to delegation of authority to recognised organisations and nominated surveyors; an annual list of ships to which an IRRC has been issued including the name of the recycling company and the location of the ship recycling facility; the annual list of ships recycled and those deregistered including the name of the recycling company and location; and information on violations and actions taken. *Ibid.*

[165] *Ibid.*, art. 7.

by them." Some of the peaceful means identified therein include: enquiry, mediation, conciliation, arbitration, judicial settlement, or resort to regional agencies or arrangements.

13. *Summary of the Ship Recycling Process*

In applying these provisions, the following picture emerges: Once a shipowner decides to recycle the ship, s/he will have to follow two parallel paths. The first is commercial, which culminates in the selection of an appropriate recycling facility; the second relates to fulfilling certain procedural requirements. The shipowner notifies the flag administration of his/her intent to recycle the ship, thereby setting in motion the survey and certification requirements.[166] However, long before this decision is made, the survey process in relation to recycling may have been initiated. Under the Ship Recycling Convention, all ships new or existing are to have an IHM, thereby subjecting them to a series of surveys.[167] New ships, before being pressed into service, as well as existing ships have to undergo an initial survey and subsequent renewal surveys at specified intervals to verify whether Part I of the IHM conforms to the requirements of the convention.[168] On the successful completion of the initial and renewal surveys, the International Certificate on Inventory of Hazardous Materials is issued by the administration of the flag state.[169]

Once an IHM is issued, the focus shifts to the ship recycling facility housed in the ship recycling state. The ship recycling facility will have to develop a ship-specific SRP, prior to the recycling, taking into account the information provided by the shipowner.[170] The shipowner will also have to provide the recycling facility with the IHM, since the ship recycling facility is required to notify its competent authority(ies) of its intent to receive the ship and one of the main pieces of information that has to be included in this notification is the IHM.[171] Also, the recycling facility will have to provide in the ship-specific SRP, information on how the hazardous materials identified in the IHM will be managed.[172] The notification submitted to the competent authority(ies) also has to include a draft SRP which the authority(ies) will have to approve.[173] Once approval is secured, the action shifts back to the flag state. The ship is again subject to a final survey before it is taken out of service.[174] This survey verifies whether the IHM (which is

[166] *Ibid.*, reg. 24(1).
[167] *Ibid.*, reg. 5(1), (2).
[168] *Ibid.*, reg. 10(1)(1), (2).
[169] *Ibid.*, reg. 11(1).
[170] *Ibid.*, reg. 9(1).
[171] *Ibid.*, reg. 24(2)(11).
[172] *Ibid.*, reg. 9(3).
[173] *Ibid.*, reg. 24(2)(12).
[174] *Ibid.*, reg. 10(4)(1), (2).

now an updated document comprised of three parts) complies with the convention. It also examines whether the SRP (now approved by the competent authority(ies)) properly reflects the IHM and contains information on safe-for-entry and safe-for-hot-work; and information on whether the ship recycling facility is duly authorised.[175] Once the final survey is successfully completed, the administration issues the IRRC.[176] Thereafter, the action reverts back to the recycling state with the recycling facility reporting to its competent authority(ies) the planned start date of the recycling process together with the copy of the IRRC.[177] Recycling starts only after the submission of this report.[178] Once the recycling is complete, the ship recycling facility issues a statement of completion, and reports this fact to its competent authority(ies), who in turn sends a copy of the same to the administration that issued the IRRC.[179]

III. The Ship Recycling Convention, IMO Guidelines, Basel Convention and Dilution of the Principle of Equivalency

The Ship Recycling Convention has been hailed by several countries, United Nations bodies, and inter-governmental organisations.[180] For the United States, it is a "unique Convention … that clearly address[es] the environmental issues related to ship recycling."[181] Denmark considers the convention as "a big step … with ambitious requirements for future safe and environmentally sound recycling of ships."[182] Bangladesh believes that, "we now have something which we needed so much for so long to have a smoother sailing of ship recycling".[183] Japan feels that the Ship Recycling Convention "would be the most important international instrument in terms of promoting Safe and Environmentally Sound Recycling of Ships" and that "… 2009 will undoubtedly be remembered so far as enhancing international awareness in the field of Ship Recycling".[184] For Kenya, the convention is an "important milestone" that will "go a long way in alleviating the suffering of thousands of workers in the ship recycling industry. It will also play a very

---

175 *Ibid.*, reg. 10(4).
176 *Ibid.*, reg. 11(11).
177 *Ibid.*, reg. 24(3).
178 *Ibid.*
179 *Ibid.*, reg. 25.
180 See International Conference on the Safe and Environmentally Sound Recycling of Ships: Hong Kong, China, 11–15 May 2009, Statements by Delegations and Observers at the Closing Session of the Conference, Friday, 15 May 2009, SR/CONF/INF. 8, 21 May 2009, online: SJÖFARTSVERKET <http://www.sjofartsverket.se/pages/19514/sr-conf-inf8.pdf> [Statements by Delegations and Observers].
181 *Ibid.*, at 11.
182 *Ibid.*, at 3.
183 *Ibid.*, at 2.
184 *Ibid.*, at 6.

important role in reducing the negative impact of the industry to the marine environment".[185] Malta holds the view that

> [t]he adoption of this Convention is an important milestone that would ensure great improvements in the ship recycling industry to the benefit particularly of those who work in it and to the environment. It might not be the perfect instrument but nothing is perfect. The implementation of the Hong Kong International Convention for the Safe and Environmentally Sound Recycling of Ships would certainly be a giant step forward to the benefit of all.[186]

The International Labour Organization in its statement stresses the need for the different agencies to work together to develop relevant guidelines and to cooperate technically to ensure that the health and safety of workers and the general public are protected.[187] On the other hand, the UNEP Basel Convention drew attention to the decision taken at the ninth COP regarding the establishment of equivalent level of controls as contemplated under the Basel Convention and the need to avoid duplication of regulatory instruments having the same objective.[188] Welcoming the adoption of the convention, the International Chamber of Shipping points out that, the Ship Recycling Convention "will improve recycling practices worldwide and safeguard the health, welfare and safety of recycling workers".[189]

In sharp contrast to these views hailing the Ship Recycling Convention as the panacea to the ills plaguing the ship recycling industry, the Friends of the Earth International considers that the global response has been "a profound disappointment for we have sadly missed an opportunity."[190] The Ship Recycling Convention, they opine, is "an obsolete relic that ignores long standing environmental and social principles."[191] "When the world stood up and called for real change to prevent environmental injustice and exploitation of human rights … it was given instead – an inventory, a plan and some guidelines… [A] green rubber stamp on a disastrous status quo."[192] They further point out:

> This IMO Convention in the eyes of civil society stakeholders now must be deemed a failure. It has failing marks in the five most vital areas needed for true reform.
>
> First, it has failed to provide an equivalent level of control to that of the Basel Convention.
>
> Second, it has failed the test of environmental justice as it does nothing to prevent the disproportionate burdening of developing countries from toxic waste from ships.

---

185 *Ibid.*, at 7.
186 *Ibid.*, at 8.
187 *Ibid.*, at 12.
188 *Ibid.*, at 13.
189 *Ibid.*
190 *Ibid.*
191 *Ibid.*, at 14.
192 *Ibid.*

> Third, it has failed the test of cost internalization as it has avoided attaching financial responsibility to the beneficiaries of the useful life of a ship, nor to the generators of the risks and liabilities and has instead pushed all costs downstream.
>
> Fourth, it has failed to create an incentive or a mandate for toxic-free more safely recyclable ships and thus has failed the test of prevention through design.
>
> And, finally this week, it has failed the last test of environmentally sound management as it refuses to draw a line in the sand to bar the fatally flawed beaching method – the cause of death and pollution of the marine environment.
>
> Instead of these things we have been given an inventory, a plan and some guidelines. When the world cried for help, we threw paper rather than a life ring.[193]

Against this backdrop of bouquets and brickbats, it is useful to consider how comprehensively the Ship Recycling Convention addresses the longstanding problems noted in this industry. It is also appropriate to examine how this convention compares with the Basel Convention and the IMO Guidelines on Ship Recycling in going beyond the measures established under their auspices.

On a first look, the Ship Recycling Convention seems to have adopted a holistic approach to issues relating to heath, safety and environmental protection with the primary objective of ensuring "safe and environmentally sound recycling of ships". To accomplish this broad goal, it introduces comprehensive systems of control and enforcement to ensure that ships destined for their last rites are certified and that they head only to authorised ship recycling facilities.[194] Among the notable features, the Ship Recycling Convention identifies certain actors and assigns them responsibilities. It also establishes certain control and enforcement mechanisms: recycling facility authorisation, documentary control procedures, detection of violations, an audit scheme to ensure greater transparency, prohibition or restriction on the use of hazardous materials on ships, the no-more favourable treatment clause, and ship survey and certification rules. In addition, it stresses worker safety and training, and emphasises upstream and downstream management of hazardous materials.

The Ship Recycling Convention also provides that hazardous materials on board for the IHM be determined based on threshold levels to be specified in the guidelines. This is a practical approach to the management of hazardous materials given that without threshold levels, hazardous materials would have to be listed from the smallest known or detectable concentrations onwards, including those where the concentrations do not have the potential to endanger human health.[195] The concept of approved ship-specific SRPs is also noteworthy.

---

[193] *Ibid.*, at 15.

[194] EU Comments, *supra* note 21 ¶ 46.

[195] Recycling of Ships: Utilization of Applicable Threshold Levels Relevant for the Preparation of Inventory of Hazardous Materials and Related Documentation, Submitted by Germany, IMO/MEPC 58/3/10, 30 July 2008 (KR-CON).

The requirement that ship recycling facilities provide parties and the IMO with the relevant information on the basis of which authorisation was granted to a particular yard are further significant elements.[196] The Ship Recycling Convention also addresses the vital issue of "green shipbuilding and design", which otherwise is only addressed by the non-binding IMO Guidelines on Ship Recycling. The Ship Recycling Convention establishes a system of controls at every stage in the life of a ship – from the time it is built, continuing through its operations, culminating in its decommissioning and deconstruction. In fact, it recognises recycling to be an integral part of a ship's life cycle, with recycling requirements being factored in right from the time of planning its construction to the subsequent fitting-out of the ship.[197]

Perhaps the most important feature of the Ship Recycling Convention is the prohibition that it imposes on parties from using the recycling facilities of non-parties. This stands in sharp contrast to the earlier draft version of the Ship Recycling Convention which permitted ship recycling at non-party facilities provided that the administration of the ship to be recycled, or any organisation recognised by it, after a site inspection of the non-party recycling yard issued a "document of compliance" (DOC) to the effect that the recycling facility "fully meets" the requirements of the Draft Ship Recycling Convention.[198] Many countries, including Australia and the United States, argued for the retention of similar provisions, vehemently opposing trade protectionism.[199] For them, the prohibition on parties using the recycling facilities of non-parties, regardless of the capability of those facilities to recycle a ship in a safe and environmentally sound manner, could jeopardise the implementation of the Ship Recycling Convention, particularly since there is a strong possibility that there will be insufficient recycling capacity at the time when it enters into force. Restricting access to safe and environmentally sound facilities available to party ships could prompt end-of-life ships flying the flags of parties to change to non-party flags. The possibility for this potential exodus from the registries of parties to non-party states could discourage prominent flag states from ratifying the Ship Recycling Convention

[196] Recycling of Ships: The Way Ahead – Practical and Pragmatic, Submitted by ICS, BIMCO, INTERCARGO, INTERTANKO, IPTA & WNTI, IMO/MEPC 53/3/5, 13 May 2005, ¶ 22 (KR-CON).

[197] Recycling of Ships: European Parliament Resolution of 26 March 2009 on an EU Strategy for Better Ship Dismantling, Note by the Secretariat, IMO/MEPC 59/3/6, 22 May 2009, ¶ E(14) (KR-CON) [EU Strategy for Better Ship Dismantling].

[198] Regulation 7bis of the Draft Ship Recycling Convention. For the text of the Draft Convention see *Recycling of Ships: Report of the Third Intersessional Meeting of the Working Group on Ship Recycling*, IMO/MEPC 57/3, 25 January 2008, annex 1 (KR-CON) [Draft Convention].

[199] See International Conference on the Safe and Environmentally Sound Recycling of Ships: Agenda Item 6, Consideration of the Draft International Convention for the Safe and Environmentally Sound Recycling of Ships: Non-Party Recycling Facilities, Submitted by Australia, SR/CONF/9, 6 February 2009, ¶ 6, online: SJÖFARTSVERKET <http://www.sjofartsverket.se/pages/19514/sr-conf-9.pdf>.

thereby proving fatal to its success. Moreover, any such prohibition could also fall afoul of international trade rules such as the most-favoured-nation-treatment obligation under GATT.

Two proposals were forwarded to facilitate the recycling of ships in non-party states. The United States, in line with article 11 of the Basel Convention, wanted ships that are registered with state parties to be recycled in non-party states pursuant to a bilateral, multilateral, or regional agreement or arrangement to which both the recycling state and flag state are parties. This agreement or arrangement should incorporate provisions that are at least as protective of human health and the environment as those provided for in the Ship Recycling Convention.[200] For Norway, such recycling could be possible only if the administration of the ship destined to be recycled assumed full responsibility to the effect that the selected recycling facility meets the conditions mentioned therein and that a DOC was issued to demonstrate this compliance.[201] In contrast, countries like India (India had earlier pleaded for recycling ships in non-party facilities provided that such facilities were capable of recycling ships in a safe and environmentally sound manner[202]) opposed the retention of such a provision in any form whatsoever.[203] India argued that even though the World Trade Organization requirements advocate non-discriminatory treatment to nations, exemptions are possible when issues relating to health and environmental hazards come in the way.[204] Since the very bedrock of the IMO's Ship Recycling Convention is to ensure that no compromise is made on any matter pertaining to health and environment protection, not permitting non-parties to the convention to participate in the trade of ship-recycling does not violate the tenets of the WTO.[205]

Even while acknowledging the possibility of insufficient recycling capacity when the convention enters into force that could hamper its success, given the dynamics of the ship scrapping industry and the lopsided nature of its concentration in just three to four countries, any attempt to recognise the trade between parties and non-parties without appropriate safeguards such as those that have been incorporated in the Basel Convention can have major repercussions.

Let us consider a hypothetical situation involving Bangladesh and India (major shipbreaking nations that together command more than two-thirds of the ship

[200] Recycling of Ships: Proposal for Recycling of Ships at Safe and Environmentally Sound Ship Recycling Facilities in States that are not Parties to the Convention, Submitted by the United States, IMO/MEPC 56/3/18, 18 May 2007 (KR-CON).

[201] Use of Ship Recycling Facilities under the Jurisdiction of Non-parties, *supra* note 135.

[202] Recycling of Ships: Further Development of the Draft Convention – Non-party Recycling Facilities, Submitted by India, IMO/MEPC 56/3/14, 17 May 2007, ¶ 7 (KR-CON) [Non-party Recycling Facilities].

[203] Statements by Delegations and Observers, *supra* note 180 at 4.

[204] *Ibid.*

[205] *Ibid.*

recycling business) *vis-à-vis* the safeguards that were incorporated in the draft text of the Ship Recycling Convention permitting trade with non-party ship recycling facilities. Suppose Bangladesh refuses to join the Ship Recycling Convention, while India on the other hand ratifies the convention. Since Bangladesh virtually commands the market and has a distinct edge over India in respect of market share, the convention could in due course lose relevance. Ships from party flag states (if they find the regulations which India enacts under the convention as being too stringent) and non-party states could flow to the ship-breaking yards in Bangladesh primarily based on the strength of the DOC. For Bangladesh, not subscribing to the Ship Recycling Convention would make sense, since under the safeguards envisaged, the duty to ensure that a particular ship recycling facility in Bangladesh complies with the requirements of the Draft Ship Recycling Convention, lies on the administration of the ship to be recycled.[206] The Government of Bangladesh could easily wash its hands of investing in resources and making sure that ship recycling facilities within its territory acts in a manner consistent with its constitutional goals and the larger objective of ensuring safe and environmentally sound ship recycling. Given the poor state of enforcement of the different IMO regulations by FOCs, it is doubtful whether the recycling yards functioning on the basis of the DOC would in actuality be functioning according to the mandates of the Draft Ship Recycling Convention. In such a scenario, India having joined the Draft Ship Recycling Convention would be clearly at a disadvantage, and in due course prompting it to denounce this instrument.[207]

The prohibition of the trade with non-party states under the Ship Recycling Convention means that ship recycling facilities operating largely in the international market must be located in a party state. If the ship recycling facilities in a state are in a position to comply with the requirements necessary for authorisation, it is to be expected that they would, together with ENGOs and the public at large, pressure their government to ratify the convention. It is unlikely that this pressure would come about if a recycling facility could keep operating in the market even if the state in which it is operating has not ratified the convention. The non-party recycling state would not experience the downside of not being a party as it would continue to receive ships from both parties as well as non-parties without assuming the responsibilities as laid out in the convention. This would considerably weaken the convention. In short, given that the drafter could not envisage appropriate safeguards to permit the recycling of ships in non-party state facilities, by not endorsing party-non-party ship recycling, an anomalous situation has been avoided.

---

[206] Draft Convention, *supra* note 198 reg., 7bis.

[207] *Ibid.*, art. 19. See also *Ship Recycling Convention*, *supra* note 4 art. 19.

Incorporation of rules relating to safe-for-hot-work and safe-for-entry is also laudable. India has long been canvassing for inclusion of regulations relating to the gas-free-for-hot-work certification in the text of the Ship Recycling Convention. Under the Indian national law, shipowners are to ensure that cargo oil tanks, slop tanks, and cargo piping are gas-freed for man-entry and hot-work before the ship is delivered to the recycling yard. Even though these laws had the unintended effect of transferring a large portion of the business to neighbouring Bangladesh, they had the desired impact on safety and pollution prevention in the recycling yards and considerably reduced the number of accidents.[208] The Ship Recycling Convention however goes a step further by recognising the need to maintain safe-for-entry and safe-for-hot-work conditions throughout the recycling process. During the recycling process, workers can be harmed by not only by toxic gases but also by oxygen-displacing gases, and gases which create flammable or explosive environments.[209] Rather than relating hot work only to gas-free status, the convention amalgamates protection from harmful gases into the requirements for safe-for-hot-work and safe-for-entry conditions, which are more comprehensive concepts.

From the above synopsis, it can be inferred that the principles and measures set out in the Ship Recycling Convention will re-engineer existing shipbreaking practices. However, as good as it is, the convention also has its share of weaknesses. The primary *raison d'être* behind this instrument, as stated in its preamble, is to reduce human rights abuses and environmental injustice consequent to this trade. From this perspective, it is doubtful that the Ship Recycling Convention has comprehensively addressed the ills that plague this industry.

One of the primary reasons for the weakness inherent in the Ship Recycling Convention is that it introduces an entirely new set of actors to manage dying ships. Simultaneously, it allows the real polluters and those who have benefitted from the ship during its operational life to evade responsibility for environmentally sound recycling of decommissioned ships. The Ship Recycling Convention

---

[208] International Conference on the Safe and Environmentally Sound Recycling of Ships: Agenda Item 6, Consideration of the Draft International Convention for the Safe and Environmentally Sound Recycling of Ships: Preparation of Oil Tankers for Ship Recycling, Submitted by India, SR/CONF/ 26, 2 April 2009, online: SJÖFARTSVERKET <http://www.sjofartsverket.se/pages/19514/sr-conf-26.pdf>.

[209] International Conference on the Safe and Environmentally Sound Recycling of Ships: Agenda Item 6, Consideration of the Draft International Convention for the Safe and Environmentally Sound Recycling of Ships: Proposal for Providing Clarity in the Convention and Reaffirming the Importance of Safe-for-Hot Work and Safe-for-Entry Conditions during Recycling, Submitted by the United States, SR/CONF/33, 29 April 2009, online: SJÖFARTSVERKET <http://www.sjofartsverket.se/pages/19514/sr-conf-33.pdf>. See also Recycling of Ships: Comments on the Report of the Correspondence Group on Ship Recycling, and a Proposal for the Framework of Guidance on Safe-for-Entry and Safe-for-Hot-Work Conditions in Safe and Environmentally Sound Ship Recycling, Submitted by United States, IMO/MEPC 59/3/7, 22 May 2009 (KR-CON).

places the primary responsibility for ship recycling on flag states and on shipbreaking states. In the shipbreaking business, these are the entities that derive the least economic benefit during the ships' commercially useful life. They also have the least technological, infrastructure or legal resources to ensure the environmentally sound management of hazardous wastes. By placing an inequitable burden on these states, which are also in the least advantageous position to ensure compliance, the convention drafters seem to have ignored the polluter pays principle. By doing so, they have sowed the seeds for its failure. To quote Greenpeace, the Ship Recycling Convention reflects a "polluter profits principle".[210]

The Ship Recycling Convention fails to take into account the fact that nearly 90 percent of shipbreaking happens on the Indian subcontinent. While winds of change have begun to blow over the Indian ship recycling yards, the situation in Bangladesh and Pakistan continues to be dismal. Given the overall economic conditions in these three countries and in light of other compelling priorities, it may prove difficult for their governments and for the recycling yards themselves to augment the infrastructure and improve standards necessary to obtain authorisation under the Ship Recycling Convention. The convention ignores this vital aspect even though it refers to technical assistance and cooperation. Resolution 3, entitled "Promotion of Technical Co-operation and Assistance", appeals to member states to support other states that request technical assistance, which can take the form of assessment of implications relating to operationalising the convention, development of national legislation and institutional arrangements to give effect to the convention, safe and environmentally sound hazardous materials management, training of personnel, and research concerning implementation of the convention.[211] Despite this, the Ship Recycling Convention still

---

[210] Recycling of Ships: Comments on the Report of the Correspondence Group, Submitted by Greenpeace International, IMO/MEPC 48/3/3, 16 August 2002, ¶ 11 (KR-CON) [Comments by Greenpeace International].

[211] See also International Conference on the Safe and Environmentally Sound Recycling of Ships: Agenda Item 6, Consideration of the Draft International Convention for the Safe and Environmentally Sound Recycling of Ships: Proposed Amendments to the Draft International Convention for the Safe and Environmentally Sound Recycling of Ships, Submitted by Bangladesh, SR/CONF/37, 4 May 2009, online: SJÖFARTSVERKET <http://www.sjofartsverket.se/pages/19514/sr-conf-37.pdf> (proposing an amendment to article 13 entitled "technical assistance and co-operation" to include "infrastructure"). It seems that there was an attempt to introduce two resolutions entitled "Draft Conference Resolution on an Implementation Mechanism for Providing Technical Assistance with the Aim to Direct Funds Toward Phasing-out of the Beaching Method" and "Draft Conference Resolution Calling for the Creation of a Fund for the Pre-cleaning and the Safe and Environmentally Sound Recycling of Ships" respectively. However, the attempt was unsuccessful. For more details see International Conference on the Safe and Environmentally Sound Recycling of Ships: Hong Kong, China, 11–15 May 2009, Committee of the Whole Agenda Items 4 and 5, Consideration of the Draft Text of the International Convention for the Safe and Environmentally Sound Recycling of Ships and Consideration of the Draft Text of Conference Resolutions and Recommendations: Overview of Submissions on Articles, Regulations, Appendices, or Proposed Conference Resolutions, SR/CONF/CW/WP/1, 11 May 2009, online: SJÖFARTSVERKET, <http://www.sjofartsverket.se/pages/19514/sr-conf-cw-wp1.pdf>.

lacks a mechanism to encourage and ensure that these countries in particular are provided with the technical assistance needed to meet the terms of the convention.

A significant drawback of the Ship Recycling Convention is that it evades the most contentious issue in shipbreaking, *i.e.*, the prior removal of hazardous wastes before the ship is sent for recycling. It only states that ships destined to be recycled shall conduct operations in the period prior to entering the ship recycling facility to minimise the amount of cargo residues, fuel oil, and wastes remaining on board. "Minimise" is too vague a concept to offer any guidance as to what degree the ship should really be cleaned. In effect, the entire responsibility of managing the inbuilt hazardous wastes falls onto the recycling states, and by doing so, the Ship Recycling Convention seems to fortify "the injustice of cost externalization".[212]

A. *The Ship Recycling Convention Compared to the Basel Convention*

Before probing other gaps that may hamper the efficacy of the Ship Recycling Convention, it is important to compare the levels of control that it introduces *vis-à-vis* the Basel Convention. Although article 11 of the Basel Convention calls for adoption of an equivalent level of controls, the Ship Recycling Convention as will be seen falls short of this requirement.[213] Some of these aspects are highlighted next.

1. *Illegal Traffic as a Criminal Act*

From the perspective of punitive sanctions, the Ship Recycling Convention leaves room for abuse. While the Basel Convention characterises "illegal traffic" as a criminal act,[214] the Ship Recycling Convention does not address this issue,

---

[212] Basel Action Network, Briefing Paper, "Legal Shipwreck: IMO Convention Legalizes Toxic Ship Dumping Running from Basel to Turn Back the Clock" (May 2009), online: Basel Action Network <http://www.ban.org/Library/BP05_May_2009.pdf>. See also Recycling of Ships: Prior Removal of Hazardous Materials, Submitted by France and the United Kingdom, IMO/MEPC 58/3/6, 22 July 2008 (KR-CON) (arguing for the incorporation of a specific provision for the prior removal of hazardous materials before the final survey if necessary).

[213] An analysis of the draft IMO convention by a coalition of environmental, labour and human rights organisations concluded that out of the 24 widely identified elements needed to redress the ship scrap crisis, the IMO Draft Convention adequately addresses only two of these while the Basel Convention covered almost 20 elements. Basel Action Network on Behalf of Global NGO Platform on Shipbreaking, *Critique of the Draft I.M.O. "International Convention for Safe and Environmentally Sound Recycling of Ships"* (15 March 2006), online: Basel Action Network <http://www.ban.org/Library/IMO_Draft_Convention_CritiqueFINAL.pdf>. See also Recycling of Ships: European Parliament on the Green Paper on Better Ship Dismantling, Note by the Secretariat, IMO/MEPC 58/INF.16, 31 July 2008, annex, online: SJÖFARTSVERKET <http://www.sjofartsverket.se/pages/14228/58-INF16.pdf> (noting that the Ship Recycling Convention in its draft version has not put in place an equivalent level of control as under the Basel Convention).

[214] *Basel Convention*, *supra* note 5 art. 4(3).

leaving it to be determined by the national laws of state parties. This provides a loophole for shipowners who may want to recycle their ships in countries that have lighter sanctions, as a precaution in case they find themselves to be on the wrong side of the law.

2. *Express Recognition of the Sovereign Right to Prohibit an Import*

An important feature of the Basel Convention is that it expressly recognises the sovereign right of a state to ban the entry or disposal of foreign hazardous wastes and other wastes in its territory.[215] The intricacies of this right are illuminated in article 4, which subjects it to certain procedural requirements. In contrast, the Ship Recycling Convention does not expressly recognise the right to ban the entry of specific waste streams/materials. Of course, this is an inherent right in each state as a necessary incident of its sovereign powers. Consequently, an express affirmation of this right may be unnecessary. However, the lack of a provision akin to article 4 of the Basel Convention could lead to ambiguities and disputes since parties are deprived of well-articulated norms on which they can base their actions.

3. *National Waste Management Capacity*

The Basel Convention calls upon parties to adopt "appropriate measures" to develop in-house capacity for the "environmentally sound management of hazardous wastes".[216] The Ship Recycling Convention, however, offers no incentives nor does it mandate the creation of ship decontamination facilities and green recycling capacity in countries that engage in the export of these wastes.

4. *Prior Decontamination of Ships*

The Ship Recycling Convention is weak from the perspective of environmental justice. To attain the overarching goal of minimising the transboundary movement of hazardous wastes,[217] one major step suggested by ENGOs is the decontamination or pre-cleaning of ships to the fullest extent possible prior to the final journey of the ship.[218] The Basel Ban (though not in force), which has been incorporated into European Union law, implies such a course of action. However, as mentioned earlier, the Ship Recycling Convention does not address the issue of pre-cleaning of ships. This could impair its effectiveness in the long run.

---

[215] *Ibid.*, preamble.
[216] *Ibid.*, art. 4(2)(b).
[217] *Ibid.*, art. 4(2)(d).
[218] *Alternate or Additional Instrument to the Basel Convention, supra* note 6 ¶ 3.5.4.

5. *State of Export v. the Flag State and Flags of Convenience (FOC) Regime*

One of the primary reasons for the effectiveness of the Basel Convention is that before a transboundary movement of hazardous waste is initiated, the convention requires all decisions in relation to an approval for an export, import or transit of hazardous wastes to be scrutinised and consented to by national governments, and not by private actors.[219] State of export is defined under the Basel Convention as "a party from which a transboundary movement of hazardous wastes or other wastes is planned to be initiated or is initiated".[220] When applied to a transboundary movement of wastes arising from a land-based activity this definition makes it easy to identify the state of export. However, this is not the case in shipping, where during the life of a ship, ownership and flag could have changed hands several times.[221] By substituting the state of export with the flag state, the Ship Recycling Convention dilutes the requirement to maintain an "equivalent level of control", since in many cases there is an absence of a genuine link between the flag state and the ship.[222] The flag state may not have the same kind of control over the shipowner compared to what an export state can exercise.[223] This is particularly the case where ownership of a vessel has been transferred to an entity such as a cash buyer.

6. *Proximity Principle v. Not in My Back Yard Syndrome (NIMBY)*

An important feature of the Basel Convention framework is that it requires treatment and disposal of wastes to take place in the state of generation[224] or as close

---

[219] *Basel Convention*, *supra* note 5 art. 6. See also *ibid.*, ¶ 3.5.7.

[220] *Basel Convention*, *ibid.*, art. 2(10).

[221] See Greenpeace, *Playing Hide and Seek* (December 2003), online: Greenpeace Nederland <http://www.greenpeaceweb.org/shipbreak/rapport_DEF_versie_2.pdf> (provides an overview on how the shipping industry, protected by flags of convenience, dumps toxic waste on shipbreaking beaches). Following engine failure off Dunkirk, the Liberian flagged 'Sea Beirut' was abandoned. As it was not possible to trace the actual shipowner, the Dunkirk port authorities became responsible for the ship under the European Waste Shipment Regulation. Knowing very well that the ship contained asbestos, the Dunkirk port authorities auctioned the ship. Finally, the 'Sea Beirut' left Dunkirk for Turkey. However, the ship was refused permission to enter Turkish waters/ports with Turkey insisting that the ship to return to its country of origin. Under the Basel Convention, France was the state of export. However, France refused to take back the 'Sea Beirut', which led Turkey to inform the Basel Convention Secretariat of this matter. Selected Cases of Decommissioning of Vessels, *supra* note 3.

[222] *United Nations Convention on the Law of the Sea*, 10 December 1982, 21 I.L.M. 1261 (entered into force 16 November 1994), art. 91.

[223] In this context, it has to be noted that there was a proposal for an interim solution submitted by Canada, Denmark, and the Netherlands before the Open-ended Working Group of the Basel Convention, to equate the flag state to the state of export under article 2(10) of the Basel Convention. By doing so, the Basel Convention obligations imposed on a state of export could easily be made applicable to the flag state. Recycling of Ships: Outcome of the Basel Convention Open-ended Working Group Meeting on the Issue of Ship Recycling, Submitted by the Secretariat of the Basel Convention, IMO/MEPC 52/3/7, 6 August 2004, ¶ 6 (KR-CON).

[224] *Basel Convention*, *supra* note 5 preamble.

as possible to the point of production depending on technical and environmental factors. Thus the Basel Convention has the principle of proximity as one of its central tenets. This flows from the obligation of states to manage hazardous and other wastes in an environmentally sound manner which cannot, under any circumstance, be transferred to the states of import or transit.[225] Parties are also to undertake periodic reviews to examine the possibility of reducing the amount of hazardous and other wastes that are exported to other states, particularly in the developing world.[226] The Ship Recycling Convention, in comparison, by facilitating the transfer of responsibility to the flag state and to parties where authorised ship recycling facilities operate, seems to fortify the NIMBY syndrome where the intention is solely to ensure that decommissioned ships no longer rust in the ports and waters of developed countries (the major exporters of ships).

7. *The Principle of Prior Informed Consent*

The Basel Convention introduces a system of control measures for the transboundary movement of hazardous wastes based on the principle of prior informed consent.[227] Article 6 requires consent to be sought by the state of export,[228] and the state of import has to be provided with certain information[229] to enable it to make an informed decision on whether or not to accept or reject an import of wastes. Even the transit state is required to give its prior informed consent. In contrast, the Ship Recycling Convention only requires notification between the shipowner and the flag state (administration) and between the recycling facility and the recycling state (competent authorities). There is no provision for either direct state-to-state reporting, e.g., notification between the flag state and the recycling state, or for reporting to transit or port states.[230] The only instances where there is some kind of official communication between the competent authority of the recycling state and the administration in the flag state is in respect of operationalising the explicit or tacit approval procedures for the ship-specific SRP and for reporting upon completion of the recycling. However, the provisions relating to explicit or tacit approval of the ship-specific SRP under the Ship Recycling Convention falls short of the prior informed consent procedures contemplated under the Basel Convention primarily because this information flows from the importing (recycling state) to the exporting state (flag state). As well,

---

[225] *Ibid.,* art. 4(10).

[226] *Ibid.,* art. 4(13).

[227] For further details on the principle of prior informed consent, see Chapter Four entitled "Contemporary International Law and Ship Recycling". See also *ibid.,* arts. 4(1)(c), 6, annex VA.

[228] Recycling of Ships: Control System of the Basel Convention on the Control of Transboundary Movements of Hazardous Wastes and their Disposal, Submitted by the Secretariat of the Basel Convention, IMO/MEPC 53/INF.19, 27 May 2005, ¶ 15 (KR-CON).

[229] *Basel Convention, supra* note 5 annex VA.

[230] EU Comments, *supra* note 21 ¶ 71.

the reporting that occurs upon the completion of the recycling process takes place in a completely different factual situation (the end of the transaction). What is envisaged for the principle of prior informed consent under the Basel Convention are measures taken prior to the initiation of the transaction.

8. *Trade between Parties and Non-Parties*

The Ship Recycling Convention departs from the standard practice embodied in some of the recent environment-related agreements not to restrict trade with non-parties subject to the incorporation of provisions designed to ensure that the environmental objectives as embodied in the agreement are complied with. Even though the Ship Recycling Convention does not impose any express prohibition, it practically proscribes the trade in obsolete ships with non-party states. Provisions, in an earlier draft version of the Ship Recycling Convention, recognised trade with non-party states subject to the issuance of a document of compliance. However, the contemplated safeguards were not adequate, and since no consensus could be achieved, these provisions were finally deleted. The Basel Convention also proscribes the pernicious trade between parties and non-parties. Article 4(5) states that "[a] Party shall not permit hazardous wastes or other wastes to be exported to a non-Party or to be imported from a non-Party". Still an exception has been provided in article 11(1), under which parties can enter into "bilateral, multilateral, or regional agreements or arrangements" with non-parties for the transboundary movement of hazardous or other wastes provided that such agreements or arrangements do not derogate from the requirement to ensure environmentally sound management of hazardous and other wastes. These agreements or arrangements are to stipulate conditions that are not less environmentally sound than those provided for by the Basel Convention and have to take "into account the interests of developing countries".

The peculiar features of the ship recycling industry may warrant a blanket ban on the recycling of party ships in non-party facilities. However, when there are serious concerns regarding the possible non-availability of authorised ship recycling facilities when the Ship Recycling Convention enters into force, much could have been achieved had provisions been included in the Ship Recycling Convention akin to the Basel Convention permitting trade with non-party state facilities provided such agreements or arrangements stipulate standards that are not less environmentally sound than those provided for by the Ship Recycling Convention or the more stringent Basel Convention and in particular taking into account the "interests of developing countries".[231]

---

[231] *Basel Convention*, *supra* note 5 art. 11(1).

9. *Applicability of Basel Convention in Light of Article 15 and Regulation 3 of the Ship Recycling Convention*

It might be expected that once the Ship Recycling Convention comes into force, the Basel Convention will no longer be applicable to ships covered by it. However, article 15(2) of the Ship Recycling Convention, which outlines its relationship with other international agreements, including the Basel Convention, states that:

> [n]othing in this Convention shall prejudice the rights and responsibilities of Parties under other relevant and applicable international agreements.

Again, article 1(2) provides that:

> No provision of this Convention shall be interpreted as preventing a Party from taking, individually or jointly, more stringent measures consistent with international law, with respect to safe and environmentally sound recycling of ships in order to prevent, reduce or minimize any adverse effects on human health and the environment

These provisions assume relevance for two reasons. A plain reading implies that as a natural consequence, ships that are not covered by the Ship Recycling Convention, e.g., warships and government ships, will still be subject to the provisions of the Basel Convention.[232] More importantly, merely because the Ship Recycling Convention has appeared on the scene, it does not preclude the rights and responsibilities of parties under other international agreements, including the Basel Convention. Regulation 3 of the Ship Recycling Convention provides that in implementing the requirements of "Regulations for Safe and Environmentally Sound Recycling of Ships", parties are to take into account "... relevant and applicable standards, recommendations and guidance developed by the International Labour Organization and ... under the Basel Convention ..." Given the divergence between the Basel Convention and the Ship Recycling Convention and in light of the language of articles 15(2), 1(2), and regulation 3 of the Ship Recycling Convention and article 11 of the Basel Convention, which spells out the principle of equivalency, parties may be subject to two sets of mutually inconsistent obligations thereby creating a complicated scenario for the international legal shipbreaking regime.

B. *The IMO Guidelines on Ship Recycling and the Ship Recycling Convention*

The IMOGSR is only a voluntary code; its implementation is left to the goodwill of shipowners. Unfortunately, most ship recycling operations are carried out without regard for its stipulations, which has necessitated drafting of a binding legal instrument on the subject. It is surprising, however, that the Ship Recycling

---

[232] Recycling of Ships: Further Development of the Draft Convention, Submitted by Denmark, IMO/MEPC 56/3/12, 17 May 2007, ¶ 4, 5 (KR-CON).

Convention ignores several beneficial provisions in the IMOGSR. For instance, the IMOGSR identifies a range of stakeholders in the shipbreaking business and clearly defines their roles and responsibilities. Given the inequities in power relations that exist between the different players, this is an important step that can ensure greater accountability. In contrast, the Ship Recycling Convention identifies two players, the flag state and the recycling state, and apportions the majority of the responsibility between them. The Ship Recycling Convention exonerates the shipowner from any responsibilities. In contrast, the IMOGSR requires the shipowner to assume several obligations, beginning with the duty to maintain the accuracy of the green passport, through to providing insurance and contingency arrangements for the final journey. The shipowner is also expected, *inter alia*, to specify the method of recycling, to monitor the recycling process, to provide incentive payments to ensure that the recycling is carried out in accordance with relevant guidelines, and to ensure that the hazardous wastes are removed and are properly disposed of in cases where the recycling facility lacks the capacity to manage the wastes.

Another omission in the Ship Recycling Convention is that it does not mention the need for a standard contract. To streamline the processes and transactions related to ship recycling, there should be a legally-binding ship recycling contract between the shipowner and the recycling facility that clearly defines their roles and responsibilities, which will also facilitate dispute resolution.[233] Under the Basel Convention, one of the primary conditions for the transboundary movement of wastes is the presence of a contract. The IMOGSR also recommends a contract at "the end of a ship's operating life" covering the sale and purchase of the ship for recycling. In this regard, the IMOGSR even refers to the DEMOLISHCON, the BIMCO standard contract on ship recycling. Unfortunately, the Ship Recycling Convention ignores both the Basel Convention mandate and the IMOGSR, and does not espouse a legally acceptable demolition contract.[234]

The IHM is to serve as a reference point during the recycling process. The safety of the recycling to a large extent depends upon its accuracy in relation to the quantity of hazardous substances and their exact location on the ship. The Ship Recycling Convention is, however, silent on the issue of liability in cases where the information provided in the IHM is inaccurate. While the IMOGSR calls upon successive owners of a ship to maintain the accuracy of the IHM, the Ship Recycling Convention imposes no such obligation.[235] Additionally, the

---

[233] Non-party Recycling Facilities, *supra* note 202 ¶ 3.

[234] Recycling of Ships: Development of the Draft International Convention for the Safe and Environmentally Sound Recycling of Ships, Submitted by India, IMO/MEPC 55/3/12, 17 August 2006, ¶ 1 (KR-CON). See also IMOGSR, *supra* note 3 ss. 9.2.2, 9.4.3.3 (stating that flag states and recycling states should promote the use of a ship recycling sale and purchase contract such as DEMOLISHCON).

[235] IMOGSR, *ibid.*, s. 5.1.

IMOGSR expressly recognises the principle of partial decontamination; in the Ship Recycling Convention this principle finds only an implied mention.

The narrowing of the requirements mandated by the IMOGSR, which are central to a strategy to ensure sustainable shipbreaking, raises serious doubts as to whether the Ship Recycling Convention will contribute to the development of sustainable ship recycling.

C. *Other Deficiencies in the Ship Recycling Convention*

The most obvious weakness of the Ship Recycling Convention is that it comes too late in the day. The convention was adopted on 19 May 2009 and will be opened for signature from 1 September 2009. At the earliest, it will come into force in 2011.[236] This weakness is tacitly evident in resolution 5 adopted by the Hong Kong Ship Recycling Conference. The Resolution calls upon parties to apply the technical standards contained in the annex to the Ship Recycling Convention on a voluntary basis and to ship recycling facilities in order that the benefits of the convention can accrue due to their early application.[237]

A treaty is not a magic wand that can simply be waved to set things right. To transcend the black letter and to encourage the ship recycling business to adopt a sustainable model, a treaty has to go through several stages. The Ship Recycling Convention incorporates in its entry into force provisions a capacity criterion based on the historic volumes of recycled ships. Since the known recycling capacities of major recycling states far exceed the volume of ships that need recycling, it can be concluded that the Ship Recycling Convention, by ensuring that a critical mass of the dilapidated world tonnage will be ensured of recycling services, definitely overcomes any obstacles as regards its coming into force.[238] However, this is unlikely to solve the problem of lack of availability of authorised recycling capacity. Given the poor conditions in most of the recycling facilities on the Indian subcontinent, and the large capital investments needed to improve the situation, it is doubtful as to how many units will qualify to be *authorised*.[239] This will significantly impact the ship recycling

---

[236] See *Ship Recycling Convention, supra* note 4 art. 17. See also Recycling of Ships: Conditions on Entry into Force, Submitted by Japan, IMO/MEPC 57/3/13, 8 February 2008 (KR-CON).

[237] Work of the Conference, *supra* note 9 res. 5 (entitled "Early Implementation of the Technical Standards of the Hong Kong International Convention for the Safe and Environmentally Sound Recycling of Ships, 2009").

[238] Recycling of Ships: Recycling Capacity for Entry into Force Criteria for the International Convention for the Safe and Environmentally Sound Recycling of Ships, Submitted by India, IMO/MEPC 58/3/14, 13 August 2008 (KR-CON).

[239] See generally International Conference on the Safe and Environmentally Sound Recycling of Ships: Agenda Item 6, Consideration of the Draft International Convention for the Safe and Environmentally Sound Recycling of Ships: Entry-into-Force Provision, Submitted by Denmark, SR/CONF/ 30, 7 April 2009, ¶ 4, online: SJÖFARTSVERKET <http://www.sjofartsverket.se/pages/19514/sr-conf-30.pdf>.

industry. Many units may close down leading to shrinkage in the global ship recycling capacity, even when there will be an overwhelming demand for additional recycling facilities. If a state lacks the means to upgrade its recycling facilities so as to be convention compliant, it will be uninterested in ratifying the Ship Recycling Convention even if it has a large number of recycling facilities. The Ship Recycling Convention does not contain any concrete incentives or mechanisms to support such recycling facilities during the phase-in period.[240] This is a serious lacuna that needs to be remedied if the Ship Recycling Convention is to produce any tangible results in transforming the shipbreaking industry.

The Ship Recycling Convention reflects some facets of the ship recycling business. It does not mention any precautions that need to be taken when a ship is on its last journey, particularly when it is being towed to a recycling facility. This is a serious drawback since many end-of-life ships undertake their last journey to the recycling yards in leaking condition. It also does not deal with recycling wrecked ships. Given their environmental impact, wrecked ships will have to be removed, dismantled and sometimes recycled. However, it may prove difficult to meet the ideal conditions for ship recycling work as envisaged by the Ship Recycling Convention when dealing with wrecks.[241]

The main mode of ship recycling in South Asia is the beaching of ships. This has been described by the Indian government as "a practical and cost effective solution to ship recycling".[242] In a submission by India regarding the environmental comparison of the beaching method with other methods of recycling of ships, it has been observed that

[240] Recycling of Ships: Proposal for a Mechanism to Facilitate Implementation of the Convention for the Safe and Environmentally Sound Recycling of Ships, Submitted by the United Kingdom, MEPC 56/3/21, 18 May 2007 (KR-CON) (pointing to the need for an implementation compliance mechanism to facilitate the development of ship recycling facilities to the standards required by the Draft Convention). See also Recycling of Ships, Progress Report on the Work Related to the Requirements for Ship Recycling Facilities including the Trial Ship Recycling Project and the Proposal for Future Work Plan, Submitted by Japan, IMO/MEPC 57/3/6, 25 January 2008, ¶ 14, 15 (KR-CON) (suggesting that during the interim period, *i.e.*, the period between the adoption of the convention and its entry, it is necessary that potential parties take measures to achieve the overall objectives of the convention, for instance, implementing the guidelines and the relevant requirements under the Draft Convention to the extent practicable).

[241] See Recycling of Ships: Comments on the Proposal for a New Legally-binding Instrument on Recycling of Ships (MEPC 54/3), Submitted by Brazil, IMO/MEPC 54/3/4, 13 January 2006, ¶ 9.4 (KR-CON). See also Recycling of Ships: Proposed Text for the Draft International Convention for the Safe and Environmentally Sound Recycling of Ships (MEPC 58/3), Submitted by Bangladesh, IMO/MEPC 58/3/21, 1 August 2008 (KR-CON) (proposing for the inclusion of certain circumstances, e.g., cases where the vessel is abandoned, ordered by a court of law to be scrapped, declared to be a total constructive loss, wrecked, subject to force majeure, unregistered, driven to the shore with or without cargo by storm or other circumstances beyond control, etc.).

[242] Recycling of Ships: Comments on the Report of the Correspondence Group to MEPC 49, Submitted by India, IMO/MEPC 49/3/3, 13 May 2003, ¶ 1.10 (KR-CON) [Comments by India].

> … the beaching method has been in successful existence for several years for ship recycling in India at Alang and this has posed no threat to the environment. Furthermore, this method is the most efficient and cost effective method for recycling of ships from the smallest to the largest sizes.[243]

Despite these claims, it is common knowledge that among the different processes for ship recycling, beaching poses maximum obstacles to environmentally sound recycling.[244] Beaching makes it practically impossible to contain the oils and toxic contaminants from entering the fragile intertidal coastal zone and to bring lifting cranes alongside ships to lift heavy cut pieces or to rescue workers and bring emergency equipment.[245] The European Parliament Resolution on an "EU Strategy for Better Ship Dismantling" observes that "… 'beaching' … has been globally condemned as being incapable of delivering safety for workers and of providing adequate protection of the marine environment from ship-borne pollutants."[246] It further

> calls for an explicit prohibition on 'beaching' …., and considers that any technical assistance to South Asian countries within an EU framework should … aim at … phasing out of this grossly unsustainable and seriously flawed breaking method.[247]

Under United Kingdom law, grounding of vessels on a beach or a river bank for recycling will not be given the requisite permissions due to the dangers inherent in this process.[248] The Ship Recycling Convention, while adopting measures to control environmental dangers during ship recycling, ignores the need to phase-out the practice of breaking ships on beachheads.[249] If ship recycling in South Asia is to be safe, new and modern methods such as dry docking must replace the

---

243 International Conference on the Safe and Environmentally Sound Recycling of Ships: Agenda Item 6, Consideration of the Draft International Convention for the Safe and Environmentally Sound Recycling of Ships: Environmental Comparison of the Beaching Method with Other Methods of Recycling of Ships, Submitted by India, SR/CONF/36, 1 May 2009, online: SJÖFARTSVERKET <http://www.sjofartsve rket.se/pages/19514/SR-CONF-36.pdf>.

244 See also *Recycling of Ships: Report of the Correspondence Group*, Submitted by the Co-ordinator of the Correspondence Group, IMO/MEPC 46/7, 18 January 2001 (KR-CON) [*Report of the Correspondence Group*].

245 International Conference on the Safe and Environmentally Sound Recycling of Ships, Hong Kong, China, 11–15 May 2009: Committee of the Whole, Agenda Item 6, Consideration of the Draft International Convention for the Safe and Environmentally Sound Recycling of Ships: Statement of Concern, Submitted by Greenpeace International and FOEI, SR/CONF/CW/INF.1, 14 May 2009, online: SJÖFARTSVERKET <http://www.sjofartsverket.se/pages/19514/sr-conf-cw-inf1.pdf>.

246 EU Strategy for Better Ship Dismantling, *supra* note 197 ¶ E.

247 *Ibid.*, ¶ 6.

248 See also U.K., Department for Environment, Food and Rural Affairs, *Overview of Ship Recycling in the UK: Guidance* (London: Crown Copyright, Department for Environment, Food and Rural Affairs, 2007) at 14, online: Defra <http://www.defra.gov.uk/environment/waste/strategy/pdf/shiprecycle-strategy-overview.pdf>.

249 See also *Report of the Correspondence Group*, *supra* note 244 at 8 (pointing out that beaching is the "most difficult" mode under which to ensure environmentally friendly scrapping).

present shipbreaking practices. However, the Ship Recycling Convention sidesteps this important issue.

As a result of the poor state of regulation by many FOCs, certain IMO legal regimes envisage increased port state control.[250] However, in the Ship Recycling Convention, port state control is limited mainly to verifying the IHM.[251] This is inadequate given the potential of end-of-life ships to contaminate the local environment. It may not be possible to know the actual chemical composition of substances housed in different parts and equipment of the ship without actually subjecting them to further testing. Since this is time consuming, and the Ship Recycling Convention provides for compensation claims in cases of delay, the verification of the ship may be reduced to a mere visual inspection.[252]

The Ship Recycling Convention also does not address the practical difficulties a ship faces in furnishing an up-to-date IHM as required under Part II for operationally-generated wastes and Part III for storage. These wastes keep on changing until the ship reaches the recycling yard unless the ship is towed to the facility.

The shipowner remains an enigma under the Ship Recycling Convention. Interestingly, regulation 8 (4) provides that "[s]hips destined to be recycled shall: … (4) provide to the Ship Recycling Facility all available information relating to the ship for the development for the Ship Recycling Plan required by regulation 9." Obviously, the drafters would not have intended an inanimate ship to provide information for the SRP. Regulation 9 clarifies the matter by providing that, "[t]he Ship Recycling Plan shall: 1. be developed taking into account information provided by the shipowner." An instance of sloppy drafting, it is rather strange that the Ship Recycling Convention uses convoluted language to camouflage the responsibilities of the shipowner rather than articulate them in clear precise terms. Furthermore, only very few responsibilities are imposed on the shipowner, some of which may even seem irrelevant. For instance, the Ship Recycling Convention provides in express terms that according to the circumstance, additional surveys can be carried out at the request of the shipowner. In actual practice, it is doubtful whether shipowners will initiate additional survey given the cost involved.

The Ship Recycling Convention defines the shipowner as follows:

> … the person or persons or company registered as the owner of the ship or, in the absence of registration, the person or persons or company owning the ship or any other organization or person such as the manager, or the bareboat charterer, who has assumed the

---

[250] For instance see *Memorandum of Understanding on Port State Control in Implementing Agreements on Maritime Safety and the Protection of the Marine Environment*, 26 January 1982, 21 I.L.M. 1 (entered into force 1 July 1982).

[251] Recycling of Ships: Substitution, Internalizing Costs, Third Party Auditing and Rejection of Non-party Trade Needed to Secure an Effective Convention, Submitted by Greenpeace International & Friends of the Earth International, IMO/MEPC 57/3/7, 25 January 2008, ¶ 4 (KR-CON).

[252] See *Ship Recycling Convention*, *supra* note 4 art. 11.

> responsibility for operation of the ship from the owner of the ship … *This term also includes those who have ownership of the ship for a limited period pending its sale or handing over of a ship to a recycling facility.*[253]

Currently, to escape the vagaries of the ship recycling market and the administrative hassles of having to solicit bids from a variety of recycling yards and facilities, shipowners sell their ships to cash buyers who are paper entities that are virtually impossible to identify.[254] The system of provisional registration for only one voyage by the cash buyer makes it extremely difficult for flag states to enforce their regulations.[255] The situation is rendered more complex in cases where the ship is sold for recycling by the owner to the cash buyer and is delivered to the outer anchorage of a recycling facility and from there within hours, days or weeks it is re-delivered to the ship recycling facility during which time the asset is not re-flagged by the cash buyer. Thus the beaching is often done in a flagless status.

Rather than articulating a chain of responsibility to make it impossible for the real shipowners to abrogate their responsibilities in respect of decommissioning,[256] the Ship Recycling Convention as seen from the above definition afforded to the term "shipowner", confers the status of a shipowner even to these 'fly-by-night' entities who have to carry out the responsibilities and obligations anticipated under the convention, namely, the preparation of the IHM, initiation of the surveys including requesting for additional surveys, helping the recycling facility to prepare the SRP, and negotiation with the ship recycling facility.[257] In addition, the cash buyer will also have to attend to the other administrative requirements expected of the owner such as obtaining reissuance of statutory certificates, registration of the vessel, and setting in motion all of those actions that occur when a vessel's ownership changes, most of which are outside the purview of the Ship Recycling Convention.[258] Whether the cash buyer will carry out all these responsibilities is anybody's guess. Failure to ensure accountability on the part of the real shipowner will only reinforce current practices, since responsibility can be easily transferred on to entities not necessarily subject to flag state jurisdiction.

The Ship Recycling Convention is ambiguous on several points. For instance, ship recycling in the convention has been defined to include "associated

---

[253] *Ibid.*, reg. 1(8) [emphasis added].

[254] For a further discussion on the role of cash buyers, see Chapter Two, entitled "The Global Business of Shipbreaking". See also International Conference on the Safe and Environmentally Sound Recycling of Ships: Agenda Item 6, Consideration of the Draft International Convention for the Safe and Environmentally Sound Recycling of Ships: The Role of the Cash Buyer, Submitted by the Republic of the Marshall Islands, SR/CONF/15, 27 February 2009, ¶ 2, online: SJÖFARTSVERKET <http://www.sjofartsverket.se/pages/19514/sr-conf-15.pdf> [The Role of the Cash Buyer].

[255] Comments by India, *supra* note 242 ¶ 2.6.

[256] Comments by Greenpeace International, *supra* note 210 ¶ 18.

[257] The Role of the Cash Buyer, *supra* note 254 ¶ 4.

[258] *Ibid.*

operations such as storage and treatment of components and materials on site, but not their further processing or disposal in separate facilities." Even though the delimitation of the applicability of the Ship Recycling Convention can lead to greater predictability and consistency, if the final fate of the hazardous materials is not addressed, this approach may not necessarily further the broader objective of safe and environmentally sound ship recycling.[259] Similarly, the Ship Recycling Convention applies only to ships above 500 GT. Accordingly, the controls on hazardous substances like asbestos and ozone depleting substances apply only to ships that are above 500 GT. Even though parties are called upon to adopt appropriate measures to ensure that ships less than 500 GT are recycled in a manner consistent with the Ship Recycling Convention, in restricting the application of the convention, it seems an opportunity has been lost to influence the production of safer ships irrespective of size and development of better shipbreaking practices. These interests would have been served better had the Ship Recycling Convention adopted a more flexible approach, i.e., it would also apply the controls on hazardous materials to ships below 500 GT.[260] Such an approach would be appropriate as such controls are relevant to all ships irrespective of size.

The Ship Recycling Convention does not apply to ships operating throughout their life only in waters subject to the sovereignty or jurisdiction of the state whose flag the ship is entitled to fly. Consequently, it may happen that coastal ships temporarily engaged in international voyages for their repair or for their sale, or due to certain unavoidable circumstances and ships now exclusively engaged in coastal voyages but whose past records show that such vessels had undertaken international voyages, can technically be brought under the ambit of the convention. This can create uncertainty, and it may even be unreasonable that such ships be considered as falling under the same scope of application as ships constructed to be engaged on international voyages.[261] Controversy can

---

259 Recycling of Ships: Further Development of the Draft Guidelines for Safe and Environmentally Sound Ship Recycling, Submitted by Denmark, IMO/MEPC 58/3/7, 22 July 2008, ¶ 14 (KR-CON).

260 International Conference on the Safe and Environmentally Sound Recycling of Ships: Agenda Item 6, Consideration of the Draft International Convention for the Safe and Environmentally Sound Recycling of Ships: Proposal for Amending the Application of the Draft Convention to be Drafted as Other Relevant IMO Conventions, Submitted by Norway, SR/CONF/10, 6 February 2009, online: SJÖFARTSVERKET <http://www.sjofartsverket.se/pages/19514/sr-conf-10.pdf> (noting that the MARPOL, the SOLAS, the BWM Convention, and the AFC 2001 do not have any size limitation in their articles and therefore the Ship Recycling Convention in general should apply to ships above 500 GT, as well as for ships below 500 GT for certain elements).

261 International Conference on the Safe and Environmentally Sound Recycling of Ships: Agenda Item 6, Consideration of the Draft International Convention for the Safe and Environmentally Sound Recycling of Ships: Proposed Amendments to Paragraph 3 of Article 3, Submitted by the Republic of Korea, SR/CONF/40, 7 May 2009, ¶ 4, online: SJÖFARTSVERKET <http://www.sjofartsverket.se/pages/19514/sr-conf-40.pdf>.

also arise in respect of de-commissioned warships. Once the armaments are removed from a warship they are either recycled in a yard under the jurisdiction of the state that owns the ship, or sometimes the ship is sold to an intermediate buyer for recycling in another state. In such cases, the ship no longer retains its sovereign immunity. Therefore questions may arise as to whether exclusion of such ships from the regulatory ambit of the Ship Recycling Convention can be justified or not.[262]

Although the incorporation of provisions relating to safe-for-entry and safe-for hot-work are credible, the vagueness in the relevant provision may cause concern. The Ship Recycling Convention provides that when tankers arrive at a ship recycling facility, the cargo tanks and pump rooms are to be in a condition that is "ready for certification" as safe-for-entry and safe-for-hot-work, "according to the national laws, regulations and policies of the party under whose jurisdiction the Ship Recycling Facility operates." In other words, the flag states have no responsibility to certify the tanker as safe-for-entry and safe-for-hot-work. Consequently, can a flag state and the shipowner who only have to ensure that the ship is "ready for certification" be held liable if an explosion were to occur once the tanker found its way into the recycling yard? The Ship Recycling Convention throws no light on how such situations are to be dealt with.

The dispute settlement procedures as envisaged under article 14 of the Ship Recycling Convention do not offer a larger range of choices nor do they correspond to the "state of the art" dispute settlement mechanisms that are embodied in other recent IMO conventions like the *Nairobi International Convention on the Removal of Wrecks, 2007*.[263] The term "agreed upon by them" in article 14 presupposes that a specific means of settlement is applicable only if both parties agree to it.[264] Consequently, this does not ensure that disputes will be resolved in all cases because one of the parties can render an agreement on a dispute settlement mode impossible to achieve.

It can sometimes happen that the inventories of hazardous materials for existing ships may be inaccurate, the discrepancy in certain cases not being deliberate. As well, in preparing the ship-specific SRP, the ship recycling facility has to depend upon the information provided by the shipowner. It can also turn out that the information was inaccurate and consequently, the ship recycling yard is

---

[262] Recycling of Ships: Proposals to Facilitate Implementation of the Convention, Submitted by India, IMO/MEPC 58/3/13, 12 August 2008, ¶ 6–8 (KR-CON).

[263] International Conference on the Safe and Environmentally Sound Recycling of Ships: Agenda Item 6, Consideration of the Draft International Convention for the Safe and Environmentally Sound Recycling of Ships: Consideration on Dispute Settlement, Submitted by Belgium, Denmark and the Republic of Korea, SR/CONF/17, 10 March 2009, ¶ 4, online: SJÖFARTSVERKET <http://www.sjofartsverket.se/pages/19514/sr-conf-17.pdf>.

[264] *Ibid.* at ¶ 7.

saddled with a hazardous material which it is unable to manage. In such situations, can the ship be repatriated to the flag state? Such situations are a real possibility since by the time the ship is sent for recycling, the shipowner would in most cases be a cash buyer who may not have adequate knowledge about the structural features of the ship. The Ship Recycling Convention is again silent on how such situations are to be tackled.[265]

Regulation 9(4) of the Ship Recycling Convention requires that the SRP be subject to an explicit or tacit approval procedure by the competent authority in the recycling state. For tacit approval, the regulation mentions a 14 day review period, after which, if the competent authority has not forwarded any written objection, the SRP is deemed to have been approved. In cases where a recycling state requires explicit approval of the SRP, the omission to impose obligations on the competent authority to decide the matter expeditiously can result in avoidable delays.

A primary reason for the initiation of the Ship Recycling Convention is to protect occupational safety and health of workers. The convention does contain certain measures that can help attain this goal. However, the measures contemplated are inadequate. In many of the shipbreaking yards on the subcontinent there is large-scale prevalence of child labour. The Ship Recycling Convention simply ignores this evil practice. The shipbreakers often live in shanty accommodation situated in close vicinity of the breaking yards, and usually there is no provision for basic amenities or even primary health care. Technically, many of these issues could well be beyond the scope of an IMO convention given that the shipbreakers constitute an altogether different class than seafarers. Nevertheless, given that ship recycling is not a mere maritime concern, some movement in this direction in the form of certain general principles could definitely have served the development of a culture of safety in the shipbreaking yards.

IV. Conclusion

It is too early to predict whether the Ship Recycling Convention will swim or sink. However, as the preceding discussion reveals, the regulatory framework envisaged by the Ship Recycling Convention approaches the issue of breaking defunct ships primarily from a maritime lens, taking into consideration the commercial realities of the shipping world. It does address certain environmental and occupational safety issues; however, the primary flaw is that it seeks to craft

[265] See International Conference on the Safe and Environmentally Sound Recycling of Ships: Agenda Item 6, Consideration of the Draft Conference Solutions: Draft Conference Resolution on the Processing of Incorrectly Declared Waste, Submitted by France, SR/CONF 21, 27 March 2009, online: SJÖFARTSVERKET <http://www.sjofartsverket.se/pages/19514/sr-conf-21.pdf>.

solutions to these problems mainly from the shipping perspective, ignoring the complexity of the ship recycling business and significant issues of environmental justice and human rights. The future effectiveness of the Ship Recycling Convention will rely on the flag state administration and on the recycling states. Given that these entities are generally the weakest in the long chain of actors in the ship recycling industry, it is doubtful whether this regime will achieve the objective of safe and environmentally sound ship recycling. Although, the Ship Recycling Convention will be the most significant international instrument providing the legal framework within which the trade will have to operate, it addresses only a few of the contentious issues in shipbreaking. The solutions envisaged do not add up to produce a robust legal design. Consequently, the Ship Recycling Convention merely fortifies the status quo and is far from being the appropriate legal response to the fragmented and incomplete regime already in place.

Chapter 6

Conclusion

I. Towards a Sustainable International Legal Regime on Ship Recycling

Regimes are basically systems based on certain principles, norms, rules and decision-making procedures that have evolved out of a desire by concerned actors to meet the demands of certain expectations in a given issue area.[1] Regime theorists argue that states function in an environment beset with problems that are transnational in character. In attempting to tackle them, states have to evaluate the merits afforded by a range of solutions. Sometimes workable solutions to a problem, due to its complexity and permanency, call for a high degree of interdependence between the concerned actors. The cooperation demanded ultimately leads to the development of a regime as a managerial response to the problem.[2] In developing regimes as responses to particular issues under their jurisdiction, states may have to accept a circumscribed sovereignty to foster cooperation. Otherwise, their expectations, as reflected in the objectives of their regime, may not be met. This limitation on sovereignty stands to benefit each participant state since regime-governed behaviour secures sustained benefits based on maximised possible utility for all.[3]

The ship scrapping industry has the dubious distinction of being one of the most dangerous and exploitative occupations in the world. Hundreds of decommissioned ships are being taken apart annually around the world for their valuable scrap metal. However, older ships, particularly those built before the 1980s contain hazardous materials as part of their structure. Shipowners transport these toxic ships to South Asian beaches to be broken down under extremely dangerous and primitive conditions that do not conform to minimum occupational safety and environmental standards. Lack of health, safety and environmental protection measures means that there is no possibility for containing the toxins that are released into the coastal and the marine environment. Moreover, hundreds of

---

[1] Stuart B. Kaye, *International Fisheries Management: A Comparative Analysis of Legal Approaches to Management in the Context of Polar Fisheries Regimes* (J.S.D. Thesis, Dalhousie Law School, 1999) [unpublished] at 25–33.

[2] For instance, see Aldo Chircop, *Cooperative Regimes in Ocean Management: A Study in Mediterranean Regionalism* (J.S.D. Thesis, Dalhousie Law School, Vol. 1, 1998) [unpublished] at 230.

[3] Jamison E. Colburn, "Turbot Wars: Straddling Stocks, Regime Theory, and a New U.N. Agreement" (1997) 6 J. Transnat'l L. & Pol'y 323 at 357 (HeinOnline).

workers die or are seriously injured every year, either due to explosions, or by asphyxiation when they descend into un-cleaned cargo holds. The workers are also crushed under falling steel. Many of those who survive the ordeal, long after they leave the yards, die painful deaths from cancers caused by exposure to carcinogens released from the ships during their breaking. They endure all this suffering for a daily wage that is usually less than USD2. Though many of these issues, by themselves, are technically not transboundary in nature, they constitute an offshoot to the international trade in obsolete ships and involve a variety of transnational players. Clearly, the operating conditions of this industry call for the development of an international regime to govern this sector.

Since market mechanisms do not articulate the fair rules necessary for operation of this industry, government regulation is necessary at the national level to ensure a sustainable shipbreaking industry. India's unilateral efforts to regulate the shipbreaking industry demonstrate this scenario. Even though there are several general and specific laws to regulate this industry, the Supreme Court of India was called upon to formulate guidelines for sustainable ship scrapping. Despite having a normative framework that could act as a springboard to streamline the activities in the ship demolition yards, there was a growing disconnect between laws and daily practice. Rather than attracting more ships, the business in India dwindled. Shipowners found an alternative destination in Bangladesh, where lack of laws and lax law enforcement ensured that profit margins were not affected. This lead to resistance in India by those responsible for administering the legal regime, as was evident when high profile toxic European ships reached Indian shores for their last rites. Thus we have the curious situation of a race amongst nations competing to emerge as the most preferred destination for end-of-life ships, even if this entails huge costs in terms of human life and welfare. Clearly, domestic regimes on their own are incapable of containing the centripetal global forces that underlie the functioning of this industry. There is a need for an internationally "negotiated order"[4] on ship recycling lest the industry migrate to new locations where they can continue to exploit human and environmental resources.

Building a regime for ship recycling is a complicated task. A fine balance has to be struck between the competing demands of the economics of ship demolition and the need to protect the environment and safeguard basic human rights. To be sustainable, there has to be a balance between economic, social and ecological objectives.[5] Thus, solutions to the problems posed by shipbreaking can emerge only if the governance framework is guided by sustainability principles.

---

[4] Oran R. Young, "Regime Dynamics: The Rise and Fall of International Regimes" in Stephen D. Krasner, ed., *International Regimes* (London: Cornell University Press, 1983) 93 at 99.

[5] See generally Ismail Serageldin & Andrew Steer, eds. with the collaboration of Michael M. Cernea *et al.*, *Making Development Sustainable: From Concepts to Action*, Environmentally Sustainable Development Occasional Paper Series No. 2 (Washington D.C.: The International Bank for Reconstruction and Development/The World Bank, 1994).

Since early history, humankind has interacted with nature and depended on the environment to meet basic necessities.[6] The rise in population and the march of civilisation led to intense competition between the different players for control over financial and natural resources.[7] Development, in a sense, led to increased pollution.[8] It was to address these concerns at the international level that the United Nations convened in 1972 the Stockholm Conference. This conference was revolutionary because it emphasised incorporating some of the "prevailing notions of ecological science into legislation".[9]

Even after the Stockholm Conference and the development of legal mechanisms to address pollution at the international level, deteriorating environmental trends continued.[10] Development and environment were viewed as incompatible.[11] The World Commission on Environment and Development, chaired by a former Prime Minister of Norway, Mrs. Gro Harlem Brundtland, was convened by the United Nations to make recommendations for action to transcend this view. The result was the monumental work entitled *Our Common Future*.[12] Overcoming the manifest incongruities between environment and developmental considerations, the Brundtland Report laid out a conceptual framework for the pursuit of sustainable development. Development and environment were viewed not as two opposite poles, but as being mutually inclusive of each other.[13]

The brooding omnipresence of this concept has led to the conclusion of international and regional treaties, conventions, declarations, judicial decisions, and policy pronouncements by governments all over the world. However, its practical

---

[6] *For an exposition of this interplay in terms of sustainable development see separate opinion of the* Vice-President in *Case Concerning the Gabcikovo-Nagymaros Project* (*Hungary/Slovakia*) (1997), [1997] I.C.J. 7 at 88, Christopher Gregory Weeramantry J.

[7] *Ibid*. See also H.E. Judge C.G. Weeramantry, "Achieving Sustainable Justice through International Law" in Marie-Claire Cordonier Segger & Judge C.G. Weeramantry, eds., *Sustainable Justice: Reconciling Economic, Social and Environmental Law* (Boston: Martinus Nijhoff Publishers, 2005) 15.

[8] James P. Karp, "Sustainable Development: Toward a New Vision" (1994) 13 Va. Envtl. L.J. 239 at 241 (WLeC) (highlighting the conflict between technosphere and ecosphere).

[9] Pedro Tarak, "Introduction" (Seminar on the Law of Sustainable Development, Argentina, 19 April 1995), (1996) 13 Pace. Envtl. L. Rev. 561 (WLeC).

[10] For further details see Andronico O. Adede, "The Treaty System from Stockholm (1972) to Rio de Janeiro (1992)" (Statement at the Rio Environmental Treaties Colloquium, Pace Law School Center for Environmental Legal Studies, 30 October 1994), (1995) 13 Pace. Envtl. L. Rev. 33 at 37. During the 1970s, the attention of the international community was on the "first generation" environmental problems which soon gave way to "second generation" environmental problems in the mid-1980s namely, global warming, depletion of stratospheric ozone, climate change, desertification, habitat protection, environmental protection in times of armed conflict, and the transboundary trade in toxic chemicals and hazardous wastes. *Ibid*.

[11] *Ibid*. at 38.

[12] See generally World Commission on Environment & Development, *Our Common Future* (Oxford: Oxford University Press, 1987).

[13] MaryPat Williams Silveira, "International Legal Instruments and Sustainable Development: Principles, Requirements, and Restructuring" (1995) 31 Willamette L. Rev. 239 (HeinOnline).

success has been the subject of intense debate.[14] One major difficulty with the concept is what its precise contours are. The Brundtland Commission offered a simple formulation: sustainable development seeks a symbiotic balance between developmental and environmental considerations while undertaking economic activities.[15] Putting this into practice has proven to be difficult and contentious in many industrial sectors.

What constitutes sustainable development in ship recycling? Shipbreaking enables recycling of obsolete tonnage, with almost 95 percent of a dead ship being put to more productive uses. The activity also provides employment and livelihood opportunities to thousands, revenue to Third World governments, affordable high quality steel while conserving huge amounts of energy and reduced mining of ore that, in turn, results in fewer environmental problems. In this sense, this industry can lay claim to being a truly sustainable activity. But behind this veil of economic development, there is a false economy.[16] The shipbreaking process contaminates aquatic and terrestrial environments, affects ecosystem integrity, and leads to a sustained destruction of the commons. It also creates another tragedy:[17] a large number of people suffer in order to benefit a few. Thus, while the principle of ship recycling may apparently be sound, the practices utilised and the environmental standards that prevail are far from being sustainable.

If we are to move towards a sustainable international ship recycling legal regime we must strike a balance between the need to protect the environment and the human element without being unduly restrictive of the trade. While the issue of ship recycling has been on the international agenda since the 1990s, after nearly two decades, efforts at the international level to establish a regulatory regime on ship recycling has finally fructified. Prior to this development, we had the Basel Convention,[18] which though relevant to shipbreaking, could not further sustainable ship disposal as its application to the ship recycling industry was strongly opposed by certain powerful shipping interests. The International Maritime Organization's (IMO) efforts to develop a legal framework to counter the ills plaguing the shipbreaking business were also not successful. While some progress

---

[14] Michael McCloskey, "The Emperor has No Clothes: The Conundrum of Sustainable Development" (1999) 9 Duke Envtl. L. & Pol'y F. 153 at 159 (WLeC) (noting that sustainability will "prove to be no more than a boon to publicists who will paste new labels on old bottles and claim that every project that makes their clients rich is sustainable").

[15] "Sustainable development is development that meets the needs of the present without compromising the ability of future generations to meet their own needs." World Commission on Environment & Development, *supra* note 12 at 43.

[16] Dawn A. Russell & Moira L. McConnell, "Preface" (1995) 18:1 Dal. L.J. 5 at 9 (HeinOnline).

[17] *Ibid.*

[18] *Basel Convention on the Control of Transboundary Movements of Hazardous Wastes and their Disposal*, 22 March 1989, 1673 U.N.T.S 126, 28 I.L.M. 657 (entered into force 5 May 1992) [*Basel Convention*].

towards developing a sustainable model of ship recycling was made in the form of the IMO Guidelines on Ship Recycling (IMOGSR), they are non-binding. Their implementation depends on the goodwill of shipowners, and not a single ship that has sailed to the Indian subcontinent has complied with the IMOGSR in its entirety.

The recently adopted *Hong Kong International Convention for the Safe and Environmentally Sound Recycling of Ships, 2009* (Ship Recycling Convention)[19], which is intended to replace the supposedly rigid Basel Convention *vis-à-vis* shipbreaking, views the entire issue as a maritime problem, ignoring the other important dimensions that need to be taken into account if there is to be a sustainable face to shipbreaking. While profit maximisation and investment returns remain the central goal of all corporate endeavours, including shipbreaking, there is a need to fundamentally re-examine these objectives in the light of growing number of casualties in the recycling yards and the systematic destruction of the marine environment. However, the Ship Recycling Convention ignores the Basel Convention as well as some of the stipulations contained in the IMOGSR. Notwithstanding the overall strength of the Ship Recycling Convention, it could have been considerably enhanced if measures introduced in both of these documents had been taken into consideration. The Ship Recycling Convention cloaks the real polluter, usually the last or previous commercial operator of the ship, from any sort of liability. Further, it introduces new actors who are generally unconnected with the ship until it is put up for sale in the recycling market to manage the recycling process. It imposes obligations on these new entities in a bid to ensure sustainable shipbreaking, which they are incapable of assuming.

There are several gaps, ambiguities, contradictions, overlaps, and open-ended issues in the current applicable international regimes including the as yet in force Ship Recycling Convention rendering them insufficient to ensure sustainable shipbreaking. It seems that the Ship Recycling Convention was a political compromise and is not the final solution. As noted in the Introduction to this book, sustainable development has been reduced to an oxymoron. Merely coining new terms such as "ship recycling", which in itself implies an element of sustainability, without addressing the fundamental underlying issues could prove counterproductive to the stability of the regime itself.

[19] For the text of the convention, see International Conference on the Safe and Environmentally Sound Recycling of Ships: Agenda Item 8, Adoption of the Final Act and Any Instruments, Recommendations and Resolutions Resulting from the Work of the Conference: Hong Kong International Convention for the Safe and Environmentally Sound Recycling of Ships, 2009, Text Adopted by the Conference, IMO/SR/CONF/45, 19 May 2009, online: SJÖFARTSVERKET <http://www.sjofartsverket.se/pages/19514/sr-conf-45.pdf> (opened for signature 1 September 2009).

II. The Way Forward ...

If sustainability is to be attained in the ship recycling business, there must an international legal regime that harmonises its commercial interests, minimising market distortions, while safeguarding human rights and environmental considerations. Such a legal framework should be built on four elements, namely: (1) economic viability, (2) social progress, (3) equity, and (4) environmental justice. It should incorporate measures to augment infrastructure, ensure appropriate disposal of hazardous wastes and substances, encourage technology transfer and support to developing nations, adequately train and equip personnel, and provide for a proper plan for dismantling to be approved by national regulators in consultation with relevant local authorities. It should also contain provisions in relation to preparation of an inventory of hazardous materials, identifying exact locations where the hazards are present, prior decontamination, the setting up of health care facilities, compulsory medical insurance, and safe work conditions including provision for basic needs of the workers.

While some of these measures are already embodied in the Ship Recycling Convention what follows are 11 suggestions that can help to secure sustainability in this enterprise. These prescriptions do not constitute a complete code on sustainable ship recycling. Rather, they bring together points that have emerged through the debate, and which, if adequately reflected in the implementation of the Ship Recycling Convention could go a long way to help improve the situation.

A. *Developing "Green Capacity": Replacing Beaching by More Acceptable Methods*

As of now, the Indian subcontinent remains the prime ship recycling destination, and it seems that this model will not be replicated elsewhere. However, the calibre of the facilities where the ships are broken and the practices that are followed in these recycling yards leaves room for improvement. Upgrading the dismantling facilities in this region to an acceptable green recycling standard is one of the core elements in the pursuit of a sustainable ship recycling strategy.

The introduction of a green recycling standard can take many forms. One major area that requires change on the Indian subcontinent is the present practice of beaching ships by driving them onto tidal flats at high tide, usually under a ship's own power. Even though Third World governments that matter hold on to *a priori* assumption that beaching is practical and cost effective, it is the primary reason for the perpetuation of environmental and human right violations as the locus of the activity affords little opportunity for safe cutting of the ship. Whatever be the law, it makes economic sense for unscrupulous scrappers to dump the hazardous substances from the ship into the oceans, rather than having them carted off to appropriate recycling sites or landfills.[20]

Finding a viable alternative to beaching has become the most contentious issue in the ship recycling debate. While it may be technically possible to switch over to environmentally sound dismantling techniques, e.g., dry docking, where breakers have greater control over the processes, it is the cost and the vagaries of the ship recycling market that deter developing economies from investing in them.[21] Moreover, given the constraints of time, it may not be possible to introduce dry docking before the phase-out of single-hulls. However, as a long-term solution, it would be advisable if programmes and strategies for setting up dry dock infrastructure were initiated. Introduction of dry docking would necessarily lead to greater mechanisation in processes, streamline procedures, increase recycling output,[22] create better waste containment, and necessitate the presence of a trained and organised work force to run recycling operations on the Indian subcontinent.

B. *Need for a Ship Recycling Contract*

In all cases, there should be a clear ship recycling contract, which can serve as an essential reference point in case of a dispute. A contract allocates rights and

---

[20] V.K. Shashikumar, "Poisoned Lives – IX Alang the Waste Coast" *Tehelka [of India]* (13 August 2005), online: Tehelka – The People's Paper <http://www.tehelka.com/story_main13.asp?filename=Ne081305alang_the.asp> (reporting that even though Indian law proscribes dumping of hazardous materials from ships into the oceans, workers at Alang are instructed to throw jute sacks full of waste into the sea).

[21] The cost of establishing a green recycling facility in Europe in 2000 was estimated by *Det Norske Veritas* to be around USD96 million with about 70% of these costs incurred for establishing dry dock and dockside areas. The total investment involved in establishing the Pipavav recycling facility in India with an annual capacity of 400,000 LDT is reported to be approximately USD90 million. The setting up of the S.T.O.P. recycling facility in the Netherlands cost about USD63 million. EC, Directorate-General Energy & Transport, *Oil Tanker Phase Out and the Ship Scrapping Industry: A Study on the Implications of the Accelerated Phase out Scheme of Single Hull Tankers Proposed by the EU for the World Ship Scrapping and Recycling Industry* (EC, 2004) at 139–40, online: European Commission <http://ec.europa.eu/transport/maritime/safety/doc/prestige/2004_06_scrapping_study_en.pdf>.

[22] France, Interdepartmental Committee on the Dismantling of Civilian and Military End-of-Life Ships, *Le rapport de la Mission Interministérielle portant sur le Démantèlement des Navires civils et militaires en fin de vie*, at 13 (27 March 2007, Chair: Xavier de la Gorce) online: SGMer <http://www.sgmer.gouv.fr/IMG/pdf/2007-06-18_-_Rapport_MIDN_english_version.pdf>. But see International Conference on the Safe and Environmentally Sound Recycling of Ships: Agenda Item 6, Consideration of the Draft International Convention for the Safe and Environmentally Sound Recycling of Ships: Environmental Comparison of the Beaching Method with Other Methods of Recycling of Ships, Submitted by India, SR/CONF/36, 1 May 2009, ¶ 4, online: SJÖFARTSVERKET <http://www.sjofartsverket.se/pages/19514/SR-CONF-36.pdf>. Based on a study conducted by the Indian Institute of Technology, Mumbai, India concludes that the capability of the beaching method at Alang to recycle wastes is far superior than the dry dock method. Moreover, the infrastructure required to practice the dry dock method which utilises items like cement, concrete, steel, etc. to construct the docks and also the power required to operate them, add to the carbon footprint of the dry dock method. *Ibid.*

responsibilities between the parties. It provides clarity, consistency and predictability in the process. Currently, due to economic reasons, there is growing alienation between shipowners and the final act of dismantling a ship. The presence of a standard ship recycling contract could rectify this situation to a great extent. Moreover, shipowners prefer to sell their ships to cash buyers who are usually the party that make final decisions regarding recycling. This complicates the whole process because the identities of cash buyers are not known and, generally, they share no relationship with the flag state. In disputes raised by the end-buyers, the cash buyer often passes the dispute on to the original owner, sometimes even appropriating a portion of the compensation.[23] If environmentally and socially responsible ship recycling practices are to develop worldwide, it is necessary that the shipowners exercise greater control over the last stages of a ship through a contract. The IMOGSR addresses this point, but it has been omitted in the Ship Recycling Convention. The BIMCO standard contract DEMOLISHCON must be modified to bring it in line with the requirements of the Ship Recycling Convention to address some of the concerns highlighted here. As well, copies of the recycling contract should be sent to the flag state and the recycling state before initiating the transfer.

C. *Prior Decontamination of Hazardous Materials*

Given the poor infrastructure facilities available in most of the recycling yards for hazardous wastes processing, pre-cleaning of ships should be made the general rule for all end-of-life ships before they embark on their final voyage. As far as the extent of decontamination is concerned, two considerations are important. Firstly, decontamination should not be carried out to such an extent that a ship is rendered dead, unable to ply on its own power. This is necessary since overly-stringent removal would result in the towing of hulls posing considerable risks and possibly even rendering the business commercially unviable.[24] Moreover, there is also the risk of non-availability of tugs in sufficient numbers and the possibility that towing in large numbers through busy sealanes would render normal sea traffic risky.[25] Secondly, while preparing the ship recycling plan, it should be

---

[23] Recycling of Ships: General Comments on Document, Submitted by India, IMO/MEPC 48/3/1, 9 August 2002, ¶ 2.7 (KR-CON).

[24] The towing of 'Otapan' from Amsterdam to Turkey cost about €205,000 (approx. USD265,544, exchange rate as on 30 January 2007 for USD1 = €0.7720), while the return trip cost yet another €350,000 (approx. USD453,368). Towing a ship from Europe to the Indian subcontinent through the Suez Canal reduces the costs marginally compared to circumventing the Cape of Good Hope. EC, Directorate General Environment, *Ship Dismantling and Pre-cleaning of Ships: Final Report* (EC, 2007) at 118, online: European Commission <http://ec.europa.eu/environment/waste/ships/pdf/ship_dismantling_report.pdf>.

[25] Recycling of Ships: Comments on the Report of the Correspondence Group to MEPC 49, Submitted by India, IMO/MEPC 49/3/3, 13 May 2003, ¶ 2.5 (KR-CON).

required that there be an assessment of the capacities of the dismantling yard. If it appears that the recycling yard is not in a position to manage the waste, the shipowner should ensure that there is proper decontamination elsewhere, or that alternative arrangements are made to dispose of the wastes in facilities that possess the necessary capacity. There should also be mandatory certification detailing the extent of prior decontamination undertaken.

D. *Responsibility of Organisation for Economic Co-operation and Development (OECD) Countries*

Although it has often been argued that the sole responsibility for worker health, safety and environment protection in the ship recycling industry falls on the recycling states, the fact is that the OECD countries have a responsibility to ensure that the ships are safely dismantled in the recycling state. This view is consistent with the principle of environmental justice. OECD countries should provide technical and financial assistance, or even soft loans, to set up adequate numbers of dry dock facilities in developing countries. They can also play a role in transferring relevant hazardous wastes management technology and in matters relating to improving worker competence and help train local authorities in the nuances of ship recycling management. The OECD countries should also adopt measures to closely monitor ships that have reached the end-of-life stage. For instance, despite the presence of a strong legal framework, the European Union continues to be a major exporter of toxic ships as a result of its ownership of 25 percent of the worldwide tonnage and 40 percent of all tonnage owned by European Union domiciled companies.[26] However, due to the considerable reflagging of end-of-life ships, only a miniscule number of merchant ships are left under the operational ambit of the European Union Waste Shipment Regulation. As of now, there are no legal obstacles to the transfer of end-of-life ships. Yet, given that the timetable for the phase-out of single-hull tankers is fast approaching, it would be ideal if, at some point, policy makers in Europe treated a certain reflagging rate and sale volume of former European flagged single-hull oil tankers or other ships as a *de facto* evasion of the relevant European Union law.[27] Steps should be initiated to proscribe such transfers. Any subsidies to the shipping industry should be provided only if they comply with relevant waste management laws. The OECD countries should also initiate steps to set up more clean ship recycling facilities within their boundaries.

---

[26] Thomas Ormond, "Developments in Europe: The European Commission's Green Paper on Better Ship Recycling" (PowerPoint presented to the Preparatory Discussions on Promoting Sustainable Ship Recycling through the Global Programme, Dhaka, Bangladesh, 13 January 2008), online: Basel Convention <http://www.basel.int/ships/gpssr/index.html>.

[27] Directorate-General Energy & Transport, *supra* note 21 at 141.

E. *Responsibilities of Third World Governments*

Since its establishment in 1983, the shipbreaking industry at Alang has been viewed by the ship recyclers, the Gujarat State Government and the Government of India as the proverbial goose that lays golden eggs. It is estimated that between 1994 and 2004, when the activity in Alang was at its peak, the Gujarat Maritime Board earned INR468 crores (approx. USD103 million)[28] by renting out shipbreaking plots; the State Government over INR1,000 crores (approx. USD220 million)[29] in sales tax; and the Central Government about INR8,000 crores (approx. USD1.7 billion)[30] in customs duties.[31] Individual plot owners also made good profits, not to mention the huge windfall reaped by the owners of these derelict ships. However, not a penny was ploughed back by any of them to better the infrastructure in Alang.[32] The situation is no different in Bangladesh, where it is believed that the national government earns 700 crores taka (approx. USD102 million)[33] each year by way of import duties and taxes. If it were not for the hue and cry raised by the environmental non-governmental organisations (ENGOs), the industry would have continued to function at a primitive level.

Governments have the primary responsibility to ensure that recycling operations within their national boundaries are conducted in a safe and sound manner. Time and time again, governments in the major shipbreaking countries, the recyclers, and shipowners have taken refuge behind a hollow proposition that since more than half a million people from this region depend on ship scrapping for a living, introduction of new technologies would be catastrophic, throwing thousands out of employment. This argument belies the fact that shipbreaking is not the sole solution to the endemic unemployment problem. Governments must come up with innovative strategies to combat unemployment, and in particular, be more concerned with enhancing human capabilities.

Since the Ship Recycling Convention virtually bans recycling in non-party states, it is in the best interests of the recycling states to accede to the convention. Once accession is complete, states will have to create domestic legal regimes that meet the requirements of the Ship Recycling Convention. Therefore stocktaking

---

28 Manas Dasgupta, "Slow death for ship-breaking industry" *The Hindu [of India]* (6 March 2006), online: The Hindu <http://www.thehindu.com/>. This conversion uses the exchange rate USD1 = INR45.3005 as on 30 January 2004.

29 *Ibid.*

30 *Ibid.*

31 *Ibid.*

32 *Ibid.*

33 Md. M. Maruf Hossain & Mohammad Mahmudul Islam, *Ship Breaking Activities and Its Impact on the Coastal Zone of Chittagong, Bangladesh: Towards Sustainable Development*, (Chittagong: Advocacy & Publication Unit, YPSA, 2006), online: Young Power in Social Action <http://www.ypsa.org/publications/Impact.pdf>. This conversion uses the exchange rate, USD1 = BDT68.500 as on 13 August 2008.

of their recycling capacity is crucial to give these countries an idea of what more they need to do. Governments will also have to make necessary investments in infrastructure and provide incentives to ship recyclers to adopt new recycling practices. Financial support in the form of subsidies and tax benefits could be provided to recyclers to undertake programmes to improve their yards. To protect their interests better, the major ship recycling nations on the Indian subcontinent could even form regional agreements consistent with the terms of the Ship Recycling Convention, and initiate programmes to protect the Arabian Sea and the Bay of Bengal, from further environmental degradation due to ship recycling activities.

F. *Promoting Safety in the Recycling Yards*

All recycling states are to legislate on worker safety, health and welfare in accordance with the International Labour Organization guidelines. There should be provision for compulsory health insurance for workers to take care of their medical needs after the worker leaves the demolition yard. All workers should be issued with photo-identity cards with personal details. Adequate residential facilities with basic civic amenities should also be provided. There should be a mechanism to ensure that the working hours of labourers are strictly regulated and that workers are given weekly holidays. Night-work should be regulated in order to avoid worker fatigue. Proper lighting arrangements must be provided in the yards. Hospitals with emergency health care and anti-burn care facilities must be set up. Permission for establishing recycling yards should be conditional upon the recycler providing adequate firefighting equipment. Child labour in the yards must be strictly prohibited. Minimum wages should be established and illegal deductions from salary should be penalised. A worker must be provided with adequate training for the work s/he is supposed to perform and must be given protective gear. Such training should also include raising awareness regarding their entitlements, and the need to maintain general cleanliness. Though a male-dominated profession, tasks which may employ women labourers must be identified, and special protections should be provided for them.

G. *Promoting a Green Culture*

In constructing a ship, the primary consideration of the shipbuilder is making sure that the ship proves to be economically advantageous to the owner. Accordingly, shipbuilding yards rarely consider decommissioning while designing and constructing a ship. This is the primary causative factor for the present state of affairs in the shipbreaking yards. With growing awareness of the hazards that emanate from the process of breaking of a ship, measures should be considered at all stages of ship design and construction to minimise the environmental and other hazards at the end of the vessel's life.[34] Ships should be green and "breaker-friendly", but of course not at the cost of compromising stability and safety. To the extent practicable and possible, materials used in ship construction should be non-hazardous.[35] As well, each time a ship is repaired, the standard of repair should retain the green standard of the ship.[36] Shipowners should also be motivated to divert their ships only to green ship recycling yards. A culture of green ships can be promoted in the shipping industry if charterers stipulate appropriate norms while hiring a ship, like requiring the ship to be green, and forcing owners to build, operate and maintain them green throughout their life.[37] Insurance companies can also play an important role in this regard by lowering the insurance premium if the ship is certified as a green ship.[38]

The culture of green should extend to the recycling yards as well. Landfills must be secured, and there must be constant monitoring to ensure that there is no groundwater contamination though leachate. Pollution prevention laws should be strictly enforced, not only in the recycling yards, but also in the surrounding steel foundries.

H. *Transparency*

Perhaps the most important issue for consideration in the context of ship recycling is transparency. As discussed earlier, unscrupulous shipowners adopt several ingenious techniques to cloak their identities and change flags to circumvent inconvenient flag state regulations. To combat this ploy, there must be an international register of vessels that are likely to be decommissioned (Registry of Vessels Destined for Disposal) that is maintained by the IMO in conjunction with individual flag states.[39] Since shipowners have knowledge about the likely decommissioning schedules of their ships, the cost of adding information concerning probable timeframes for withdrawing ships from service to the already existing

---

[34] See Recycling of Ships: Proposals for Consideration by the Committee on the Development of Internationally Applicable Guidelines on Recycling of Ships, Submitted by India, IMO/MEPC 47/3/1, 6 November 2001, ¶ 3 (KR-CON) [Committee on the Development of Internationally Applicable Guidelines].

[35] Recycling of Ships: Urgent Considerations for Correcting Deficiencies in the Draft Convention on the Safe and Environmentally Sound Recycling of Ships, Submitted by Greenpeace, IMO/MEPC 55/3/7, 10 August 2006, ¶ 18, 19 (KR-CON). See also Recycling of Ships: Substitution, Internalizing Costs, Third Party Auditing and Rejection of Non-party Trade Needed to Secure an Effective Convention, Submitted by Greenpeace International & Friends of the Earth International, IMO/MEPC 57/3/7, 25 January 2008, ¶ 9 (KR-CON).

[36] See Committee on the Development of Internationally Applicable Guidelines, *supra* note 34 ¶ 6.

[37] *Ibid.*, ¶ 5.

[38] *Ibid.*, ¶ 7.

[39] See generally Recycling of Ships: Proposals for the Further Development of the List of Future Work Items on Ship Recycling as Approved by MEPC 49, Submitted by Greenpeace International, IMO/MEPC 51/3/6, 6 February 2004 (KR-CON).

reporting requirements would be minimal.[40] In addition, there could also be a "ship equipment database" maintained by the flag states in conjunction with shipowners. This would contain details regarding the nature and quantity of material included in specific equipment onboard the ships that are accessible to all stakeholders.[41] Such tracking systems could greatly reduce abandonment of ships by shipowners and the usage of substandard hazardous equipment in ships to decrease costs. In the ship recycling yards there always exists a dichotomy between official figures on causalities, the quantity of wastes generated, the number of workers employed, wage rates, etc., and those uncovered by ENGOs. Therefore, it is essential that there be an effective system for health, environment and death-related documentation for workers and monitoring to facilitate effective redressal of grievances.

I. *Greater Flag State Responsibilities*

Under the IMO-sponsored Ship Recycling Convention, flag states have onerous responsibilities. These include preparing the IHMs, carrying out periodical surveys, issuing the ready for recycling certificate, etc. Flag states should take the whole process of decommissioning more seriously, and should not consider it as a mere means to take in more registration fees. They should carefully scrutinise applications for registration of end-of-life ships under their flag and discourage cases where subscription to the new flag is solely with the objective of deflecting legal responsibility.

J. *Application of the Polluter Pays Principle and Creation of a Ship Recycling Fund*

Ship recycling is a commercial operation and not public works.[42] The ship recycling business is estimated to have a total market turnover of about USD1.5 billion annually, rendering it one of the largest recycling activities in the world.[43] It is not the case that the developing economies in South Asia are getting rusty

[40] Recycling of Ships: Comments on the Report of the Correspondence Group, Submitted by Greenpeace International, IMO/MEPC 47/3/5, 11 January 2002, ¶ 16 (KR-CON). See also *Recycling of Ships: Report of the Correspondence Group*, Submitted by the Co-ordinator of the Correspondence Group, IMO/MEPC 47/3/3, 30 November 2001, ¶ 4.4. (KR-CON) (this aspect was considered by the correspondence group which, however, rejected the same).

[41] Recycling of Ships: Comments on the Report of the Correspondence Group, Submitted by Japan, IMO/MEPC 49/3/5, 23 May 2003, ¶ 2.4.2.2. (KR-CON) (also noting that Japan's Ministry of Land, Infrastructure and Transport has earmarked a budget for research on ship recycling including the development of a database for the financial year 2002).

[42] *Ibid.*, ¶ 2.4.2.3.

[43] Department for Environment, Food & Rural Affairs, "UK Ship Recycling Strategy: Consultation Paper, March 2006", online: Basel Convention <http://cop8.basel.int/ships/docs/16e.pdf>.

ships for free. Therefore, as *quid pro quo*, they should bear the costs involved in cleaning up the hazardous by-products. While shipowners are being paid the market price for the scrap steel, they do not pay anything for toxic waste disposal. If a shipowner was to have a ship recycled in the West, s/he would have to put aside the thought of profiting from the scrap. They could even incur costs for complying with environmental and safety requirements. On the Indian subcontinent, the situation is the reverse. To date, no environmentally responsible procedures for waste disposal have been followed. The wastes are simply dumped into the oceans or landfills. As well, there is a huge labour force that is willing to perform the dangerous work without any protections. This explains why the Indian subcontinent is able to offer huge prices for the scrap. In fact, the ship scrapping phenomenon represents a classical case of negative externalities where the consequences of scrapping activities result not only in destruction of the commons, but more importantly, it affects human life and well being. Thus, while the benefits are enjoyed by the recyclers, the shipowners, other intermediaries in the chain, and even by governments, it is the labourers and the poor communities who live in and around the scrap yards that are forced to bear the burdens ensuing from degradation of their water, air, local fisheries and opportunities for livelihood and, more importantly, their right to life and health.

A balanced economic approach must be pursued to set right present distortions in this industry, whereby the costs in managing the hazardous wastes and ensuring occupational safety and health are internalised by the polluters. The same has to be reflected in the current market prices offered for the scrap tonnage on the Indian subcontinent. Application of the polluter pays principle, which has been sidelined in the debate, could definitely reverse the situation and force shipowners to pay the costs for the necessary clean-up. Although it can be argued that the industry can become uncompetitive, the fact is that asking shipowners to pay for pre-cleaning services could lead to a reduction of only 15 to 30 percent of the scrapping price.[44] If adequate infrastructure and a sound system for environmentally sound management of hazardous wastes can be established in the developing world, the costs involved in effecting a clean-up would still be much cheaper than doing the same in industrialised countries. For instance, in 2003, it was estimated that the costs for pre-cleaning in Asia were between USD25 to USD50 per light displacement tonnage (LDT), when the average scrap price was only USD170 per LDT.[45] At the same time, it was estimated that to pre-clean the

---

[44] ECORYS Transport, "The Ship Recycling Fund: Financing Environmentally Sound Scrapping and Recycling of Sea-going Ships, Final", (1 February 2005), 26 online: GREENPEACE International <http://www.greenpeace.org/raw/content/india/press/reports/the-ship-breaking-fund.pdf>.

[45] *Ibid.*

'Sandrien'[46] in Europe would cost about €800,000 (approx. USD996,388) to €900,000 (approx. USD1,120,937)[47] equalling approximately €100 (approx. USD125)[48] per LDT.[49]

The major impediment to setting up the necessary infrastructure, as mentioned above, is the cost involved. As part of funding measures to establish green infrastructure, a ship recycling fund should be established, into which shipping companies pay levies.[50] The amount to be paid would be determined by IMO according to the class and category of the ship. The contribution would be collected each time the ship is registered with a flag state and the proceeds would be placed into this fund, which would be administered by IMO. The fund should be utilised to assist national governments and recyclers in the Third World to establish adequate green ship recycling measures and facilities.

K. *Responsibility of the IMO*

The dismantling of end-of-life ships is primarily a land-based activity and has been dealt with by national authorities. However, the global nature of the business has proved to be the major obstacle to the success of national regulatory regimes which have failed to discipline the industry and ensure responsible execution of regulations. At the same time, it is clear that ship recycling is a major cog in the smooth sailing of the shipping industry. Thus, IMO was called upon to develop an international convention to streamline this industry's sustainable

---

[46] For further information on the controversy regarding the recycling of this vessel, see Chapter Three entitled "Limitations of a National Response to Regulate the Global Ship Breaking Industry: A Study of the Indian Experience".

[47] Directorate General Environment, *supra* note 24 at 116, 117 (also noting that the costs involved in cleaning two United States Navy ships for artificial reefing, the USS 'Spiegel Grove' (a landing ship dock weighing 7,000 LDT) and 'USS Oriskany' (an aircraft carrier of 32,000 LDT) were USD625,000 and USD8 million requiring a clean-up time of seven and 12 months respectively). The exchange rate used is USD1 = €0.8029 as on 30 January 2004. The year 2004 is used since the data regarding the pre-cleaning cost of 'Sandrien' is based on an earlier study in 2004 entitled *Oil Tanker Phase Out and the Ship Scrapping Industry: A Study on the Implications of the Accelerated Phase out Scheme of Single Hull Tankers Proposed by the EU for the World Ship Scrapping and Recycling Industry*.

[48] *Ibid.*

[49] *Ibid.*

[50] See Recycling of Ships: European Parliament Resolution of 26 March 2009 on an EU Strategy for Better Ship Dismantling, Note by the Secretariat, IMO/MEPC 59/3/6, 22 May 2009, ¶ E(14) (KR-CON) (calling for a funding mechanism based on mandatory contributions from the shipping industry in line with the producer liability principle). See also International Conference on the Safe and Environmentally Sound Recycling of Ships: Agenda Item 6, Consideration of the Draft International Convention for the Safe and Environmentally Sound Recycling of Ships: Ensuring Sustainable Green and Safe and Ship Dismantling – Concerning Beaching and the Establishment of a Mandatory Fund, Submitted by Greenpeace and FOEI, SR/CONF/14, 9 February 2009, ¶ 7–10, online: Basel Action Network <http://www.ban.org/Library/SR-CONF-14.pdf> (pleading for an indirect financial mechanism in which financing is raised at the new-built stage or during the lifetime of the ship).

operation. For the IMO, shipbreaking has been the second most important marine environmental challenge that it has had to face after the amendments to annex VI of the MARPOL Convention.[51] While an IMO convention is welcome, the fact is that the promise of equivalent levels of control, as required by the Basel Convention, is not reflected in the terms of the Ship Recycling Convention. The Ship Recycling Convention dilutes the principle of prior informed consent, does not criminalise illegal traffic of ships, turns a blind eye to the demand for pre-cleaning of ships, replaces the state of export with the flag state, and sidesteps the proximity principle.[52] These are all essential requirements for sustainable transboundary waste management under the Basel Convention. In addition, the parties to the Ship Recycling Convention will also have to address issues not included in this convention such as wreck recycling, special procedures for the towing of end-of-life ships and balancing the right of innocent passage with the interests of the coastal states to protect their marine environment, and removal of anti-fouling paints with tributyltin additives from end-of-life ships.

However, the greatest challenge for the IMO will be to convince recycling states to initiate steps to develop their low-standard ship recycling industries so as to conform to the parameters which the IMO is required to develop. The IMO will also have to ensure that flag states seriously implement their obligations. Despite the protracted ratification process for the Ship Recycling Convention, it may take considerable time for it to come into force. This interim, however, is most crucial, given that the accelerated phase-out of single-hull oil tankers will be at its peak during that time. It is expected that in 2010 alone, 16.7 million LDT will be available for breaking.[53] There is a strong possibility that there will be no corresponding legal framework or system in place to manage the deluge of time expired tonnage. This calls for interim measures to be adopted as soon as possible in order to ensure the smooth phase-out of obsolete ships.

At this point in time, the international legal regime on shipbreaking is perched at a virtual precipice on its journey to effective normative maturity. While the ship recycling yards on the Indian subcontinent have adequate demolition capabilities, the lack of green capacity remains a major obstacle to creating a sustainable ship recycling culture. The OECD countries have green recycling capacity, but these are limited. On the subcontinent, shipbreaking provides huge economic benefits, particularly for Bangladesh and Pakistan as a secondary source of steel,

---

[51] See International Conference on the Safe and Environmentally Sound Recycling of Ships: Hong Kong, China, 11–15 May 2009, Statements by Delegations and Observers at the Closing Session of the Conference, Friday, May 15 2009, SR/CONF/INF.8, 21 May 2009, at 4, online: SJÖFARTSVERKET <http://www.sjofartsverket.se/pages/19514/sr-conf-inf8.pdf>.

[52] For a pointed analysis of the Ship Recycling Convention, see Chapter Five entitled, "Deciphering the Ship Recycling Convention".

[53] Directorate-General Energy & Transport, *supra* note 21 at 99.

since both countries do not have any iron ore resources. However, ship scrapping comes at the costs of significant environmental degradation and human casualties. For shipowners, the removal of outdated tonnage and recycling on the subcontinent provides them with a final opportunity to dispose of obsolete fleets and earn some funds from the sale of decrepit ships at minimal costs to them. If this recycling gateway is blocked, they may resort to unscrupulous means to get rid of their ships. The OECD countries have an interest in ensuring the removal of decaying, hazardous ships lest their shores and ports be forced to harbour them. Thus, ship recycling is in the best interests of all, and there is an opportunity for all the players to benefit from ensuring that a sustainable regime for undertaking the task is established. Otherwise, this opportunity will become a recipe for a disaster.

Clearly, greater cooperation among the players is the key to a sustainable regime for ship recycling. The recently adopted Ship Recycling Convention, supplemented by other IMO initiatives, provides a platform to facilitate this. More importantly, the need to eradicate irresponsible scrapping practices by operationalising at least some of the measures suggested in this study to develop a new culture for sustainable ship recycling is all the more urgent. Even though ship recycling plays a vital role in ridding the oceans of obsolete and hazardous tonnage, it does not have to come at the cost of human lives and environmental degradation. Let the dead ships rest in peace and not torment the living …

INTERNATIONAL MARITIME ORGANIZATION

IMO

E

INTERNATIONAL CONFERENCE ON THE SAFE AND ENVIRONMENTALLY SOUND RECYCLING OF SHIPS
Agenda item 8

SR/CONF/45
19 May 2009
Original: ENGLISH

ADOPTION OF THE FINAL ACT AND ANY INSTRUMENTS, RECOMMENDATIONS AND RESOLUTIONS RESULTING FROM THE WORK OF THE CONFERENCE

HONG KONG INTERNATIONAL CONVENTION FOR THE SAFE AND ENVIRONMENTALLY SOUND RECYCLING OF SHIPS, 2009

Text adopted by the Conference

1 As a result of its deliberations, as recorded in the Record of Decisions of the Plenary (SR/CONF/RD/2) and the Final Act of the Conference (SR/CONF/46), the Conference adopted the Hong Kong International Convention for the Safe and Environmentally Sound Recycling of Ships, 2009.

2 The above-mentioned Convention, as adopted by the Conference, is annexed hereto.

***

ANNEX

HONG KONG INTERNATIONAL CONVENTION FOR THE SAFE AND ENVIRONMENTALLY SOUND RECYCLING OF SHIPS, 2009

THE PARTIES TO THIS CONVENTION,

NOTING the growing concerns about safety, health, the environment and welfare matters in the ship recycling industry,

RECOGNIZING that recycling of ships contributes to sustainable development and, as such, is the best option for ships that have reached the end of their operating life,

RECALLING resolution A.962(23), adopted by the Assembly of the International Maritime Organization (Guidelines on Ship Recycling); amendments to the Guidelines adopted by resolution A.980(24); Decision VI/24 of the Sixth Meeting of the Conference of the Parties to the Basel Convention on the Control of Transboundary Movements of Hazardous Wastes and their Disposal, which adopted Technical Guidelines for the Environmentally Sound Management of the Full and Partial Dismantling of Ships; and the Guidelines approved by the 289th session of the Governing Body of the International Labour Office (Safety and Health in Shipbreaking: Guidelines for Asian countries and Turkey),

RECALLING ALSO resolution A.981(24), by which the Assembly of the International Maritime Organization requested the Organization's Marine Environment Protection Committee to develop a legally-binding instrument on ship recycling,

NOTING ALSO the role of the International Labour Organization in protecting the occupational safety and health of workers involved in ship recycling,

NOTING FURTHER the role of the Basel Convention on the Control of Transboundary Movements of Hazardous Wastes and their Disposal in protecting human health and the environment against the adverse effects which may result from such wastes,

MINDFUL of the precautionary approach set out in Principle 15 of the Rio Declaration on Environment and Development and referred to in resolution MEPC.67(37), adopted by the Organization's Marine Environment Protection Committee on 15 September 1995,

MINDFUL ALSO of the need to promote the substitution of hazardous materials in the construction and maintenance of ships by less hazardous, or preferably, non-hazardous materials, without compromising the ships' safety, the safety and health of seafarers and the ships' operational efficiency,

RESOLVED to effectively address, in a legally-binding instrument, the environmental, occupational health and safety risks related to ship recycling, taking into account the particular characteristics of maritime transport and the need to secure the smooth withdrawal of ships that have reached the end of their operating lives,

CONSIDERING that these objectives may best be achieved by the conclusion of an International Convention for the Safe and Environmentally Sound Recycling of Ships,

HAVE AGREED as follows:

ARTICLE 1
General obligations

1 Each Party to this Convention undertakes to give full and complete effect to its provisions in order to prevent, reduce, minimize and, to the extent practicable, eliminate accidents, injuries and other adverse effects on human health and the environment caused by Ship Recycling, and enhance ship safety, protection of human health and the environment throughout a ship's operating life.

2 No provision of this Convention shall be interpreted as preventing a Party from taking, individually or jointly, more stringent measures consistent with international law, with respect to the safe and environmentally sound recycling of ships, in order to prevent, reduce or minimize any adverse effects on human health and the environment.

3 Parties shall endeavour to co-operate for the purpose of effective implementation of, compliance with and enforcement of this Convention.

4 The Parties undertake to encourage the continued development of technologies and practices which contribute to safe and environmentally sound Ship Recycling.

5 The Annex to this Convention forms an integral part of it. Unless expressly provided for otherwise, a reference to this Convention constitutes at the same time a reference to its Annex.

ARTICLE 2
Definitions

For the purposes of this Convention, unless expressly provided otherwise:

1 "Convention" means the Hong Kong International Convention for the Safe and Environmentally Sound Recycling of Ships, 2009.

2 "Administration" means the Government of the State whose flag the ship is entitled to fly, or under whose authority it is operating.

3 "Competent Authority(ies)" means a governmental authority or authorities designated by a Party as responsible, within specified geographical area(s) or area(s) of expertise, for duties related to Ship Recycling Facilities operating within the jurisdiction of that Party as specified in this Convention.

4 "Organization" means the International Maritime Organization.

5 "Secretary-General" means the Secretary-General of the Organization.

6 "Committee" means the Marine Environment Protection Committee of the Organization.

7 "Ship" means a vessel of any type whatsoever operating or having operated in the marine environment and includes submersibles, floating craft, floating platforms, self elevating platforms, Floating Storage Units (FSUs), and Floating Production Storage and Offloading Units (FPSOs), including a vessel stripped of equipment or being towed.

8 "Gross tonnage" means the gross tonnage (GT) calculated in accordance with the tonnage measurement regulations contained in Annex I to the International Convention on Tonnage Measurement of Ships, 1969, or any successor convention.

9 "Hazardous Material" means any material or substance which is liable to create hazards to human health and/or the environment.

10 "Ship Recycling" means the activity of complete or partial dismantling of a ship at a Ship Recycling Facility in order to recover components and materials for reprocessing and re-use, whilst taking care of hazardous and other materials, and includes associated operations such as storage and treatment of components and materials on site, but not their further processing or disposal in separate facilities.

11 "Ship Recycling Facility" means a defined area that is a site, yard or facility used for the recycling of ships.

12 "Recycling Company" means the owner of the Ship Recycling Facility or any other organization or person who has assumed the responsibility for operation of the Ship Recycling activity from the owner of the Ship Recycling Facility and who on assuming such responsibility has agreed to take over all duties and responsibilities imposed by this Convention

ARTICLE 3
Application

1 Unless expressly provided otherwise in this Convention, this Convention shall apply to:

.1 ships entitled to fly the flag of a Party or operating under its authority;

.2 Ship Recycling Facilities operating under the jurisdiction of a Party.

2 This Convention shall not apply to any warships, naval auxiliary, or other ships owned or operated by a Party and used, for the time being, only on government non-commercial service. However, each Party shall ensure, by the adoption of appropriate measures not impairing operations or operational capabilities of such ships owned or operated by it, that such ships act in a manner consistent with this Convention, so far as is reasonable and practicable.

3 This Convention shall not apply to ships of less than 500 GT or to ships operating throughout their life only in waters subject to the sovereignty or jurisdiction of the State whose flag the ship is entitled to fly. However, each Party shall ensure, by the adoption of appropriate measures, that such ships act in a manner consistent with this Convention, so far as is reasonable and practicable.

4 With respect to ships entitled to fly the flag of non-Parties to this Convention, Parties shall apply the requirements of this Convention as may be necessary to ensure that no more favourable treatment is given to such ships.

ARTICLE 4
Controls related to Ship Recycling

1 Each Party shall require that ships entitled to fly its flag or operating under its authority comply with the requirements set forth in this Convention and shall take effective measures to ensure such compliance.

2 Each Party shall require that Ship Recycling Facilities under its jurisdiction comply with the requirements set forth in this Convention and shall take effective measures to ensure such compliance.

ARTICLE 5
Survey and certification of ships

Each Party shall ensure that ships flying its flag or operating under its authority and subject to survey and certification are surveyed and certified in accordance with the regulations in the Annex.

ARTICLE 6
Authorization of Ship Recycling Facilities

Each Party shall ensure that Ship Recycling Facilities that operate under its jurisdiction and that recycle ships to which this Convention applies, or ships treated similarly pursuant to Article 3.4 of this Convention, are authorized in accordance with the regulations in the Annex.

ARTICLE 7
Exchange of information

For the Ship Recycling Facilities authorized by a Party, such Party shall provide to the Organization, if requested, and to those Parties which request it, relevant information, in regard to this Convention, on which its decision for authorization was based. The information shall be exchanged in a swift and timely manner.

ARTICLE 8
Inspection of ships

1 A ship to which this Convention applies may, in any port or offshore terminal of another Party, be subject to inspection by officers duly authorized by that Party for the purpose of determining whether the ship is in compliance with this Convention. Except as provided in paragraph 2, any such inspection is limited to verifying that there is on board either an International Certificate on Inventory of Hazardous Materials or an International Ready for Recycling Certificate, which, if valid, shall be accepted.

2 Where a ship does not carry a valid certificate or there are clear grounds for believing that:

.1 the condition of the ship or its equipment does not correspond substantially with the particulars of the certificate, and/or Part I of the Inventory of Hazardous Materials; or

.2 there is no procedure implemented on board the ship for the maintenance of Part I of the Inventory of Hazardous Materials;

a detailed inspection may be carried out taking into account guidelines developed by the Organization.

ARTICLE 9
Detection of violations

1 Parties shall co-operate in the detection of violations and the enforcement of the provisions of this Convention.

2 When there is sufficient evidence that a ship is operating, has operated or is about to operate in violation of any provision in this Convention, a Party holding the evidence may request an investigation of this ship when it enters the ports or offshore terminals under the jurisdiction of another Party. The report of such an investigation shall be sent to the Party requesting it, to the Administration of the ship concerned and to the Organization, so that action may be taken as appropriate.

3 If the ship is detected to be in violation of this Convention, the Party carrying out the inspection may take steps to warn, detain, dismiss, or exclude the ship from its ports. A Party taking such action shall immediately inform the Administration of the ship concerned and the Organization.

4 If a request for an investigation is received from any Party, together with sufficient evidence that a Ship Recycling Facility is operating, has operated or is about to operate in violation of any provision of this Convention, a Party should investigate this Ship Recycling Facility operating under its jurisdiction and make a report. The report of any such investigation shall be sent to the Party requesting it, including information on action taken or to be taken, if any, and to the Organization for appropriate action.

ARTICLE 10
Violations

1 Any violation of the requirements of this Convention shall be prohibited by national laws and:

.1 in the case of a ship, sanctions shall be established under the law of the Administration, wherever the violation occurs. If the Administration is informed of such a violation by a Party, it shall investigate the matter and may request the reporting Party to furnish additional evidence of the alleged violation. If the

Administration is satisfied that sufficient evidence is available to enable proceedings to be brought in respect of the alleged violation, it shall cause such proceedings to be taken as soon as possible, in accordance with its law. The Administration shall promptly inform the Party that reported the alleged violation, as well as the Organization, of any action taken. If the Administration has not taken any action within one year after receiving the information, it shall inform the Party which reported the alleged violation, and the Organization, of the reasons why no action has been taken;

.2 in the case of a Ship Recycling Facility, sanctions shall be established under the law of the Party having jurisdiction over the Ship Recycling Facility. If the Party is informed of such a violation by another Party, it shall investigate the matter and may request the reporting Party to furnish additional evidence of the alleged violation. If the Party is satisfied that sufficient evidence is available to enable proceedings to be brought in respect of the alleged violation, it shall cause such proceedings to be taken as soon as possible, in accordance with its law. The Party shall promptly inform the Party that reported the alleged violation, as well as the Organization, of any action taken. If the Party has not taken any action within one year after receiving the information, it shall inform the Party which reported the alleged violation, and the Organization, of the reasons why no action has been taken.

2 Any violation of the requirements of this Convention within the jurisdiction of any Party shall be prohibited and sanctions shall be established under the law of that Party. Whenever such a violation occurs, that Party shall either:

.1 cause proceedings to be taken in accordance with its law; or

.2 furnish to the Administration of the ship such information and evidence as may be in its possession that a violation has occurred.

3 The sanctions provided for by the laws of a Party pursuant to this Article shall be adequate in severity to discourage violations of this Convention wherever they occur.

ARTICLE 11
Undue delay or detention of ships

1 All possible efforts shall be made to avoid a ship being unduly detained or delayed under Article 8, 9 or 10 of this Convention.

2 When a ship is unduly detained or delayed under Article 8, 9 or 10 of this Convention, it shall be entitled to compensation for any loss or damage suffered.

ARTICLE 12
Communication of information

Each Party shall report to the Organization and the Organization shall disseminate, as appropriate, the following information:

.1 a list of Ship Recycling Facilities authorized in accordance with this Convention and operating under the jurisdiction of that Party;

.2 contact details for the Competent Authority(ies), including a single contact point, for that Party;

.3 a list of the recognized organizations and nominated surveyors which are authorized to act on behalf of that Party in the administration of matters relating to the control of Ship Recycling in accordance with this Convention, and the specific responsibilities and conditions of the authority delegated to the recognized organizations or nominated surveyors;

.4 an annual list of ships flying the flag of that Party to which an International Ready for Recycling Certificate has been issued, including the name of the Recycling Company and location of the Ship Recycling Facility as shown on the certificate;

.5 an annual list of ships recycled within the jurisdiction of that Party;

.6 information concerning violations of this Convention; and

.7 actions taken towards ships and Ship Recycling Facilities under the jurisdiction of that Party.

ARTICLE 13
Technical assistance and co-operation

1 Parties undertake, directly or through the Organization and other international bodies, as appropriate, in respect of the safe and environmentally sound recycling of ships, to provide support for those Parties which request technical assistance:

.1 to train personnel;

.2 to ensure the availability of relevant technology, equipment and facilities;

.3 to initiate joint research and development programmes; and

.4 to undertake other actions aimed at the effective implementation of this Convention and of guidelines developed by the Organization related thereto.

2 Parties undertake to co-operate actively, subject to their national laws, regulations and policies, in the transfer of management systems and technology in respect of the safe and environmentally sound recycling of ships.

ARTICLE 14
Dispute settlement

Parties shall settle any dispute between them concerning the interpretation or application of this Convention by negotiation or any other peaceful means agreed upon by them, which may include enquiry, mediation, conciliation, arbitration, judicial settlement, or resort to regional agencies or arrangements.

ARTICLE 15
Relationship with international law and other international agreements

1 Nothing in this Convention shall prejudice the rights and obligations of any State under the United Nations Convention on the Law of the Sea, 1982, and under the customary international law of the sea.

2 Nothing in this Convention shall prejudice the rights and obligations of Parties under other relevant and applicable international agreements.

ARTICLE 16
Signature, ratification, acceptance, approval and accession

1 This Convention shall be open for signature by any State at the Headquarters of the Organization from 1 September 2009 to 31 August 2010 and shall thereafter remain open for accession by any State.

2 States may become Parties to this Convention by:

.1 signature not subject to ratification, acceptance, or approval; or

.2 signature subject to ratification, acceptance, or approval, followed by ratification, acceptance or approval; or

.3 accession.

3 Ratification, acceptance, approval or accession shall be effected by the deposit of an instrument to that effect with the Secretary-General.

4 If a State comprises two or more territorial units in which different systems of law are applicable in relation to matters dealt with in this Convention, it may at the time of signature, ratification, acceptance, approval, or accession declare that this Convention shall extend to all its territorial units or only to one or more of them and may modify this declaration by submitting another declaration at any time.

5 A declaration under paragraph 4 shall be notified to the Secretary-General in writing and shall state expressly the territorial unit or units to which this Convention applies.

6 A State at the time it expresses its consent to be bound by this Convention shall declare whether it requires explicit or tacit approval of the Ship Recycling Plan before a ship may be recycled in its authorized Ship Recycling Facility(ies). This declaration may be revised thereafter by notification to the Secretary-General. Such revision shall specify the effective date of the revision.

ARTICLE 17
Entry into force

1 This Convention shall enter into force 24 months after the date on which the following conditions are met:

.1 not less than 15 States have either signed it without reservation as to ratification, acceptance or approval, or have deposited the requisite instrument of ratification, acceptance, approval or accession in accordance with Article 16;

.2 the combined merchant fleets of the States mentioned in paragraph 1.1 constitute not less than 40 per cent of the gross tonnage of the world's merchant shipping; and

.3 the combined maximum annual ship recycling volume of the States mentioned in paragraph 1.1 during the preceding 10 years constitutes not less than 3 per cent of the gross tonnage of the combined merchant shipping of the same States.

2 For States which have deposited an instrument of ratification, acceptance, approval or accession in respect of this Convention after the requirements for entry into force thereof have been met, but prior to the date of entry into force, the ratification, acceptance, approval or accession shall take effect on the date of entry into force of this Convention, or three months after the date of deposit of the instrument, whichever is the later date.

3 Any instrument of ratification, acceptance, approval or accession deposited after the date on which this Convention enters into force shall take effect three months after the date of deposit.

4 After the date on which an amendment to this Convention is deemed to have been accepted under Article 18, any instrument of ratification, acceptance, approval or accession deposited shall apply to the Convention, as amended.

ARTICLE 18
Amendments

1 This Convention may be amended by either of the procedures specified in the following paragraphs.

2 Amendments after consideration within the Organization:

.1 Any Party may propose an amendment to this Convention. A proposed amendment shall be submitted to the Secretary-General, who shall then circulate it to the Parties and Members of the Organization at least six months prior to its consideration.

.2 An amendment proposed and circulated as above shall be referred to the Committee for consideration. Parties, whether or not Members of the Organization, shall be entitled to participate in the proceedings of the Committee for consideration and adoption of the amendment.

.3 Amendments shall be adopted by a two-thirds majority of the Parties present and voting in the Committee, on condition that at least one-third of the Parties shall be present at the time of voting.

.4 Amendments adopted in accordance with subparagraph 3 shall be communicated by the Secretary-General to the Parties for acceptance.

.5 An amendment shall be deemed to have been accepted in the following circumstances:

.5.1 An amendment to an article of this Convention shall be deemed to have been accepted on the date on which two-thirds of the Parties have notified the Secretary-General of their acceptance of it.

.5.2 An amendment to the Annex shall be deemed to have been accepted at the end of a period to be determined by the Committee at the time of its adoption, which period shall not be less than ten months after the date of adoption. However, if by that date more than one-third of the Parties notify the Secretary-General that they object to the amendment, it shall be deemed not to have been accepted.

.6 An amendment shall enter into force under the following conditions:

.6.1 An amendment to an article of this Convention shall enter into force, for those Parties that have declared that they have accepted it, six months after the date on which it is deemed to have been accepted in accordance with subparagraph .5.1.

.6.2 An amendment to the Annex shall enter into force with respect to all Parties six months after the date on which it is deemed to have been accepted, except for any Party that has:

.6.2.1 notified its objection to the amendment in accordance with subparagraph .5.2 and that has not withdrawn such objection; or

.6.2.2 notified the Secretary-General, prior to the entry into force of such amendment, that the amendment shall enter into force for it only after a subsequent notification of its acceptance.

.6.3 A Party that has notified an objection under subparagraph .6.2.1 may subsequently notify the Secretary-General that it accepts the amendment. Such amendment shall enter into force for such Party six months after the date of its notification of acceptance, or the date on which the amendment enters into force, whichever is the later date.

.6.4 If a Party that has made a notification referred to in subparagraph .6.2.2 notifies the Secretary-General of its acceptance with respect to an amendment, such amendment shall enter into force for such Party six months after the date of its notification of acceptance, or the date on which the amendment enters into force, whichever is the later date.

3 Amendment by a Conference:

.1 Upon the request of a Party concurred in by at least one-third of the Parties, the Organization shall convene a Conference of Parties to consider amendments to this Convention.

.2 An amendment adopted by such a Conference by a two-thirds majority of the Parties present and voting shall be communicated by the Secretary-General to all Parties for acceptance.

.3 Unless the Conference decides otherwise, the amendment shall be deemed to have been accepted and shall enter into force in accordance with the procedures specified in paragraphs 2.5 and 2.6 respectively.

4 Any Party that has declined to accept an amendment to the Annex shall be treated as a non-Party only for the purpose of application of that amendment.

5 Any notification under this Article shall be made in writing to the Secretary-General.

6 The Secretary-General shall inform the Parties and Members of the Organization of:

.1 any amendment that enters into force and the date of its entry into force generally and for each Party; and

.2 any notification made under this Article.

ARTICLE 19
Denunciation

1 This Convention may be denounced by any Party at any time after the expiry of two years from the date on which this Convention enters into force for that Party.

2 Denunciation shall be effected by written notification to the Secretary-General, to take effect one year after receipt or such longer period as may be specified in that notification.

ARTICLE 20
Depositary

1 This Convention shall be deposited with the Secretary-General, who shall transmit certified copies of this Convention to all States which have signed this Convention or acceded thereto.

2 In addition to the functions specified elsewhere in this Convention, the Secretary-General shall:

.1 inform all States that have signed this Convention, or acceded thereto, of:

.1.1 each new signature or deposit of an instrument of ratification, acceptance, approval or accession, together with the date thereof;

.1.2 the date of entry into force of this Convention;

.1.3 the deposit of any instrument of denunciation from this Convention, together with the date on which it was received and the date on which the denunciation takes effect; and

.1.4 other declarations and notifications received pursuant to this Convention; and

.2 as soon as this Convention enters into force, transmit the text thereof to the Secretariat of the United Nations, for registration and publication in accordance with Article 102 of the Charter of the United Nations.

ARTICLE 21
Languages

This Convention is established in a single original in the Arabic, Chinese, English, French, Russian and Spanish languages, each text being equally authentic.

DONE AT HONG KONG, CHINA, this fifteenth day of May, two thousand and nine.

IN WITNESS WHEREOF the undersigned, being duly authorized by their respective Governments for that purpose, have signed this Convention.

* * *

ANNEX

REGULATIONS FOR SAFE AND ENVIRONMENTALLY SOUND RECYCLING OF SHIPS

CHAPTER 1 – GENERAL PROVISIONS

Regulation 1 – Definitions

For the purposes of this Annex:

1 "Competent person" means a person with suitable qualifications, training, and sufficient knowledge, experience and skill, for the performance of the specific work. Specifically, a Competent person may be a trained worker or a managerial employee capable of recognizing and evaluating occupational hazards, risks, and employee exposure to potentially Hazardous Materials or unsafe conditions in a Ship Recycling Facility, and who is capable of specifying the necessary protection and precautions to be taken to eliminate or reduce those hazards, risks, or exposures. The Competent Authority may define appropriate criteria for the designation of such persons and may determine the duties to be assigned to them.

2 "Employer" means a natural or legal person that employs one or more workers engaged in Ship Recycling.

3 "Existing ship" means a ship which is not a new ship.

4 "New ship" means a ship:

.1 for which the building contract is placed on or after the entry into force of this Convention; or

.2 in the absence of a building contract, the keel of which is laid or which is at a similar stage of construction on or after six months after the entry into force of this Convention; or

.3 the delivery of which is on or after 30 months after the entry into force of this Convention.

5 "New installation" means the installation of systems, equipment, insulation, or other material on a ship after the date on which this Convention enters into force.

6 "Safe-for-entry" means a space that meets the following criteria:

.1 the oxygen content of the atmosphere and the concentration of flammable vapours are within safe limits;

.2 any toxic materials in the atmosphere are within permissible concentrations; and

.3 any residues or materials associated with the work authorized by the Competent person will not produce uncontrolled release of toxic materials or an unsafe concentration of flammable vapours under existing atmospheric conditions while maintained as directed.

7 Safe-for-hot work means a space that meets the following criteria:

.1 a safe, non-explosive condition, including gas-free status, exists for the use of electric arc or gas welding equipment, cutting or burning equipment or other forms of naked flame, as well as heating, grinding, or spark generating operations;

.2 Safe-for-entry requirements of regulation 1.6 are met;

.3 existing atmospheric conditions will not change as a result of the hot work; and

.4 all adjacent spaces have been cleaned, or inerted, or treated sufficiently to prevent the start or spread of fire.

8 "Shipowner" means the person or persons or company registered as the owner of the ship or, in the absence of registration, the person or persons or company owning the ship or any other organization or person such as the manager, or the bareboat charterer, who has assumed the responsibility for operation of the ship from the owner of the ship. However, in the case of a ship owned by a State and operated by a company which in that State is registered as the ship's operator, "owner" shall mean such company. This term also includes those who have ownership of the ship for a limited period pending its sale or handing over to a Ship Recycling Facility.

9 "Site inspection" means an inspection of the Ship Recycling Facility confirming the condition described by the verified documentation.

10 "Statement of Completion" means a confirmatory statement issued by the Ship Recycling Facility that the Ship Recycling has been completed in accordance with this Convention.

11 "Tanker" means an oil tanker as defined in MARPOL Annex I or an NLS tanker as defined in MARPOL Annex II.

12 "Worker" means any person who performs work, either regularly or temporarily, in the context of an employment relationship including contractor personnel.

Regulation 2 – General applicability

Unless expressly provided otherwise, the design, construction, survey, certification, operation and recycling of ships shall be conducted in accordance with the provisions of this Annex.

Regulation 3 – Relationship with other standards, recommendations and guidance

Parties shall take measures to implement the requirements of the regulations of this Annex, taking into account relevant and applicable standards, recommendations and guidance developed by the International Labour Organization and the relevant and applicable technical standards, recommendations and guidance developed under the Basel Convention on the Control of Transboundary Movements of Hazardous Wastes and their Disposal.

CHAPTER 2 – REQUIREMENTS FOR SHIPS

Part A – Design, construction, operation and maintenance of ships

Regulation 4 – Controls of ships' Hazardous Materials

In accordance with the requirements specified in Appendix 1 to this Convention each Party:

.1 shall prohibit and/or restrict the installation or use of Hazardous Materials listed in Appendix 1 on ships entitled to fly its flag or operating under its authority; and

.2 shall prohibit and/or restrict the installation or use of such materials on ships, whilst in its ports, shipyards, ship repair yards, or offshore terminals,

and shall take effective measures to ensure that such ships comply with those requirements.

Regulation 5 – Inventory of Hazardous Materials

1 Each new ship shall have on board an Inventory of Hazardous Materials. The Inventory shall be verified either by the Administration or by any person or organization authorized by it taking into account guidelines, including any threshold values and exemptions contained in those guidelines, developed by the Organization. The Inventory of Hazardous Materials shall be specific to each ship and shall at least:

.1 identify as Part I, Hazardous Materials listed in Appendices 1 and 2 to this Convention and contained in ship's structure or equipment, their location and approximate quantities; and

.2 clarify that the ship complies with regulation 4.

2 Existing ships shall comply as far as practicable with paragraph 1 not later than 5 years after the entry into force of this Convention, or before going for recycling if this is earlier, taking into account the guidelines developed by the Organization and the Organization's Harmonized System of Survey and Certification. The Hazardous Materials listed in Appendix 1, at least, shall be identified when the Inventory is developed. For existing ships a plan shall be prepared describing the visual/sampling check by which the Inventory of Hazardous Materials is developed, taking into account the guidelines developed by the Organization.

3 Part I of the Inventory of Hazardous Materials shall be properly maintained and updated throughout the operational life of the ship, reflecting new installations containing Hazardous Materials listed in Appendix 2 and relevant changes in ship structure and equipment, taking into account the guidelines developed by the Organization.

4 Prior to recycling the Inventory shall, in addition to the properly maintained and updated Part I, incorporate Part II for operationally generated wastes and Part III for stores, and be verified either by the Administration or by any person or organization authorized by it, taking into account the guidelines developed by the Organization.

Regulation 6 – Procedure for proposing amendments to Appendices 1 and 2

1 Any Party may propose an amendment to Appendix 1 and/or Appendix 2 in accordance with this regulation. The proposed amendment shall be considered within the Organization under Article 18 paragraph 2 and this regulation.

2 When the Organization receives a proposal, it shall also bring the proposal to the attention of the United Nations and its Specialized Agencies, intergovernmental organizations having agreements with the Organization and non-governmental organizations in consultative status with the Organization and shall make it available to them.

3 The Committee shall establish a technical group in accordance with regulation 7 to review proposals submitted in accordance with paragraph 1 of this regulation.

4 The technical group shall review the proposal along with any additional data, including decisions adopted by other international bodies regarding their lists of materials or hazardous substances, submitted by any interested entity, and shall evaluate and report to the Committee whether the Hazardous Material in question is likely, in the context of this Convention, to lead to significant adverse effects on human health or the environment such that the amendment of Appendix 1 or Appendix 2 is warranted. In this regard:

.1 The technical group's review shall include:

.1.1 an evaluation of the association between the Hazardous Material in question and the likelihood, in the context of this Convention, that it will lead to significant adverse effects on human health or the environment based on the submitted data or other relevant data brought to the attention of the group;

.1.2 an evaluation of the potential risk reduction attributable to the proposed control measures and any other control measures that may be considered by the technical group;

.1.3 consideration of available information on the technical feasibility of control measures;

.1.4 consideration of available information on other effects arising from the introduction of such control measures relating to:

- the environment;
- human health and safety including that of seafarers and workers; and
- the cost to international shipping and other relevant sectors.

.1.5 consideration of the availability of suitable alternatives to the Hazardous Material to be controlled, including a consideration of the potential risks of alternatives;

.1.6 consideration of the risks posed by the Hazardous Material during the recycling process; and

.1.7 consideration of suitable threshold values and any useful or necessary exemptions.

.2 If the technical group finds that the Hazardous Material in question is likely, in the context of this Convention, to lead to significant adverse effects on human health or the environment, lack of full scientific certainty shall not be used as a reason to prevent the group from proceeding with an evaluation of the proposal.

.3 The technical group's report shall be in writing and shall take into account each of the evaluations and considerations referred to in subparagraph .1, except that the technical group may decide not to proceed with the evaluations and considerations described in subparagraphs .1.2 to .1.7 if it determines after the evaluation in subparagraph .1.1 that the proposal does not warrant further consideration.

.4 The technical group's report shall include, *inter alia*, a recommendation on whether international controls pursuant to this Convention are warranted on the Hazardous Material in question, on the suitability of the specific control measures suggested in the comprehensive proposal, or on other control measures which it believes to be more suitable.

5 The Committee shall decide whether to approve any proposal to amend Appendix 1 or Appendix 2, and any modifications thereto, if appropriate, taking into account the technical group's report. Any proposed amendment shall specify the application of the amendment for ships certified in accordance with this Convention before the entry into force of the amendment. If the report finds that the Hazardous Material in question is likely, in the context of this Convention, to lead to significant adverse effects on human health or the environment, lack of full scientific certainty shall not be used as a reason to prevent a decision from being taken to list a Hazardous Material in Appendix 1 or Appendix 2. A decision not to approve the proposal shall not preclude future submission of a new proposal with respect to a particular Hazardous Material if new information comes to light.

Regulation 7 – Technical Groups

1 The Committee may establish one or more technical groups pursuant to regulation 6 as needed. The technical group may comprise representatives of the Parties, Members of the Organization, the United Nations and its Specialized Agencies, intergovernmental organizations having agreements with the Organization, and non-governmental organizations in consultative status with the Organization, which should preferably include representatives of institutions and laboratories with expertise in environmental fate and effects of substances, toxicological effects, marine biology, human health, economic analysis, risk management, shipbuilding, international shipping, occupational health and safety or other fields of expertise necessary to objectively review the technical merits of a proposal.

2 The Committee shall decide on the terms of reference, organization, participation and operation of the technical groups. Such terms shall provide for protection of any confidential information that may be submitted. Technical groups may hold such meetings as required, but shall endeavour to conduct their work through written or electronic correspondence or other media as appropriate.

3 Only the representatives of Parties may participate in formulating any recommendation to the Committee pursuant to regulation 6. A technical group shall endeavour to achieve unanimity among the representatives of the Parties. If unanimity is not possible, the technical group shall communicate any minority views of such representatives.

Part B – Preparation for Ship Recycling

Regulation 8 – General requirements

Ships destined to be recycled shall:

.1 only be recycled at Ship Recycling Facilities that are:

.1 authorized in accordance with this Convention; and

.2 fully authorized to undertake all the ship recycling which the Ship Recycling Plan specifies to be conducted by the identified Ship Recycling Facility(ies);

.2 conduct operations in the period prior to entering the Ship Recycling Facility in order to minimize the amount of cargo residues, remaining fuel oil, and wastes remaining on board;

.3 in the case of a tanker, arrive at the Ship Recycling Facility with cargo tanks and pump room(s) in a condition that is ready for certification as Safe-for-entry, or Safe-for-hot work, or both, according to national laws, regulations and policies of the Party under whose jurisdiction the Ship Recycling Facility operates;

.4 provide to the Ship Recycling Facility all available information relating to the ship for the development of the Ship Recycling Plan required by regulation 9;

.5 complete the Inventory required by regulation 5; and

.6 be certified as ready for recycling by the Administration or organization recognized by it, prior to any recycling activity taking place.

Regulation 9 – Ship Recycling Plan

A ship-specific Ship Recycling Plan shall be developed by the Ship Recycling Facility(ies) prior to any recycling of a ship, taking into account the guidelines developed by the Organization. The Ship Recycling Plan shall:

.1 be developed taking into account information provided by the shipowner;

.2 be developed in the language accepted by the Party authorizing the Ship Recycling Facility, and if the language used is not English, French or Spanish, the Ship Recycling Plan shall be translated into one of these languages, except where the Administration is satisfied that this is not necessary;

.3 include information concerning *inter alia*, the establishment, maintenance, and monitoring of Safe-for-entry and Safe-for-hot work conditions and how the type and amount of materials including those identified in the Inventory of Hazardous Materials will be managed;

.4 in accordance with the declaration deposited pursuant to Article 16.6, be either explicitly or tacitly approved by the Competent Authority authorizing the Ship Recycling Facility. The Competent Authority shall send written acknowledgement of receipt of the Ship Recycling Plan to the Ship Recycling Facility, Ship Owner and Administration within three (3) working days of its receipt in accordance with regulation 24. Thereafter:

.1 where a Party requires explicit approval of the Ship Recycling Plan, the Competent Authority shall send written notification of its decision to approve or deny the Ship Recycling Plan to the Ship Recycling Facility, Ship Owner and Administration; and

.2 where a Party requires tacit approval of the Ship Recycling Plan, the acknowledgment of receipt shall specify the end date of a 14-day review period. The Competent Authority shall notify any written objection to the Ship Recycling Plan to the Ship Recycling Facility, Ship Owner and Administration within this 14-day review period. Where no such written objection has been notified, the Ship Recycling Plan shall be deemed to be approved.

.5 once approved in accordance with paragraph .4, be made available for inspection by the Administration, or any nominated surveyors or organization recognized by it; and

.6 where more than one Ship Recycling Facility is used, identify the Ship Recycling Facilities to be used and specify the recycling activities and the order in which they occur at each authorized Ship Recycling Facility.

Part C – Surveys and certification

Regulation 10 – Surveys

1 Ships to which this Convention applies shall be subject to the surveys specified below:

.1 an initial survey before the ship is put in service, or before the International Certificate on Inventory of Hazardous Materials is issued. This survey shall verify that Part I of the Inventory required by regulation 5 is in accordance with the requirements of this Convention;

.2 a renewal survey at intervals specified by the Administration, but not exceeding five years. This survey shall verify that Part I of the Inventory of Hazardous Materials required by regulation 5 complies with the requirements of this Convention;

.3 an additional survey, either general or partial, according to the circumstances, may be made at the request of the shipowner after a change, replacement, or significant repair of the structure, equipment, systems, fittings, arrangements and material. The survey shall be such as to ensure that any such change, replacement, or significant repair has been made in the way that the ship continues to comply with the requirements of this Convention, and that Part I of the Inventory is amended as necessary; and

.4 a final survey prior to the ship being taken out of service and before the recycling of the ship has started. This survey shall verify:

.1 that the Inventory of Hazardous Materials as required by regulation 5.4 is in accordance with the requirements of this Convention taking into account the guidelines developed by the Organization;

.2 that the Ship Recycling Plan, as required by regulation 9, properly reflects the information contained in the Inventory of Hazardous Materials as required by regulation 5.4 and contains information concerning the establishment, maintenance and monitoring of Safe-for-entry and Safe-for-hot work conditions; and

.3 that the Ship Recycling Facility(ies) where the ship is to be recycled holds a valid authorization in accordance with this Convention.

2 Surveys of ships for the purpose of enforcement of the provisions of this Convention shall be carried out by officers of the Administration, taking into account the guidelines developed by the Organization. The Administration may, however, entrust the surveys either to surveyors nominated for the purpose or to organizations recognized by it.

3 An Administration nominating surveyors or recognizing organizations to conduct surveys, as described in paragraph 2 shall, as a minimum, empower such nominated surveyors or recognized organizations to:

.1 require a ship that they survey to comply with the provisions of this Convention; and

.2 carry out surveys and inspections if requested by the appropriate authorities of a port State that is a Party.

4 In every case, the Administration concerned shall be responsible to ensure the completeness and efficiency of the survey and shall undertake to ensure the necessary arrangements to satisfy this obligation.

5 The initial and renewal surveys should be harmonized with the surveys required by other applicable statutory instruments of the Organization.

Regulation 11 – Issuance and endorsement of certificates

1 An International Certificate on Inventory of Hazardous Materials shall be issued either by the Administration or by any person or organization authorized by it after successful completion of an initial or renewal survey conducted in accordance with regulation 10, to any ships to which regulation 10 applies, except for existing ships for which both an initial survey and a final survey are conducted at the same time, taking into account the guidelines developed by the Organization.

2 The International Certificate on Inventory of Hazardous Materials issued under paragraph 1, at the request of the shipowner, shall be endorsed either by the Administration or by any person or organization authorized by it after successful completion of an additional survey conducted in accordance with regulation 10.

3 Notwithstanding regulation 14.2 and the requirements of regulation 10.1.2, when the renewal survey is completed within three months before the expiry date of the existing certificate, the new certificate shall be valid from the date of completion of the renewal survey to a date not exceeding five years from the date of expiry of the existing certificate.

4 When the renewal survey is completed after the expiry date of the existing certificate, the new certificate shall be valid from the date of completion of the renewal survey to a date not exceeding five years from the date of expiry of the existing certificate.

5 When the renewal survey is completed more than three months before the expiry date of the existing certificate, the new certificate shall be valid from the date of completion of the renewal survey to a date not exceeding five years from the date of completion of the renewal survey.

6 If a certificate is issued for a period of less than five years, the Administration may extend the validity of the certificate beyond the expiry date to the maximum period specified in regulation 10.1.2.

7 If a renewal survey has been completed and a new certificate cannot be issued or placed on board the ship before the expiry date of the existing certificate, the person or organization authorized by the Administration may endorse the existing certificate and such a certificate shall be accepted as valid for a further period which shall not exceed five months from the expiry date.

8 If a ship at the time when a certificate expires is not in a port in which it is to be surveyed, the Administration may extend the period of validity of the certificate but this extension shall be granted only for the purpose of allowing the ship to complete its voyage to the port in which it is to be surveyed and then only in cases where it appears proper and reasonable to do so. No certificate shall be extended for a period longer than three months, and a ship to which an extension is granted shall not, on its arrival in the port in which it is to be surveyed, be entitled by virtue of such extension to leave that port without having a new certificate. When the renewal survey is completed, the new certificate shall be valid to a date not exceeding five years from the date of expiry of the existing certificate before the extension was granted.

9 A certificate issued to a ship engaged on short voyages which has not been extended under the foregoing provisions of this regulation may be extended by the Administration for a period of grace of up to one month from the date of expiry stated on it. When the renewal survey

is completed, the new certificate shall be valid to a date not exceeding five years from the date of expiry of the existing certificate before the extension was granted.

10 In special circumstances, as determined by the Administration, a new certificate need not be dated from the date of expiry of the existing certificate as required by paragraph 4, 8 or 9 of this regulation. In these special circumstances, the new certificate shall be valid to a date not exceeding five years from the date of completion of the renewal survey.

11 An International Ready for Recycling Certificate shall be issued either by the Administration or by any person or organization authorized by it, after successful completion of a final survey in accordance with the provisions of regulation 10, to any ships to which regulation 10 applies, taking into account the authorization of the Ship Recycling Facility and the guidelines developed by the Organization.

12 A certificate issued under the authority of a Party shall be accepted by the other Parties and regarded for all purposes covered by this Convention as having the same validity as a certificate issued by them. Certificates shall be issued or endorsed either by the Administration or by any person or organization duly authorized by it. In every case, the Administration assumes full responsibility for the certificate.

Regulation 12 – Issuance or endorsement of a certificate by another Party

1 At the request of the Administration, another Party may cause a ship to be surveyed and, if satisfied that the provisions of this Convention are complied with, shall issue or authorize the issuance of a certificate to the ship, and where appropriate, endorse or authorize the endorsement of that certificate on the ship, in accordance with this Annex.

2 A copy of the certificate and a copy of the survey report shall be transmitted as soon as possible to the requesting Administration.

3 A certificate so issued shall contain a statement to the effect that it has been issued at the request of the Administration and it shall have the same force and receive the same recognition as a certificate issued by the Administration.

4 No certificate shall be issued to a ship entitled to fly the flag of a State which is not a Party.

Regulation 13 – Form of the certificates

The certificates shall be drawn up in an official language of the issuing Party, in the form set forth in Appendices 3 and 4. If the language used is not English, French or Spanish, the text shall include a translation into one of these languages. The Administration may, however, issue the International Certificate on Inventory of Hazardous Materials drawn up only in an official language of the issuing Party to ships not engaged in voyages to ports or offshore terminals under the jurisdiction of other Parties to this Convention and the International Ready for Recycling Certificate drawn up only in an official language of the issuing Party to ships recycled in Ship Recycling Facilities under the jurisdiction of the issuing Party.

Regulation 14 – Duration and validity of the certificates

1 An International Certificate on Inventory of Hazardous Materials issued under regulation 11 or 12 shall cease to be valid in any of the following cases:

.1 if the condition of the ship does not correspond substantially with the particulars of the certificate, including where Part I of the Inventory of Hazardous Materials is not properly maintained and updated, reflecting changes in ship structure and equipment, in accordance with the guidelines developed by the Organization;

.2 upon transfer of the ship to the flag of another State. A new certificate shall only be issued when the Party issuing the new certificate is fully satisfied that the ship is in compliance with the requirements of regulation 10. In the case of a transfer between Parties, if requested within three months after the transfer has taken place, the Party whose flag the ship was formerly entitled to fly shall, as soon as possible, transmit to the Administration copies of the certificates carried by the ship before the transfer and, if available, copies of the relevant survey reports;

.3 if the renewal survey is not completed within the periods specified under regulations 10.1 and 11; or

.4 if the certificate is not endorsed in accordance with regulation 11 or 12.

2 An International Certificate on Inventory of Hazardous Materials shall be issued for a period specified by the Administration, which shall not exceed five years.

3 An International Ready for Recycling Certificate shall be issued for a period specified by the Administration that shall not exceed three months.

4 An International Ready for Recycling Certificate issued under regulation 11 or 12 shall cease to be valid if the condition of the ship does not correspond substantially with the particulars of the certificate.

5 The International Ready for Recycling Certificate may be extended by the Administration or by any person or organization authorized by it for a single point to point voyage to the Ship Recycling Facility.

CHAPTER 3 – REQUIREMENTS FOR SHIP RECYCLING FACILITIES

Regulation 15 – Controls on Ship Recycling Facilities

1 Each Party shall establish legislation, regulations, and standards that are necessary to ensure that Ship Recycling Facilities are designed, constructed, and operated in a safe and environmentally sound manner in accordance with the regulations of this Convention.

2 Each Party shall establish a mechanism for authorizing Ship Recycling Facilities with appropriate conditions to ensure that such Ship Recycling Facilities meet the requirements of this Convention.

3 Each Party shall establish a mechanism for ensuring that Ship Recycling Facilities comply with the requirements of this chapter including the establishment and effective use of inspection, monitoring and enforcement provisions, including powers of entry and sampling. Such a mechanism may include an audit scheme to be carried out by the Competent Authority(ies) or an organization recognized by the Party, taking into account guidelines developed by the Organization, and the results of these audits should be communicated to the Organization.

4 Each Party shall designate one or more Competent Authorities and the single contact point to be used by the Organization, Parties to this Convention and other interested entities, for matters related to Ship Recycling Facilities operating within the jurisdiction of that Party.

Regulation 16 – Authorization of Ship Recycling Facilities

1 Ship Recycling Facilities which recycle ships to which this Convention applies, or ships treated similarly pursuant to Article 3.4, shall be authorized by a Party taking into account the guidelines developed by the Organization.

2 The authorization shall be carried out by the Competent Authority(ies) and shall include verification of documentation required by this Convention and a site inspection. The Competent Authority(ies) may however entrust the authorization of Ship Recycling Facilities to organizations recognized by it.

3 The Party shall notify the Organization of the specific responsibilities and conditions of the authority delegated to the recognized organizations, for circulation to Parties. In every case, the Competent Authority(ies) retains full responsibility for the authorization issued.

4 The authorization shall be drawn up in the form set forth in Appendix 5. If the language used is not English, French or Spanish, the text shall include a translation into one of these languages.

5 The authorization shall be valid for a period specified by the Party but not exceeding five years. The Party shall identify the terms for which the authorization will be issued, withdrawn, suspended, amended and renewed, and communicate these terms to the Ship Recycling Facilities. If a Ship Recycling Facility refuses inspection by the Competent Authority(ies) or the recognized organization operating on its/their behalf, the authorization shall be suspended or withdrawn.

6 If incidents or actions taken at the Ship Recycling Facility have the effect that the conditions for the authorization are no longer fulfilled, the Ship Recycling Facility shall inform the Competent Authority(ies). The Competent Authority(ies) may accordingly decide to suspend or withdraw the authorization, or require corrective actions by the Ship Recycling Facility.

Regulation 17 – General requirements

1 Ship Recycling Facilities authorized by a Party shall establish management systems, procedures and techniques which do not pose health risks to the workers concerned or to the population in the vicinity of the Ship Recycling Facility and which will prevent, reduce, minimize and to the extent practicable eliminate adverse effects on the environment caused by Ship Recycling, taking into account guidelines developed by the Organization.

2 Ship Recycling Facilities authorized by a Party shall, for ships to which this Convention applies, or ships treated similarly pursuant to Article 3.4:

.1 only accept ships that:

.1 comply with this Convention; or

.2 meet the requirements of this Convention;

.2 only accept ships which they are authorized to recycle; and

.3 have the documentation of its authorization available if such documentation is requested by a shipowner that is considering recycling a ship at that Ship Recycling Facility.

Regulation 18 – Ship Recycling Facility Plan

Ship Recycling Facilities authorized by a Party shall prepare a Ship Recycling Facility Plan. The Plan shall be adopted by the board or the appropriate governing body of the Recycling Company, and shall include:

.1 a policy ensuring workers' safety and the protection of human health and the environment, including the establishment of objectives that lead to the minimization and elimination to the extent practicable of the adverse effects on human health and the environment caused by Ship Recycling;

.2 a system for ensuring implementation of the requirements set out in this Convention, the achievement of the goals set out in the policy of the Recycling Company, and the continuous improvement of the procedures and standards used in the Ship Recycling operations;

.3 identification of roles and responsibilitilies for employers and workers when conducting Ship Recycling operations;

.4 a programme for providing appropriate information and training of workers for the safe and environmentally sound operation of the Ship Recycling Facility;

.5 an emergency preparedness and response plan;

.6 a system for monitoring the performance of Ship Recycling;

.7 a record-keeping system showing how Ship Recycling is carried out;

.8 a system for reporting discharges, emissions, incidents and accidents causing damage, or with the potential of causing damage, to workers' safety, human health and the environment; and

.9 a system for reporting occupational diseases, accidents, injuries and other adverse effects on workers' safety and human health,

taking into account guidelines developed by the Organization.

Regulation 19 – Prevention of adverse effects to human health and the environment

Ship Recycling Facilities authorized by a Party shall establish and utilize procedures to:

.1 prevent explosions, fires, and other unsafe conditions by ensuring that Safe-for-hot work conditions and procedures are established, maintained and monitored throughout Ship Recycling;

.2 prevent harm from dangerous atmospheres and other unsafe conditions by ensuring that Safe-for-entry conditions and procedures are established, maintained, and monitored in ship spaces, including confined spaces and enclosed spaces, throughout Ship Recycling;

.3 prevent other accidents, occupational diseases and injuries or other adverse effects on human health and the environment; and

.4 prevent spills or emissions throughout Ship Recycling which may cause harm to human health and/or the environment,

taking into account guidelines developed by the Organization.

Regulation 20 – Safe and environmentally sound management of Hazardous Materials

1 Ship Recycling Facilities authorized by a Party shall ensure safe and environmentally sound removal of any Hazardous Material contained in a ship certified in accordance with regulation 11 or 12. The person(s) in charge of the recycling operations and the workers shall be familiar with the requirements of this Convention relevant to their tasks and, in particular, actively use the Inventory of Hazardous Materials and the Ship Recycling Plan, prior to and during the removal of Hazardous Materials.

2 Ship Recycling Facilities authorized by a Party shall ensure that all Hazardous Materials detailed in the Inventory are identified, labelled, packaged and removed to the maximum extent possible prior to cutting by properly trained and equiped workers, taking into account the guidelines developed by the Organization, in particular:

.1 hazardous liquids, residues and sediments;

.2 substances or objects containing heavy metals such as lead, mercury, cadmium and hexavalent chromium;

.3 paints and coatings that are highly flammable and/or lead to toxic releases;

.4 asbestos and materials containing asbestos;

.5 PCB and materials containing PCBs, ensuring that heat inducing equipment is avoided during such operations;

.6 CFCs and halons; and

.7 other Hazardous Materials not listed above and that are not a part of the ship structure.

3 Ship Recycling Facilities authorized by a Party shall provide for and ensure safe and environmentally sound management of all Hazardous Materials and wastes removed from the ship recycled at that Ship Recycling Facility. Waste management and disposal sites shall be identified to provide for the further safe and environmentally sound management of materials.

4 All wastes generated from the recycling activity shall be kept separate from recyclable materials and equipment, labelled, stored in appropriate conditions that do not pose a risk to the workers, human health or the environment and only transferred to a waste management facility authorized to deal with their treatment and disposal in a safe and environmentally sound manner.

Regulation 21 – Emergency preparedness and response

Ship Recycling Facilities authorized by a Party shall establish and maintain an emergency preparedness and response plan. The plan shall be made having regard to the location and environment of the Ship Recycling Facility, and shall take into account the size and nature of activities associated with each Ship Recycling operation. The plan shall furthermore:

.1 ensure that the necessary equipment and procedures to be followed in the case of an emergency are in place, and that drills are conducted on a regular basis;

.2 ensure that the necessary information, internal communication and coordination are provided to protect all people and the environment in the event of an emergency at the Ship Recycling Facility;

.3 provide for communication with, and information to, the relevant Competent Authority(ies), the neighbourhood and emergency response services;

.4 provide for first-aid and medical assistance, fire-fighting and evacuation of all people at the Ship Recycling Facility, pollution prevention; and

.5 provide for relevant information and training to all workers of the Ship Recycling Facility, at all levels and according to their competence, including regular exercises in emergency prevention, preparedness and response procedures.

Regulation 22 – Worker safety and training

1 Ship Recycling Facilities authorized by a Party shall provide for worker safety by measures including:

.1 ensuring the availability, maintenance and use of personal protective equipment and clothing needed for all Ship Recycling operations;

.2 ensuring that training programmes are provided to enable workers to safely undertake all Ship Recycling operations they are tasked to do; and

.3 ensuring that all workers at the Ship Recycling Facility have been provided with appropriate training and familiarization prior to performing any Ship Recycling operation.

2 Ship Recycling Facilities authorized by a Party shall provide and ensure the use of personal protective equipment for operations requiring such use, including:

.1 head protection;

.2 face and eye protection;

.3 hand and foot protection;

.4 respiratory protective equipment;

.5 hearing protection;

.6 protectors against radioactive contamination;

.7 protection from falls; and

.8 appropriate clothing.

3 Ship Recycling Facilities authorized by a Party may co-operate in providing for training of workers. Taking into account the guidelines developed by the Organization, the training programmes set forth in paragraph 1.2 of this regulation shall:

.1 cover all workers including contractor personnel and employees in the Ship Recycling Facility;

.2 be conducted by Competent persons;

.3 provide for initial and refresher training at appropriate intervals;

.4 include participants' evaluation of their comprehension and retention of the training;

.5 be reviewed periodically and modified as necessary; and

.6 be documented.

Regulation 23 – Reporting on incidents, accidents, occupational diseases and chronic effects

1 Ship Recycling Facilities authorized by a Party shall report to the Competent Authority(ies) any incident, accident, occupational diseases, or chronic effects causing, or with the potential of causing, risks to workers safety, human health and the environment.

2 Reports shall contain a description of the incident, accident, occupational disease, or chronic effect, its cause, the response action taken and the consequences and corrective actions to be taken.

CHAPTER 4 – REPORTING REQUIREMENTS

Regulation 24 – Initial notification and reporting requirements

1 A shipowner shall notify the Administration in due time and in writing of the intention to recycle a ship in order to enable the Administration to prepare for the survey and certification required by this Convention.

2 A Ship Recycling Facility when preparing to receive a ship for recycling shall notify in due time and in writing its Competent Authority(ies) of the intent. The notification shall include at least the following ship details:

.1 name of the State whose flag the ship is entitled to fly;

.2 date on which the ship was registered with that State;

.3 ship's identification number (IMO number);

.4 hull number on new-building delivery;

.5 name and type of the ship;

.6 port at which the ship is registered;

.7 name and address of the Shipowner as well as the IMO registered owner identification number;

.8 name and address of the company as well as the IMO company identification number;

.9 name of all classification society(ies) with which the ship is classed;

.10 ship's main particulars (Length overall (LOA), Breadth (Moulded), Depth (Moulded), Lightweight, Gross and Net tonnage, and engine type and rating);

.11 Inventory of Hazardous Materials; and

.12 draft ship recycling plan for approval pursuant to regulation 9.

3 When the ship destined to be recycled has acquired the International Ready for Recycling Certificate, the Ship Recycling Facility shall report to its Competent Authority(ies) the planned start of the Ship Recycling. The report shall be in accordance with the reporting format in Appendix 6, and shall at least include a copy of the International Ready for Recycling Certificate. Recycling of the ship shall not start prior to the submission of the report.

Regulation 25 – Reporting upon completion

When the partial or complete recycling of a ship is completed in accordance with the requirements of this Convention, a Statement of Completion shall be issued by the Ship Recycling Facility and reported to its Competent Authority(ies). This report must be compiled as shown in appendix 7. The Competent Authority(ies) shall send a copy of the Statement to the Administration which issued the International Ready for Recycling Certificate for the ship. The Statement shall be issued within 14 days of the date of partial or completed Ship Recycling in accordance with the Ship Recycling Plan and shall include a report on incidents and accidents damaging human health and/or the environment, if any.

APPENDIX 1

CONTROLS OF HAZARDOUS MATERIALS

Hazardous Material	Definitions	Control measures
Asbestos	Materials containing asbestos	For all ships, new installation of materials which contain asbestos shall be prohibited.
Ozone-depleting substances	Ozone-depleting substances means controlled substances defined in paragraph 4 of article 1 of the Montreal Protocol on Substances that Deplete the Ozone Layer, 1987, listed in Annexes A,B,C or E to the said Protocol in force at the time of application or interpretation of this Annex. Ozone-depleting substances that may be found on board ship include, but are not limited to: Halon 1211 Bromochlorodifluoromethane Halon 1301 Bromotrifluoromethane Halon 2402 1,2-Dibromo-1,1,2,2-tetrafluoroethane (also known as Halon 114B2) CFC-11 Trichlorofluoromethane CFC-12 Dichlorodifluoromethane CFC-113 1,1,2-Trichloro-1,2,2-trifluoroethane CFC-114 1,2-Dichloro-1,1,2,2-tetrafluoroethane CFC-115 Chloropentafluoroethane	New installations which contain ozone-depleting substances shall be prohibited on all ships, except that new installations containing hydrochlorofluorocarbons (HCFCs) are permitted until 1 January 2020.
Polychlorinated biphenyls (PCB)	"Polychlorinated biphenyls" means aromatic compounds formed in such a manner that the hydrogen atoms on the biphenyl molecule (two benzene rings bonded together by a single carbon-carbon bond) may be replaced by up to ten chlorine atoms	For all ships, new installation of materials which contain Polychlorinated biphenyls shall be prohibited.
Anti-fouling compounds and systems	Anti-fouling compounds and systems regulated under Annex I to the International Convention on the Control of Harmful Anti-fouling Systems on Ships, 2001 (AFS Convention) in force at the time of application or interpretation of this Annex.	1. No ship may apply anti-fouling systems containing organotin compounds as a biocide or any other anti-fouling system whose application or use is prohibited by the AFS Convention. 2. No new ships or new installations on ships shall apply or employ anti-fouling compounds or systems in a manner inconsistent with the AFS Convention.

APPENDIX 2

MINIMUM LIST OF ITEMS FOR THE INVENTORY OF HAZARDOUS MATERIALS

Any Hazardous Materials listed in Appendix 1
Cadmium and Cadmium Compounds
Hexavalent Chromium and Hexavalent Chromium Compounds
Lead and Lead Compounds
Mercury and Mercury Compounds
Polybrominated Biphenyl (PBBs)
Polybrominated Diphenyl Ethers (PBDEs)
Polychlorinated Naphthalenes (more than 3 chlorine atoms)
Radioactive Substances
Certain Shortchain Chlorinated Paraffins (Alkanes, C10-C13, chloro)

APPENDIX 3

FORM OF THE INTERNATIONAL CERTIFICATE ON INVENTORY OF HAZARDOUS MATERIALS

INTERNATIONAL CERTIFICATE ON INVENTORY OF HAZARDOUS MATERIALS

(Note: This certificate shall be supplemented by Part I of the Inventory of Hazardous Materials)

(Official seal) *(State)*

Issued under the provisions of the Hong Kong International Convention for the Safe and Environmentally Sound Recycling of Ships, 2009 (hereinafter referred to as "the Convention") under the authority of the Government of

....

(*Full designation of the country*)

by ....

(*Full designation of the person or organization authorized under the provisions of the Convention*)

Particulars of the Ship

Name of Ship	
Distinctive number or letters	
Port of Registry	
Gross tonnage	
IMO number	
Name and address of shipowner	
IMO registered owner identification number	
IMO company identification number	
Date of Construction	

Particulars of Part I of the Inventory of Hazardous Materials

Part I of the Inventory of Hazardous Materials identification/verification number: ....

Note: Part I of the Inventory of Hazardous Materials, as required by regulation 5 of the Annex to the Convention, is an essential part of the International Certificate on Inventory of Hazardous Materials and must always accompany the International Certificate on Inventory of Hazardous Materials. Part I of the Inventory of Hazardous Materials should be compiled on the basis of the standard format shown in the guidelines developed by the Organization.

THIS IS TO CERTIFY:

1. that the ship has been surveyed in accordance with regulation 10 of the Annex to the Convention; and
2. that the survey shows that Part I of the Inventory of Hazardous Materials fully complies with the applicable requirements of the Convention.

Completion date of survey on which this certificate is based: .... (dd/mm/yyyy)

This certificate is valid until .... (dd/mm/yyyy)

Issued at ....

(*Place of issue of certificate*)

(dd/mm/yyyy) ....

(*Date of issue*) (*Signature of duly authorized official issuing the certificate*)

(*Seal or stamp of the authority, as appropriate*)

ENDORSEMENT TO EXTEND THE CERTIFICATE IF VALID FOR LESS THAN FIVE YEARS WHERE REGULATION 11.6 APPLIES*

The ship complies with the relevant provisions of the Convention, and this certificate shall, in accordance with regulation 11.6 of the Annex to the Convention, be accepted as valid until (dd/mm/yyyy): ....

Signed: ....

(Signature of duly authorized official)

Place:

Date: (dd/mm/yyyy)

(Seal or stamp of the authority, as appropriate)

ENDORSEMENT WHERE THE RENEWAL SURVEY HAS BEEN COMPLETED AND REGULATION 11.7 APPLIES*

The ship complies with the relevant provisions of the Convention, and this certificate shall, in accordance with regulation 11.7 of the Annex to the Convention, be accepted as valid until (dd/mm/yyyy): ....

Signed: ....

(Signature of duly authorized official)

Place: ....

Date: (dd/mm/yyyy) ....

(Seal or stamp of the authority, as appropriate)

* This page of the endorsement at survey shall be reproduced and added to the certificate as considered necessary by the Administration.

ENDORSEMENT TO EXTEND THE VALIDITY OF THE CERTIFICATE UNTIL REACHING THE PORT OF SURVEY OR FOR A PERIOD OF GRACE WHERE REGULATION 11.8 OR 11.9 APPLIES*

This certificate shall, in accordance with regulation 11.8 or 11.9** of the Annex to the Convention, be accepted as valid until (dd/mm/yyyy): ....

Signed: ....

(Signature of duly authorized official)

Place: ....

Date: (dd/mm/yyyy) ....

(Seal or stamp of the authority, as appropriate)

ENDORSEMENT FOR ADDITIONAL SURVEY*

At an additional survey in accordance with regulation 10 of the Annex to the Convention, the ship was found to comply with the relevant provisions of the Convention.

Signed: ....

(Signature of duly authorized official)

Place: ....

Date: (dd/mm/yyyy) ....

(Seal or stamp of the authority, as appropriate)

---

* This page of the endorsement at survey shall be reproduced and added to the certificate as considered necessary by the Administration.

** Delete as appropriate.

APPENDIX 4

FORM OF THE INTERNATIONAL READY FOR RECYCLING CERTIFICATE

INTERNATIONAL READY FOR RECYCLING CERTIFICATE

(Note: This certificate shall be supplemented by the Inventory of Hazardous Materials and the Ship Recycling Plan)

(Official seal) *(State)*

Issued under the provisions of the Hong Kong International Convention for the Safe and Environmentally Sound Recycling of Ships, 2009 (hereinafter referred to as "the Convention") under the authority of the Government of

....
(*Full designation of the country*)

by ....
(*Full designation of the person or organization authorized under the provisions of the Convention*)

Particulars of the Ship

Name of Ship	
Distinctive number or letters	
Port of Registry	
Gross tonnage	
IMO number	
Name and address of shipowner	
IMO registered owner identification number	
IMO company identification number	
Date of Construction	

Particulars of the Ship Recycling Facility(ies)

Name of Ship Recycling Facility	
Distinctive Recycling Company identity number*	
Full address	
Date of expiry of DASR	

* This number is based on the Document of Authorization to conduct Ship Recycling (DASR).

Particulars of the Inventory of Hazardous Materials

Inventory of Hazardous Materials identification/verification number: ....

Note: The Inventory of Hazardous Materials, as required by regulation 5 of the Annex to the Convention, is an essential part of the International Ready for Recycling Certificate and must always accompany the International Ready for Recycling Certificate. The Inventory of Hazardous Materials should be compiled on the basis of the standard format shown in the guidelines developed by the Organization.

Particulars of the Ship Recycling Plan

Ship Recycling Plan identification/verification number: ....

Note: The Ship Recycling Plan, as required by regulation 9 of the Annex to the Convention, is an essential part of the International Ready for Recycling Certificate and must always accompany the International Ready for Recycling Certificate.

THIS IS TO CERTIFY:

1 that the ship has been surveyed in accordance with regulation 10 of the Annex to the Convention;

2 that the ship has a valid Inventory of Hazardous Materials in accordance with regulation 5 of the Annex to the Convention;

3 that the Ship Recycling Plan, as required by regulation 9, properly reflects the information contained in the Inventory of Hazardous Materials as required by regulation 5.4 and contains information concerning the establishment, maintenance and monitoring of Safe-for-entry and Safe-for-hot work conditions; and

4 that the Ship Recycling Facility(ies) where this ship is to be recycled holds a valid authorization in accordance with the Convention.

This certificate is valid until (dd/mm/yyyy) ....
(Date)

Issued at ....
(*Place of issue of certificate*)

(dd/mm/yyyy) ....
(*Date of issue*) (*Signature of duly authorized official issuing the certificate*)

(*Seal or stamp of the authority, as appropriate*)

ENDORSEMENT TO EXTEND THE VALIDITY OF THE CERTIFICATE UNTIL REACHING THE PORT OF THE SHIP RECYCLING FACILITY FOR A PERIOD OF GRACE WHERE REGULATION 14.5 APPLIES*

This certificate shall, in accordance with regulation 14.5 of the Annex to the Convention, be accepted as valid for a single point to point voyage

from the port of: ....

to the port of: ....

Signed: ....

(Signature of duly authorized official)

Place: ....

Date: (dd/mm/yyyy) ....

(Seal or stamp of the authority, as appropriate)

* This page of the endorsement shall be reproduced and added to the certificate as considered necessary by the Administration.

APPENDIX 5

FORM OF THE AUTHORIZATION OF SHIP RECYCLING FACILITIES

Document of Authorization to conduct Ship Recycling (DASR) in accordance with the requirements of the Hong Kong International Convention for the Safe and Environmentally Sound Recycling of Ships, 2009

Issued under the provision of the Hong Kong International Convention for the Safe and Environmentally Sound Recycling of Ships, 2009 (hereinafter referred to as "the Convention") under the authority of the Government of:

....

(Full designation of the country)

by...

(Full designation of the Competent Authority under the Convention)

Name of Ship Recycling Facility	
Distinctive Recycling Company identity No.	
Full address of Ship Recycling Facility	
Primary contact person	
Phone number	
E-mail address	
Name, address, and contact information of ownership company	
Working language(s)	

This is to verify that the Ship Recycling Facility has implemented management systems, procedures and techniques in accordance with Chapters 3 and 4 to the Annex to the Convention.

This authorization is valid until .... and is subject to the limitations identified in the attached supplement.

This authorization is subject to amendment, suspension, withdrawal, or periodic renewal in accordance with regulation 16 of the Annex to the Convention.

Issued at ....

(*Place of issue of the authorization*)

(dd/mm/yyyy) ....

(*Date of issue*) (*Signature of duly authorized official issuing the authorization*)

....

(Typed name and title of duly authorized official issuing the authorization)

(*Seal or stamp of the authority, as appropriate*)

SUPPLEMENT TO:

Document of Authorization to undertake Ship Recycling (DASR) in accordance with the Hong Kong International Convention for the Safe and Environmentally Sound Recycling of Ships, 2009

Notes:
1 This record shall be permanently attached to the DASR. The DASR shall be available at the Ship Recycling Facility at all times.
2 All procedures, plans and other documents produced by the Ship Recycling Facility and required under the terms to which the DASR has been issued shall be available in the working language of the Ship Recycling Facility and in either English, French or Spanish.
3 The authorization is subject to the limitations defined by this supplement.

1 GENERAL TERMS

1.1 Requirements of the Convention

The Ship Recycling Facility meets the requirements that it be designed, constructed, and operated in a safe and environmentally sound manner in accordance with the Convention, including meeting the relevant requirements of:

Regulation 16 – Authorization of Ship Recycling Facilities
Regulation 17 – General requirements
Regulation 18 – Ship Recycling Facility Plan
Regulation 19 – Prevention of adverse effects to human health and the environment
Regulation 20 – Safe and environmentally sound management of Hazardous Materials
Regulation 21 – Emergency preparedness and response
Regulation 22 – Worker safety and training
Regulation 23 – Reporting on incidents, accidents, occupational diseases and chronic effects
Regulation 24 – Initial notification and reporting requirements
Regulation 25 – Reporting upon completion

These requirements are imposed on the Ship Recycling Facility by way of

....

(Identify the permit, licence, authorization, legal standards, or other mechanism that applies)

Ship Recycling Facility Plan identification/verification number: ....

1.2 Acceptance of ships

For ships to which the Convention applies and ships treated similarly pursuant to Article 3.4 of the Convention, the Ship Recycling Facility can only accept a ship for recycling in accordance with regulation 17 of the Annex to the Convention.

1.3 Safe-for-hot work and Safe-for-entry conditions

The Ship Recycling Facility is capable of establishing, maintaining and monitoring Safe-for-hot work and Safe-for-entry conditions throughout the Ship Recycling process.

1.4 Management of Hazardous Materials

The Ship Recycling Facility is designed, constructed, operated, and required to ensure that all Hazardous Materials' management shall be safe and environmentally sound in compliance with the Convention and with all relevant local or national regulations/requirements.

1.5 Map and location of Ship Recycling operations

A map of the boundary of the Ship Recycling Facility and the location of Ship Recycling operations within it, is attached.

2 CAPABILITY OF SHIP RECYCLING FACILITY

2.1 Size of ships

The Ship Recycling Facility is authorized to accept a ship for recycling subject to the following size limitations:

Maximum Size		Other Limitations
Length		
Breadth		
Lightweight		

2.2 Safe and Environmentally Sound Management of Hazardous Materials

The Ship Recycling Facility is authorized to accept a ship for recycling that contains Hazardous Materials as specified in the following table subject to the conditions noted below:

Hazardous Material(*4)	Management of Hazardous Materials			Authorization/Limitations
	Removal Y/N (*2)	Storage Y/N	Process (*1) Y/N (*3)	
Asbestos				
Ozone-depleting substances				
Polychlorinated biphenyls (PCB)				
Anti-fouling compounds and systems				
Cadmium and Cadmium Compounds				
Hexavalent Chromium and Hexavalent Chromium Compounds				
Lead and Lead Compounds				
Mercury and Mercury Compounds				
Polybrominated Biphenyl (PBBs)				
Polybrominated Diphenyl Ethers (PBDEs)				
Polychlorinated Naphthalenes (more than 3 chlorine atoms)				
Radioactive substances				
Certain Shortchain Chlorinated Paraffins (Alkanes, C10-C13, chloro)				
Hazardous liquids, residues and sediments				
Paints and coatings that are highly flammable and/or lead to toxic release				
Other Hazardous Materials not listed above and that are not a part of the ship structure (specify)				

Notes: [*]1 Process means the processing of Hazardous Materials in the Ship Recycling Facility, such as:

a. incineration of Hazardous Materials;
b. reclamation of Hazardous Materials; and
c. treatment of oily residues.

[*]2 If Yes (Y), indicate in the Ship Recycling Facility Plan the responsible personnel authorized to carry out the removal, with the certificate number or other relevant information.

[*]3 If No (N), describe in the Ship Recycling Plan where the Hazardous Materials are to be processed/disposed.

[*]4 These Hazardous Materials are specified in Appendices 1 and 2 and regulation 20 of the Convention.

APPENDIX 6

FORM OF REPORT OF PLANNED START OF SHIP RECYCLING

The ....
(Name of Ship Recycling Facility)

located at ....
(Full Ship Recycling Facility address)

Authorized in accordance with the requirements of the Hong Kong International Convention for the Safe and Environmentally Sound Recycling of Ships, 2009 (hereinafter referred to as "the Convention") to conduct Ship Recycling under the authority of the Government of:

....
(Full designation of country)

as indicated in the Document of Authorization to conduct Ship Recycling issued at

....
(Place of authorization)

by ....
(Full designation of the Competent Authority under the Convention)

on (dd/mm/yyyy) ....
(Date of issue)

Hereby reports that the Ship Recycling Facility is ready in every respect to start the recycling of the vessel ....
(IMO number)

The International Ready for Recycling Certificate issued under the provisions of the Convention under the authority of the Government of

....
(Full designation of country)

by ....
(Full designation of the person or organization authorized under the provisions of the Convention)

on (dd/mm/yyyy) ....
(Date of issue)

is enclosed.

Signed ....

APPENDIX 7

FORM OF THE STATEMENT OF COMPLETION OF SHIP RECYCLING

STATEMENT OF COMPLETION OF SHIP RECYCLING

This document is a statement of completion of Ship Recycling for

....
(*Name of the ship when it was received for recycling/at the point of deregistration*)

Particulars of the Ship as received for recycling

Distinctive number or letters	
Port of Registry	
Gross tonnage	
IMO number	
Name and address of shipowner	
IMO registered owner identification number	
IMO company identification number	
Date of Construction	

THIS CONFIRMS THAT:

The ship has been recycled in accordance with the Ship Recycling Plan as part of the Hong Kong International Convention for the Safe and Environmentally Sound Recycling of Ships, 2009 (hereinafter referred to as "the Convention") at

....
(Name and location of the authorized Ship Recycling Facility)

and the recycling of the ship as required by the Convention was completed on:

(dd/mm/yyyy) ....
(Date of completion)

Issued at ....
(*Place of issue of the Statement of Completion*)

(dd/mm/yyyy) ....
(*Date of issue*) (*Signature of the owner of the Ship Recycling Facility or a representative acting on behalf of the owner*)

Bibliography

International Materials

Conventions

Basel Convention on the Control of Transboundary Movements of Hazardous Wastes and their Disposal, 22 March 1989, 1673 U.N.T.S. 126, 28 I.L.M. 657 (entered into force 5 May 1992).

Convention for the Prevention of Marine Pollution from Land-Based Sources, 21 February 1974, 13 I.L.M. 352.

Convention on the Prevention of Marine Pollution by Dumping of Wastes and Other Matter, 29 December 1972, 1046 U.N.T.S. 120, 11 I.L.M. 1291 (entered into force 30 August 1975), online: International Maritime Organization <http://www.imo.org/includes/blastDataOnly.asp/data_id%3D16925/LC1972.pdf> replaced by *1996 Protocol to the Convention on the Prevention of Marine Pollution by Dumping of Wastes and Other Matter, 1972 and Resolutions Adopted by the Special Meeting,* 7 November 1996, 36 I.L.M. 1 (entered into force 11 March 2008).

Convention on the Prior Informed Consent Procedure for Certain Hazardous Chemicals and Pesticides in International Trade, 11 September 1998, 38 I.L.M. 1 (entered into force 24 February 2004).

International Convention for the Prevention of Pollution from Ships, 2 November 1973, 1340 U.N.T.S. 184, 12 I.L.M. 1319 as modified by the *Protocol of 1978 Relating to the International Convention for the Prevention of Pollution from Ships of 1973*, 17 February 1978, 1340 U.N.T.S. 61, 17 I.L.M. 546.

International Covenant on Civil and Political Rights, 19 December 1966, 999 U.N.T.S. 171, 6 I.L.M. 368 (entered into force 23 March 1976).

International Covenant on Economic, Social and Cultural Rights, 19 December 1966, 993 U.N.T.S. 3, 6 I.L.M. 360 (entered into force 3 January 1976).

International Convention on the Control of Harmful Anti-fouling Systems on Ships, 5 October 2001 (to enter into force 17 September 2008, 30 ratifications as on 30 June 2008), online: Australian Government <http://www.environment.gov.au/coasts/pollution/antifouling/pubs/convention.pdf>.

Nairobi International Convention on the Removal of Wrecks, 2007, 46 I.L.M. 697 (open for signature since 19 November 2007, not yet in force).

Protocol to the Convention on the Prevention of Marine Pollution by Dumping of Wastes and Other Matter, 1972 and Resolutions Adopted by the Special Meeting, 7 November 1996, 36 I.L.M. 1 (entered into force 11 March 2008).

United Nations Convention on the Law of the Sea, 10 December 1982, 21 I.L.M. 1261 (entered into force 1994).

United Nations Environment Programme: Stockholm Convention on Persistent Organic Pollutants, 22 May 2001, 40 I.L.M. 532 (entered into force on 17 May 2004).

Declarations

Declaration of the United Nations Conference on the Human Environment, 1972, 11 I.L.M. 1416.

UNCED, *Rio Declaration on Environment and Development*, 1992, 31 I.L.M. 874.

Universal Declaration of Human Rights, GA Res. 217(III), UN GAOR, 3d Sess., Supp. No. 13, UN Doc. A/810 (1948) 71.

Washington Declaration on Protection of the Marine Environment from Land-based Activities, 1 November 1995, online: UNEP <http://www.gpa.unep.org/documents/washington_declaration_english.pdf>.

European Union

EC, Commission, *Green Paper on Better Ship Dismantling* (Brussels: COM (2007) 269 Final), at 2, online: European Commission <http://ec.europa.eu/environment/waste/ships/pdf/com_2007_269_en.pdf>.

EC, Directorate General Environment, *Ship Dismantling and Pre-cleaning of Ships: Final Report* (EC, 2007), online: European Commission <http://ec.europa.eu/environment/waste/ships/pdf/ship_dismantling_report.pdf>.

EC, Directorate-General Energy & Transport, *Oil Tanker Phase Out and the Ship Scrapping Industry: A Study on the Implications of the Accelerated Phase out Scheme of Single Hull Tankers Proposed by the EU for the World Ship Scrapping and Recycling Industry* (EC, 2004), online: European Commission <http://ec.europa.eu/transport/maritime/safety/doc/prestige/2004_06_scrapping_study_en.pdf>.

France

France, Interdepartmental Committee on the Dismantling of Civilian and Military End-of-Life Ships, *Le rapport de la Mission Interministérielle portant sur le Démantèlement des Navires civils et militaires en fin de vie* (27 March 2007, Chair: Xavier de la Gorce), online: SGMer <http://www.sgmer.gouv.fr/IMG/pdf/2007-06-18_-_Rapport_MIDN_english_version.pdf>.

Guidelines

Guidelines and Standards for the Removal of Offshore Installations and Structures on the Continental Shelf and in the Exclusive Economic Zone, IMO Assembly Res. A.672(16), adopted on 19 October 1989 (KR-CON).

Guidelines for Maintenance and Monitoring of On-board Materials Containing Asbestos, MSC/Circ.1045, 28 May 2002 (KR-CON).

IMO Guidelines on Ship Recycling: Res. A.962(23) Adopted on 5 December 2003 (Agenda Item 19), A 23/Res.962, 4 March 2004 (KR-CON).

Safe Work Codes of Practice, online: International Labour Organization <http://www.ilo.org/public/english/protection/safework/cops/english/index.htm>.

Safety and Health in Shipbreaking: Guidelines for Asian Countries and Turkey, ILO, Interregional Tripartite Meeting of Experts on Safety and Health in Shipbreaking for Selected Asian Countries and Turkey, Bangkok, 7–14 October 2003, MESHS/2003/1, online: International Labour Organization <http://www.ilo.org/public/english/standards/relm/gb/docs/gb289/pdf/meshs-1.pdf>.

Specific Guidelines for Assessment of Vessels, 1997, online: I International Maritime Organization <http://www.imo.org/includes/blastDataOnly.asp/data_id%3D17023/4Vessels.pdf>.

The UNEP Montreal Guidelines for the Protection of the Marine Environment against Pollution from Land Based Sources, 24 May 1985, UNEP/GC. Decision 13/8/II, online: UNEP <http://www.unep.org/law/PDF/UNEPEnv-LawGuide&PrincN07.pdf>.

IMO Documentation

Adoption of Amendments to the International Convention for the Safety of Life at Sea, 1974, as Amended: Res.MSC.282(86) Adopted on 5 June 2009 (KR-CON).

Adoption of the Final Act and Any Instruments, Recommendations and Resolutions Resulting from the Work of the Conference, International Convention for the Control and Management of Ships' Ballast Water and Sediments, 2004, IMO BWM/CONF/36, 16 February 2004.

Consideration of the Work Programmes of the Pertinent Bodies of ILO, IMO and the Conference of Parties to the Basel Convention on the Issue of Ship Recycling: Abandonment of ships, Note by the IMO Secretariat, ILO/IMO/BC WG 1/2/2, 17 January 2005, online: Basel Convention <http://www.basel.int/ships/abandonment.html>.

Draft Convention on Wreck Removal, Experience of An Tai incident, Submitted by Malaysia, IMO/LEGAL COMMITTEE LEG 83/5/2, 14 September 2001 (KR-CON).

Environment Protection Committee – 43rd session: 28 June – 2 July 1999, online: International Maritime Organization <http://www.imo.org/newsroom/mainframe.asp?topic_id=109&doc_id=357>.

International Conference on the Safe and Environmentally Sound Recycling of Ships: Agenda Item 6, Consideration of the Draft International Convention for the Safe and Environmentally Sound Recycling of Ships: Non-Party Recycling Facilities, Submitted by Australia, SR/CONF/9, 6 February 2009, online: SJÖFARTSVERKET <http://www.sjofartsverket.se/pages/19514/sr-conf-9.pdf>.

International Conference on the Safe and Environmentally Sound Recycling of Ships: Agenda Item 6, Consideration of the Draft International Convention for the Safe and Environmentally Sound Recycling of Ships: Proposal for Amending the Application of the Draft Convention to be Drafted as Other Relevant IMO Conventions, Submitted by Norway, SR/CONF/10, 6 February 2009, online: SJÖFARTSVERKET <http://www.sjofartsverket.se/pages/19514/sr-conf-10.pdf>.

International Conference on the Safe and Environmentally Sound Recycling of Ships: Agenda Item 6, Consideration of the Draft International Convention for the Safe and Environmentally Sound Recycling of Ships: Ensuring Sustainable Green and Safe Ship Dismantling – Concerning the Entry into Force Provisions, Submitted by Greenpeace International and FOEI, SR/CONF/13, 9 February 2009, online: SJÖFARTSVERKET <http://www.sjofartsverket.se/pages/19514/sr-conf-13.pdf>.

International Conference on the Safe and Environmentally Sound Recycling of Ships: Agenda Item 6, Consideration of the Draft International Convention for the Safe and Environmentally Sound Recycling of Ships: Ensuring Sustainable Green and Safe and Ship Dismantling – Concerning Beaching and the Establishment of a Mandatory Fund, Submitted by Greenpeace and FOEI, SR/CONF/14, 9 February 2009, online: Basel Action Network <http://www.ban.org/Library/SR-CONF-14.pdf>.

International Conference on the Safe and Environmentally Sound Recycling of Ships: Agenda Item 6, Consideration of the Draft International Convention for the Safe and Environmentally Sound Recycling of Ships: The Role of the Cash Buyer, Submitted by the Republic of the Marshall Islands, SR/CONF/15, 27 February 2009, online: SJÖFARTSVERKET <http://www.sjofartsverket.se/pages/19514/sr-conf-15.pdf>.

International Conference on the Safe and Environmentally Sound Recycling of Ships: Agenda Item 6, Consideration of the Draft International Convention for the Safe and Environmentally Sound Recycling of Ships: Consideration on Dispute Settlement, Submitted by Belgium, Denmark and the Republic of Korea, SR/CONF/17, 10 March 2009, online: SJÖFARTSVERKET <http://www.sjofartsverket.se/pages/19514/sr-conf-17.pdf>.

International Conference on the Safe and Environmentally Sound Recycling of Ships: Agenda Item 6, Consideration of the Draft Conference Solutions: Draft Conference Resolution on the Processing of Incorrectly Declared Waste, Submitted by France, SR/CONF 21, 27 March 2009, online: SJÖFARTSVERKET <http://www.sjofartsverket.se/pages/19514/sr-conf-21.pdf>.

International Conference on the Safe and Environmentally Sound Recycling of Ships: Agenda Item 6, Consideration of the Draft International Convention for the Safe and Environmentally Sound recycling of Ships: Proposed Amendment to Draft Regulation 5 – Inventory of Hazardous Materials, Submitted by the Marshall Islands, Panama, Ukraine, ICS, BIMCO, IACS, OCIMF, INTERTANKO, INTERCARGO and IPTA, SR/CONF/23, 1 April 2009, online: SJÖFARTSVERKET <http://www.sjofartsverket.se/pages/19514/sr-conf-23.pdf>.

International Conference on the Safe and Environmentally Sound Recycling of Ships: Agenda Item 6, Consideration of the Draft International Convention for the Safe and Environmentally Sound Recycling of Ships: Preparation of Oil Tankers for Ship Recycling, Submitted by India, SR/CONF/26, 2 April 2009, online: SJÖFARTSVERKET <http://www.sjofartsverket.se/pages/19514/sr-conf-26.pdf>.

International Conference on the Safe and Environmentally Sound Recycling of Ships: Agenda Item 6, Consideration of the Draft International Convention for the Safe and Environmentally Sound Recycling of Ships: The Role of the Shipowner in the Development of the Ship Recycling Plan, Submitted by India, SR/CONF 27, 2 April 2009, online: SJÖFARTSVERKET <http://www.sjofartsverket.se/pages/19514/SR-CONF-27.pdf>.

International Conference on the Safe and Environmentally Sound Recycling of Ships: Agenda Item 6, Consideration of the Draft International Convention for the Safe and Environmentally

Sound Recycling of Ships: Proposed Amendments to the Draft Convention, Improvement in the Communication of Information (Articles 9 and 12) and Clarification on the Ship Recycling Plan Approval Process (regulation 9), Submitted by France, SR/CONF/28, 6 April 2009, online: SJÖFARTSVERKET <http://www.sjofartsverket.se/pages/19514/sr-conf-28.pdf>.

International Conference on the Safe and Environmentally Sound Recycling of Ships: Agenda Item 6, Consideration of the Draft International Convention for the Safe and Environmentally Sound Recycling of Ships: Entry-into-Force Provision, Submitted by Denmark, SR/CONF/ 30, 7 April 2009, online: SJÖFARTSVERKET <http://www.sjofartsverket.se/pages/19514/sr-conf-30.pdf>.

International Conference on the Safe and Environmentally Sound Recycling of Ships: Agenda Item 6, Consideration of the Draft International Convention for the Safe and Environmentally Sound Recycling of Ships: Proposal for Providing Clarity in the Convention and Reaffirming the Importance of Safe-for-Hot Work and Safe-for-Entry Conditions during Recycling, Submitted by the United States, SR/CONF/33, 29 April 2009, online: SJÖFARTSVERKET <http://www .sjofartsverket.se/pages/19514/sr-conf-33.pdf>.

International Conference on the Safe and Environmentally Sound Recycling of Ships: Agenda Item 6, Consideration of the Draft International Convention for the Safe and Environmentally Sound Recycling of Ships: Environmental Comparison of the Beaching Method with Other Methods of Recycling of Ships, Submitted by India, SR/CONF/36, 1 May 2009, online: SJÖFARTSVERKET <http://www.sjofartsverket.se/pages/19514/SR-CONF-36.pdf>.

International Conference on the Safe and Environmentally Sound Recycling of Ships: Agenda Item 6, Consideration of the Draft International Convention for the Safe and Environmentally Sound Recycling of Ships: Proposed Amendments to the Draft International Convention for the Safe and Environmentally Sound Recycling of Ships, Submitted by Bangladesh, SR/CONF/37, 4 May 2009, online: SJÖFARTSVERKET <http://www.sjofartsverket.se/pages/19514/sr-conf-37.pdf>

International Conference on the Safe and Environmentally Sound Recycling of Ships: Agenda Item 6, Consideration of the Draft International Convention for the Safe and Environmentally Sound Recycling of Ships: Proposed Amendments to Paragraph 3 of Article 3, Submitted by the Republic of Korea, SR/CONF/40, 7 May 2009, online: SJÖFARTSVERKET <http://www.sjofartsverket .se/pages/19514/sr-conf-40.pdf>.

International Conference on the Safe and Environmentally Sound Recycling of Ships: Agenda Item 6, Consideration of the Draft International Convention for the Safe and Environmentally Sound Recycling of Ships: Proposal on the Conditions on Entry into Force, Submitted by Japan, SR/ CONF/41, 8 May 2009, online: SJÖFARTSVERKET <http://www.sjofartsverket.se/pages/19514/ sr-conf-41.pdf>.

International Conference on the Safe and Environmentally Sound Recycling of Ships: Agenda Item 6, Consideration of the Draft International Convention for the Safe and Environmentally Sound Recycling of Ships: Proposal to Amend Article 2, Submitted by the United States, SR/CONF/43, 10 May 2009, online: SJÖFARTSVERKET <http://www.sjofartsverket.se/pages/19514/sr-conf-43 .pdf>.

International Conference on the Safe and Environmentally Sound Recycling of Ships: Agenda Item 6, Consideration of the Draft International Convention for the Safe and Environmentally Sound Recycling of Ships: Proposal to Amend Regulation 9, Submitted by the United States, SR/ CONF/44, 10 May 2009, online: SJÖFARTSVERKET <http://www.sjofartsverket.se/pages/ 19514/sr-conf-44.pdf>.

International Conference on the Safe and Environmentally Sound Recycling of Ships: Hong Kong, China, 11–15 May 2009, Committee of the Whole Agenda Items 4 and 5, Consideration of the Draft Text of the International Convention for the Safe and Environmentally Sound Recycling of Ships and Consideration of the Draft Text of Conference Resolutions and Recommendations: Overview of Submissions on Articles, Regulations, Appendices, or Proposed Conference Resolutions, SR/CONF/CW/WP/1, 11 May 2009, online: SJÖFARTSVERKET, <http://www .sjofartsverket.se/pages/19514/sr-conf-cw-wp1.pdf>.

International Conference on the Safe and Environmentally Sound Recycling of Ships, Hong Kong, China, 11–15 May 2009: Committee of the Whole, Agenda Item 6, Consideration of the Draft International Convention for the Safe and Environmentally Sound Recycling of Ships: Statement of Concern, Submitted by Greenpeace International and FOEI, SR/CONF/CW/INF.1, 14 May 2009, online: SJÖFARTSVERKET <http://www.sjofartsverket.se/pages/19514/sr-conf-cw-inf1.pdf>.

International Conference on the Safe and Environmentally Sound Recycling of Ships: Agenda Item 8, Adoption of the Final Act and Any Instruments, Recommendations and Resolutions Resulting from the Work of the Conference: Hong Kong International Convention for the Safe and Environmentally Sound Recycling of Ships, 2009, Text Adopted by the Conference, SR/CONF/45, 19 May 2009, online: SJÖFARTSVERKET <http://www.sjofartsverket.se/pages/19514/sr-conf-45.pdf>.

International Conference on the Safe and Environmentally Sound Recycling of Ships: Agenda Item 8, Adoption of the Final Act and Any Instruments, Recommendations and Resolutions Resulting from the Work of the Conference: Final Act of the International Conference on the Safe and Environmentally Sound Recycling of Ships, Text Adopted by the Conference, SR/CONF/46, 19 May 2009, online: SJÖFARTSVERKET <http://www.sjofartsverket.se/pages/19514/sr-conf-46.pdf>.

International Conference on the Safe and Environmentally Sound Recycling of Ships: Hong Kong, China, 11–15 May 2009, Statements by Delegations and Observers at the Closing Session of the Conference, Friday, May 15 2009, SR/CONF/INF.8, 21 May 2009, online: SJÖFARTSVERKET <http://www.sjofartsverket.se/pages/19514/sr-conf-inf8.pdf>.

Intersessional Meeting of the Working Group on Ship Recycling: Further Development of the Draft Convention: Consideration of the Use of Ship Recycling Facilities under the Jurisdiction of Non-parties, Submitted by Norway, MEPC-ISRWG 3/2/9, 14 December 2007, online: SJÖFARTSVERKET <http://www.sjofartsverket.se/pages/14272/3-2-9.pdf>.

Marine Environment Protection Committee – 44th session: 6–13 March 2000, online: International Maritime Organization <http://www.imo.org/newsroom/mainframe.asp?topic_id=109&doc_id=358>.

Marine Environment Protection Committee (MEPC): 46th session: 23–27 April 2001, online: International Maritime Organization <http://www.imo.org/newsroom/mainframe.asp?topic_id=109&doc_id=774#single>.

Marine Environment Protection Committee (MEPC), 47th session: 4–8 March 2002, online: International Maritime Organization <http://www.imo.org/newsroom/mainframe.asp?topic_id=109&doc_id=1753#otherissues>.

Marine Environment Protection Committee (MEPC), 48th session: 7–11 October 2002, online: International Maritime Organization <http://www.imo.org/newsroom/mainframe.asp?topic_id=109&doc_id=2543>.

Marine Environment Protection Committee (MEPC), 49th session: 14–18 July 2003, online: International Maritime Organization <http://www.imo.org/newsroom/mainframe.asp?topic_id=109&doc_id=2798>.

Recycling of Ships: Submitted by India, IMO/MEPC 46/7/1, 14 February 2001 (KR-CON).

Recycling of Ships: Environmental Contamination in Alang-Sosiya Shipbreaking Yards, Submitted by Greenpeace International, IMO/MEPC 46/INF.21, 15 February 2001 (KR-CON).

Recycling of Ships: Proposals for Consideration by the Committee on the Development of Internationally Applicable Guidelines on Recycling of Ships, Submitted by India, IMO/MEPC 47/3/1, 6 November 2001 (KR-CON).

Recycling of Ships: Information on the Environmental Guidelines for the Ship Recycling Industry in India, Submitted by India, IMO/MEPC 47/INF.3, 6 November 2001 (KR-CON).

Recycling of Ships, Submitted by the International Chamber of Shipping (ICS) on Behalf of the Industry Working Party on Ship Recycling, IMO/MEPC 47/3/2, 30 November 2001 (KR-CON).

Recycling of Ships: Comments on the Report of the Correspondence Group, Submitted by Greenpeace International, IMO/MEPC 47/3/5, 11 January 2002 (KR-CON).

Recycling of Ships: General Comments on Document, Submitted by India, IMO/MEPC 48/3/1, 9 August 2002 (KR-CON).

Recycling of Ships: Comments on Document MEPC 48/3, Submitted by India, IMO/MEPC 48/3/2, 9 August 2002 (KR-CON).

Recycling of Ships: Comments on the Report of the Correspondence Group, Submitted by Greenpeace International, IMO/MEPC 48/3/3, 16 August 2002 (KR-CON).

Recycling of Ships: Comments on the Report of the Correspondence Group: Selected Cases of Decommissioning of Vessels Indicating the Need for Mandatory Requirements, Submitted by Greenpeace International, IMO/MEPC 49/3/2, 9 May 2003 (KR-CON).

Recycling of Ships: Comments on the Report of the Correspondence Group to MEPC 49, Submitted by India, IMO/MEPC 49/3/3, 13 May 2003 (KR-CON).

Recycling of Ships: Comments on the Report of the Correspondence Group, Submitted by Japan, IMO/MEPC 49/3/5, 23 May 2003 (KR-CON).

Recycling of Ships: Proposed Technical Considerations for Preparing a Ship Recycling Plan, Submitted by Canada, IMO/MEPC 51/3/2, 22 January 2004 (KR-CON).

Recycling of Ships: Safety and Health in Shipbreaking: Guidelines for Asian Countries and Turkey, Note by the International Labour Organization (ILO), IMO/MEPC 51/INF.4, 23 January 2004 (KR-CON).

Recycling of Ships, Comments on Future Work Items on Ship Recycling: Mechanism to Promote the Implementation of the IMO Guidelines on Ship Recycling, Submitted by Japan, IMO/MEPC 51/3/4, 23 January 2004 (KR-CON).

Recycling of Ships: Analysis of Inconsistencies between the IMO Guidelines on Ship Recycling and the Basel Convention, Submitted by Greenpeace International, IMO/MEPC 51/INF.11, 6 February 2004 (KR-CON).

Recycling of Ships: Proposals for the Further Development of the List of Future Work Items on Ship Recycling as Approved by MEPC 49, Submitted by Greenpeace International, IMO/MEPC 51/3/6, 6 February 2004 (KR-CON).

Recycling of Ships, Proposals for Mandatory Guidelines, Submitted by Greenpeace International, IMO/MEPC 52/3/2, 2 August 2004 (KR-CON).

Recycling of Ships: Proposals for Review and Further Development of the IMO Guidelines, Submitted by Greenpeace International, IMO/MEPC 52/3/3, 2 August 2004 (KR-CON).

Recycling of Ships: The Application of the Basel Convention, Submitted by the International Chamber of Shipping (ICS) on behalf of the Industry Working Party on Ship Recycling, IMO/ MEPC 52/INF.12, 5 August 2004 (KR-CON).

Recycling of Ships: Proposal on a "Phased Approach to Ship Dismantling", Submitted by the Netherlands, IMO/MEPC 52/3/1, 6 August 2004 (KR-CON).

Recycling of Ships: Outcome of the Basel Convention Open-ended Working Group Meeting on the Issue of Ship Recycling, Submitted by the Secretariat of the Basel Convention, IMO/MEPC 52/3/7, 6 August 2004 (KR-CON).

Recycling of Ships: The Need to Develop a Definitive and Consolidated List of Single Hull Oil Tankers Subject to Phase out Regulations, Submitted by Greenpeace International, IMO/MEPC 53/3/3, 12 May 2005 (KR-CON).

Recycling of Ships: The Development of a Green Passport for an Existing Ship, Submitted by Canada Marine Environment Protection Committee, IMO/MEPC 53/INF.17, 13 May 2005 (KR-CON).

Recycling of Ships: The Way Ahead – Practical and Pragmatic, Submitted by ICS, BIMCO, INTERCARGO, INTERTANKO, IPTA & WNTI, IMO/MEPC 53/3/5, 13 May 2005 (KR-CON).

Recycling of Ships: Mandatory Requirements for the IMO Guidelines, Submitted by India, IMO/ MEPC 53/3/6, 13 May 2005 (KR-CON).

Recycling of Ships: Proposal for an Enforceable Legally Binding System for the Recycling of Ships, Submitted by Denmark, IMO/MEPC 53/3/7, 13 May 2005 (KR-CON).

Recycling of Ships: Control System of the Basel Convention on the Control of Transboundary Movements of Hazardous Wastes and their Disposal, Submitted by the Secretariat of the Basel Convention, IMO/MEPC 53/INF.19, 27 May 2005 (KR-CON).

Recycling of Ships: Implementation of the IMO Guidelines on Ship Recycling (Assembly Resolution A-962(23)) "Gas-Free-for-Hot-Work" Certification, IMO/MEPC/Circ.466, 25 July 2005 (KR-CON).

Recycling of Ships: Proposal for a New Legally-binding Instrument on Recycling of Ships, Submitted by Norway, IMO/MEPC 54/3, 5 December 2005 (KR-CON).

Recycling of Ships: Comments on the Proposal for a New Legally-binding Instrument on Recycling of Ships (MEPC 54/3), Submitted by Brazil, IMO/MEPC 54/3/4, 13 January 2006 (KR-CON).

Recycling of Ships: The Need to Develop a New Legally-binding Instrument that will Build and Improve upon Existing Environmental Justice Legislation, Submitted by Greenpeace International, IMO/ MEPC 54/3/5/Rev.1, 1 February 2006 (KR-CON).

Recycling of Ships: Proposal on the Drafting of a Code for Safe and Environmentally Sound Management of Ship-scrapping Facilities, Submitted by France, IMO/MEPC 55/3/4, 28 July 2006 (KR-CON).

Recycling of Ships: Urgent Considerations for Correcting Deficiencies in the Draft Convention on the Safe and Environmentally Sound Recycling of Ships, Submitted by Greenpeace, IMO/MEPC 55/3/7, 10 August 2006 (KR-CON).

Recycling of Ships: Development of the Draft International Convention for the Safe and Environmentally Sound Recycling of Ships, Submitted by India, IMO/MEPC 55/3/12, 17 August 2006 (KR-CON).

Recycling of Ships: Proposal to Clarify the Scope of the Convention with Respect to Land-based Facilities, Submitted by France, IMO/MEPC 55/3/14, 17 August 2006 (KR-CON).

Recycling of Ships: Normal Recycling Procedures, Submitted by ICS, IMO/MEPC 55/INF.12, 18 August 2006 (KR-CON).

Recycling of Ships: Comments on the Draft Convention and the Inventory of Hazardous Materials, Submitted by the Community of European Shipyards' Associations, IMO/MEPC 55/3/15, 18 August 2006 (KR-CON).

Recycling of Ships: Guidelines for the Development of the Ship Recycling Facility Management Plan, Submitted by the United States, IMO/MEPC 56/3/6, 6 April 2007 (KR-CON).

Recycling of Ships: Ready for Recycling Criteria, Submitted by ICS, BIMCO, INTERTANKO, INTERCARGO & IPTA, IMO/MEPC 56/3/9, 4 May 2007 (KR-CON).

Recycling of Ships: Comments to the Report of ISRWG 2 – Proposal to Include Three Hazardous Materials in the Draft Convention, Submitted by Norway, IMO/ MEPC 56/3/10, 16 May 2007 (KR-CON).

Recycling of Ships: Further Development of the Draft Convention, Submitted by Denmark, IMO/ MEPC 56/3/12, 17 May 2007 (KR-CON).

Recycling of Ships: Further Development of the Draft Convention – Non-party Recycling Facilities, Submitted by India, IMO/MEPC 56/3/14, 17 May 2007 (KR-CON).

Recycling of Ships: Comments on the Report of the Second Intersessional Meeting of the Working Group on Ship Recycling, Submitted by Belgium and the Netherlands, IMO/MEPC 56/3/16, 17 May 2007 (KR-CON).

Recycling of Ships: Proposal for Recycling of Ships at Safe and Environmentally Sound Ship Recycling Facilities in States that are not Parties to the Convention, Submitted by the United States, IMO/MEPC 56/3/18, 18 May 2007 (KR-CON).

Recycling of Ships: Proposal for a Mechanism to Facilitate Implementation of the Convention for the Safe and Environmentally Sound Recycling of Ships, Submitted by the United Kingdom, MEPC 56/3/21, 18 May 2007 (KR-CON).

Recycling of Ships: Examination of the Practical Application of the Draft International Convention for the Safe and Environmentally Sound Recycling of Ships – A Timeline for Recycling a Ship According to the Draft Convention's Current Regulations, Submitted by the United Kingdom, IMO/MEPC 56/3/22, 18 May 2007 (KR-CON).

Recycling of Ships: Equivalent Levels of Control Established under the Basel Convention and the Draft IMO Ship Recycling Convention, Submitted by the Secretariat of the Basel Convention, IMO/MEPC 57/3/3, 25 January 2008 (KR-CON).

Recycling of Ships: National Workshop on the Development of the International Convention for the Safe and Environmentally Sound Recycling of Ships, Submitted by India, IMO/MEPC 57/3/5, 25 January 2008 (KR-CON).

Recycling of Ships, Progress Report on the Work Related to the Requirements for Ship Recycling Facilities including the Trial Ship Recycling Project and the Proposal for Future Work Plan, Submitted by Japan, IMO/MEPC 57/3/6, 25 January 2008 (KR-CON).

Recycling of Ships: Substitution, Internalizing Costs, Third Party Auditing and Rejection of Non-party Trade Needed to Secure an Effective Convention, Submitted by Greenpeace International & Friends of the Earth International, IMO/MEPC 57/3/7, 25 January 2008 (KR-CON).

Recycling of Ships: Proposed Clarification on the Applicability of the Ship Recycling Convention, Submitted by the United States, IMO/MEPC 57/3/9, 7 February 2008 (KR-CON).

Recycling of Ships: Conditions on Entry into Force, Submitted by Japan, IMO/MEPC 57/3/13, 8 February 2008 (KR-CON).

Recycling of Ships: Further Development of the Draft Convention, Submitted by India, IMO/MEPC 57/3/16, 8 February 2008 (KR-CON).

Recycling of Ships: Consideration of the Non-Party Proposal during MEPC-ISRWG 3, Submitted by Norway, IMO/MEPC 57/3/20, 8 February 2008 (KR-CON).

Recycling of Ships: Prior Removal of Hazardous Materials, Submitted by France and the United Kingdom, IMO/MEPC 58/3/6, 22 July 2008 (KR-CON).

Recycling of Ships: Further Development of the Draft Guidelines for Safe and Environmentally Sound Ship Recycling, Submitted by Denmark, IMO/MEPC 58/3/7, 22 July 2008 (KR-CON).

Recycling of Ships: Utilization of Applicable Threshold Levels Relevant for the Preparation of Inventory of Hazardous Materials and Related Documentation, Submitted by Germany, IMO/MEPC 58/3/10, 30 July 2008 (KR-CON).

Recycling of Ships: European Parliament on the Green Paper on Better Ship Dismantling, Note by the Secretariat, IMO/MEPC 58/INF.16, 31 July 2008, online: SJÖFARTSVERKET <http://www.sjofartsverket.se/pages/14228/58-INF16.pdf>.

Recycling of Ships: Proposed Text for the Draft International Convention for the Safe and Environmentally Sound Recycling of Ships (MEPC 58/3), Submitted by Bangladesh, IMO/MEPC 58/3/21, 1 August 2008 (KR-CON).

Recycling of Ships: Proposals to Facilitate Implementation of the Convention, Submitted by India, IMO/MEPC 58/3/13, 12 August 2008 (KR-CON).

Recycling of Ships: Recycling Capacity for Entry into Force Criteria for the International Convention for the Safe and Environmentally Sound Recycling of Ships, Submitted by India, IMO/MEPC 58/3/14, 13 August 2008 (KR-CON).

Recycling of Ships: Appropriate Criteria to Carry out a Preliminary Assessment of Equivalency between the Basel Convention and the Hong Kong International Convention for the Safe and Environmentally Sound Recycling of Ships, 2009, Note by the Secretariat of the Basel Convention, IMO/MEPC 59/3/5, 8 May 2009 (KR-CON).

Recycling of Ships: Outcome of the 2009 International Conference on the Safe and Environmentally Sound Recycling of Ships, Note by the Secretariat, IMO/MEPC 59/3/3, 20 May 2009 (KR-CON).

Recycling of Ships: European Parliament Resolution of 26 March 2009 on an EU Strategy for Better Ship Dismantling, Note by the Secretariat, IMO/MEPC 59/3/6, 22 May 2009 (KR-CON).

Recycling of Ships: Comments on the Report of the Correspondence Group on Ship Recycling, and a Proposal for the Framework of Guidance on Safe-for-Entry and Safe-for-Hot-Work Conditions in Safe and Environmentally Sound Ship Recycling, Submitted by United States, IMO/MEPC 59/3/7, 22 May 2009 (KR-CON).

Recycling of Ships: Report of the Correspondence Group, Submitted by the Co-ordinator of the Correspondence Group, IMO/ MEPC 46/7, 18 January 2001 (KR-CON).

Recycling of Ships: Report of the Correspondence Group, Submitted by the Co-ordinator of the Correspondence Group, IMO/MEPC 47/3/3, 30 November 2001 (KR-CON).

Recycling of Ships: Report on the Current Status of Ship Recycling at Alang in India, Submitted by India, IMO/MEPC 53/INF.12, 13 May 2005 (KR-CON).

Recycling of Ships: Report of the Third Intersessional Meeting of the Working Group on Ship Recycling, IMO/MEPC 57/3, 25 January 2008 (KR-CON).

Recycling of Ships, Report of the Ninth Meeting of the Conference of the Parties to the Basel Convention, Note by the Secretariat of the Basel Convention, IMO/MEPC 58/3/9, 29 July 2008 (KR-CON).

Report on the 2009 International Conference on the Safe and Environmentally Sound Recycling of Ships, Note by the Secretary-General, C 102/13, 20 May 2009.

Report of the Working Group, ILO/IMO/BC WG 1/8, 18 February 2005 (KR-CON).

Revised List of Certificates and Documents Required to be Carried on Board Ships, MSC/Circ.1151, FAL.2/Circ.87, MEPC/Circ.426, 17 December 2004 (KR-CON).

Status of the London Convention and Protocol: Report of the Secretary-General on the Status of the London Convention 1972, IMO/LC 30/2, 2 October 2008, online: International Maritime Organization <http://www.imo.org/includes/blastDataOnly.asp/data_id%3D23854/2.pdf>.

OECD

OECD, Environment Policy Committee, *Barriers to the Implementation of Environmentally Sound Management Practices for Small and Medium Size Metal Recycling Enterprises, Case Study: Canadian Ship Dismantling Enterprise*, Working Group on Waste Prevention and Recycling, Doc. No. ENV/EPOC/WGWPR/RD(2002)7/FINAL (2003), online: OECD <http://www.oecd.org/dataoecd/61/35/2502052.pdf>.

Other United Nations Documents

Report of the Special Rapporteur on the Illicit Movement and Dumping of Toxic and Dangerous Products and Wastes on the Enjoyment of Human Rights, Fatma Zohra Ouhachi-Vesely, On Her Mission to Turkey (10 to 19 March 2004), UNESC, 61st Sess., E/CN.4/2005/44 (15 February 2005), online: NGO platform on shipbreaking <http://www.shipbreakingplatform.com/dmdocuments/other_documents/Special%20Rapporteur.Report.Turkey.pdf>.

Report of the United Nations Conference on Environment and Development, UN Doc. A/CONF.151/26 (Vol. II), (1992).

UNCTAD Secretariat, *Review of Maritime Transport*, UNCTAD/RMT/2007 (2007).

UNDP, *Human Development Report 2005* (New York: United Nations Development Programme, 2005), online: Human Development Reports <http://hdr.undp.org/en/media/hdr05_complete.pdf>.

UNEP, Conference of the Parties to the Basel Convention on the Control of Transboundary Movements of Hazardous Wastes and their Disposal, 9th Mtg., (23–27 June 2008) Environmentally Sound Management of Ship Dismantling and the Joint Working Group of the International Labour Organization, the International Maritime Organization and the Basel Convention on Ship Scrapping, Note by the Secretariat, UNEP/CHW.9/34*, 14 April 2008, online: Basel Convention <http://www.basel.int/meetings/cop/cop9/docs/34e.pdf>.

UNEP, Conference of the Parties to the Basel Convention on the Control of Transboundary Movements of Hazardous Wastes and their Disposal, 6th Mtg., (9–13 December 2002) Technical Guidelines for the Environmentally Sound Management of the Full and Partial Dismantling of Ships, UNEP/CHW.6/23, 8 August 2002, online: Basel Convention <http://www.basel.int/meetings/cop/cop6/cop6_23e.pdf>.

V/28. Dismantling of Ships, see UNEP, *Report of the Fifth Meeting of the Conference of the Parties to the Basel Convention – Advance Copy,* 5th Mtg., UNEP/CHW.5/29, (10 December 1999), online: Basel Convention <http://www.basel.int/meetings/cop/cop5/cop5reportfinal.pdf#v28>.

VII/25. Joint Working Group of the International Labour Organization, the International Maritime Organization and the Basel Convention on Ship Scrapping, online: Basel Convention <http://www.basel.int/ships/relev-decisions-cop-vii.pdf#VII_25>.

VII/26. Environmentally Sound Management of Ship Dismantling, online: Basel Convention <http://www.basel.int/ships/relev-decisions-cop-vii.pdf#VII_26>.

VII/27. Abandonment of Ships, online: Basel Convention <http://www.basel.int/ships/relev-decisions-cop-vii.pdf#VII_26>.

VIII/11: Environmentally Sound Management of Ship Dismantling, online: Basel Convention <http://www.basel.int/meetings/cop/cop8/docs/16e.pdf#viii11>.

VIII/12 Joint Working Group of the International Labour Organization, the International Maritime Organization and the Basel Convention on Ship Scrapping, online: Basel Convention <http://www.basel.int/meetings/cop/cop8/docs/16e.pdf#viii11>.

VIII/13: Abandonment of Ships on Land or In Ports, online: Basel Convention <http://www.basel.int/meetings/cop/cop8/docs/16e.pdf#viii11>.

Regional Instruments

Memorandum of Understanding on Port State Control in Implementing Agreements on Maritime Safety and the Protection of the Marine Environment, 26 January 1982, 21 I.L.M. 1 (entered into force 1 July 1982, adopted at Paris).

Legislation

European Union

EC, Commission, *Impact Assessment for an EU Strategy for Better Ship Dismantling* (Brussels: SEC (2008) 2846), online: European Commission <http://ec.europa.eu/environment/waste/ships/pdf/impact_assessment.pdf>.

EC, *Council Regulation 259/93 of 1 February 1993 on the supervision and control of shipments of waste within, into and out of the European Community,* [1993] O.J. L 30/1.
EC, *Regulation (EC) 1013/2006 of the European Parliament and of the Council of 14 June 2006 on shipments of waste*, [2006] O.J. L 190/1.
EC, *Council Directive 75/442/EEC of 15 July 1975 on waste*, [1975] O.J. L 194/39.
EC, *Council Regulation (EC) 120/97 of 20 January 1997amending Regulation (EC) 259/93 on the supervision and control of shipments of waste within, into and out of the European Community,* [1997] O.J. L 22/14.
EC, *Council Decision 97/640/EC of 22 September 1997 on the approval, on behalf of the Community, of the amendment to the Convention on the control of transboundary movements of hazardous wastes and their disposal (Basel Convention), as laid down in Decision III/ of the Conference of the* Parties, [1997] O.J. L 272/45.

India

Batteries (Management and Handling) Rules 2001 (S.O. 432(E)/2001 India), online: Ministry of Environment and Forests <http://envfor.nic.in/legis/hsm/leadbat.html>.
Child Labour (Prohibition and Regulation) Act 1986 (No. 61 of 1986, India), online: India Code <http://indiacode.nic.in/>.
Constitution of India 1950.
Draft Notification 2008 (Coastal Zone Management) (S.O. 1761(E)/2008 India), online: Ministry of Environment and Forests <http://envfor.nic.in/legis/crz/so-1761(e).pdf>.
Environment Protection Rules 1986 (S.O. 844(E)/1986 India), online: Ministry of Environment and Forests <http://envfor.nic.in/legis/env/env4.html>.
Gujarat Maritime Board Act 1981 (No. 30 of 1981, Gujarat State, India).
Gujarat Maritime Board Ship Recycling Regulations 2003 (No. GMB/ALANG/73/110/2003/42 Gujarat, India) (published in Vol. XLIV The Gujarat Government Gaz. Extra Ordinary, 7 July 2003, 281–1).
Hazardous Wastes (Management and Handling) Rules 1989 (S.O. 594(E)/1989 India).
Manufacture, Storage and Import of Hazardous Chemical Rules 1989 (S.O. 966(E)/1989 India), online: Ministry of Environment and Forests <http://envfor.nic.in/legis/hsm/hsm2.html>.
Noise Pollution (Regulation and Control) Rules 2000 (S.O. 123(E)/2000 India).
Ozone Depleting Substances (Regulation and Control) Rules 2000 (S.O. 670(E)/2000 India), online: Ministry of Environment and Forests <http://envfor.nic.in/legis/ods/odsrcr.html>.
Plastics Manufacture, Sale and Usage Rules 1999 (S.O. 705(F)/1999).
The Air (Prevention and Control) of Pollution Act 1981 (No. 14 of 1981, India), online: Ministry of Environment & Forests <http://envfor.nic.in/legis/air/air1.html>.
The Andhra Pradesh Water, Land and Trees Act 2002 (No. 10 of 2002, India), online: ielrc.org <http://www.ielrc.org/content/e0202.pdf>.
The Chemical Accidents Emergency Planning, Preparedness and Response Rules 1996 (G.S.R. 347(E)/1996), online: Ministry of Environment and Forests <http://envfor.nic.in/legis/hsm/gsr347.htm>.
The Coastal Regulation Zone Notification 1991 (S.O. 114 (E)/1990 India), online: Ministry of Environment and Forests <http://envfor.nic.in/legis/crz/crznew.pdf>.
The Contract Labour (Regulation and Abolition) Act 1970 (No. 37 of 1970, India), online: India Code <http://indiacode.nic.in/>.
The Disaster Management Act 2005 (No. 53 of 2005, India), online: National Institute of Disaster Management <http://www.nidm.gov.in/DM_act2005.pdf>.
The Environment (Protection) Act 1986 (No. 29 of 1986, India), online: Ministry of Environment and Forests <http://envfor.nic.in/legis/env/env1.html>.
The Environmental Impact Assessment Notification 2006 (S.O. 1533/2006 India), online: Ministry of Environment and Forests <http://envfor.nic.in/legis/eia/so1533.pdf>.
The Factories Act 1948 (No. 63 of 1948, India), online: India Code <http://indiacode.nic.in/>.
The Gujarat Maritime Board (Conditions and Procedures for Granting Permission for Utilising Ship Recycling Plots) Regulation 2006 (No. GMB/Alang/134/2004 Gujarat, India) (published in Vol. XLVII The Gujarat Government Gaz. Extra Ordinary, 4 October 2006, 767–1).

The Gujarat State Disaster Management Act 2003 (No. 20 of 2003, India), online: GSDMA <http://www.gsdma.org/dmact.pdf>.

The Inter-State Migrant Workmen (Regulation of Employment and Conditions of Service) Act 1979 (No. 30 of 1979, India), online: Office of Labour Commissioner <http://labour.delhigovt.nic.in/act/details_acts/interstatemigrnatwirkmen/index.html>.

The Karnataka Preservation of Trees Act 1976 (No. 76 of 1976, India), online: Karnataka Forest Department <http://www.karnatakaforest.gov.in/English/Acts_Rules/acts/KTP_Act_1976.pdf>.

The Kerala Ground Water (Control and Regulation) Act 2002 (No. 19 of 2002, India), online: The Official Web Portal of Govt. of Kerala <http://www.kerala.gov.in/dept_ground/annexure6.pdf>.

The Minimum Wages Act 1948 (No. 11 of 1948, India), online: India Code <http://indiacode.nic.in/>.

The Payment of Wages Act 1936 (No. 4 of 1936, India), online: India Code <http://indiacode.nic.in/>.

The Public Liability Insurance Act 1991 (No. 6 of 1991, India), online: India Code <http://indiacode.nic.in/>.

The Water (Prevention and Control of Pollution) Act 1974 (No. 6 of 1974, India), online: Ministry of Environment and Forests <http://envfor.nic.in/legis/water/wat1.html>.

The Workmen's Compensation Act 1923 (No. 8 of 1923, India), online: India Code <http://indiacode.nic.in/>.

United States

Bob Stump National Defense Authorization Act of Fiscal Year 2003 Pub. L. No. 107–314, 116 Stat. 2458 (2002).

Federal Property and Administrative Services Act of 1949, 40 U.S.C.A. § 101.

Merchant Ship Sales Act of 1946, Pub. L. No. 321, § 11, 60 Stat. 41 (1946) (codified as amended at 50 U.S.C. App. 1735–46 (2000)).

National Maritime Heritage Act of 1994, Pub. L. 103–451, 108 Stat. 4769 (1994).

Toxic Substances Control Act 15 U.S.C.A. § 2601 (West 2004).

Jurisprudence

National Cases

Bangladesh

Bangladesh Environmental Lawyers Association v. Bangladesh, [2003] Writ Petition No. 2911, online: Bangladesh Environmental Lawyers Association <http://www.belabangla.org/html/pil.htm>.

France

Conseil d'État, 15 February 2006, *Association ban asbestos* France, No. 288801–288811, online: Le Conseil d'État <http://www.conseil-etat.fr/ce/jurispd/index_ac_ld0607.shtml>.

India

A.D.M. Jabalpur v. Shivakant Shukla (1976), [1976] 2 S.C.C. 521(India S.C.).

A.P. Pollution Control Board v. Prof. M.V. Nayudu I (1999), [1999] A.I.R. S.C. 812 (India S.C.).

A.P. Pollution Control Board v. Prof. M.V. Nayudu II (2000), [2001] 2 S.C.C. 62.

BALCO v. Union of India (2001), [2002] 2 S.C.C. 333 (India S.C.).

Bangalore Water Supply and Sewerage Board v. A. Rajappa (1978), [1978] 2 S.C.C. 213 (India S.C.).

Bipinchandra J. Diwan v. State of Gujarat (2001), [2002] A.I.R. Guj. 99 (Gujarat H.C.).

Charanlal Sahu v. Union of India (1989), [1990] 1 S.C.C. 613 (India S.C.).

Consumer Education and Research Centre v. Union of India (1995), [1995] A.I.R. S.C. 922 (India S.C.).

Dahanu Taluka Environment Protection Group v. Bombay Suburban Electricity Supply Co. Ltd. (1991), [1991] 2 S.C.C. 539 (India S.C.).
Dr. Shivarao Shantaram Wagle v. Union of India (1988), [1988] 2 S.C.C. 115 (India S.C.).
Essar Oil Ltd. v. Halar Utkarsh Samiti (2004), [2004] 2 S.C.C. 392 (India S.C.).
F.K. Hussain v. Union of India (1990), [1990] A.I.R. Ker. 321 (Kerala H.C.).
Francis Coraile Mullin v. Administrator, Union Territory of Delhi (1981), [1981] 1 S.C.C. 608 (India S.C.).
Hinch Lal Tiwari v. Kamala Devi (2001), [2001] 6 S.C.C. 496 (India S.C.).
Hindustan Lever Employees' Union v. Hindustan Lever Limited (1994), [1995] A.I.R. S.C. 470 (India S.C.).
Indian Council for Enviro-legal Action v. Union of India (1996), [1996] 5 S.C.C. 281 (India S.C.).
Indian Council for Enviro-legal Action v. Union of India (1996), [1996] 3 S.C.C. 212 (India S.C.).
K.M. Chinnappa Applicant in T.N. Godavarman Thirumalpad v. Union of India, [2002] S.C. 1386 (indlaw.com), [2002] 10 S.C.C. 606.
L.K. Koolwal v. State of Rajasthan (1986), [1988] A.I.R. Raj. 2 (Rajasthan H.C.).
M. C. Mehta (II) v. Union of India (1988), [1988] 1 S.C.C. 471 (India S.C.).
M.C. Mehta (Calcutta Tanneries Matter) v. Union of India (1996), [1997] 2 S.C.C. 411 (India S.C.).
M.C. Mehta (Taj Trapezium Matter) v. Union of India (1996), [1997] 2 S.C.C. 353 (India S.C.).
M.C. Mehta v. Kamal Nath (1996), [1997] 1 S.C.C. 388 (India S.C.).
M.C. Mehta v. Kamal Nath (2000), [2000] 6 S.C.C. 213 (India S.C.).
M.C. Mehta v. Union of India (1986), [1987] A.I.R. S.C. 1086 (India S.C.).
M.C. Mehta v. Union of India (2002), [2002] 4 S.C.C. 356 (India S.C.).
Manchegowda v. State of Karnataka (1984), [1984] 3 S.C.C. 301 (India S.C.).
Maneka Gandhi v. Union of India (1978), [1978] A.I.R. S.C. 597 (India S.C.).
Murali S. Deora v. Union of India (2001), [2001] 8 S.C.C. 765 (India S.C.).
N.D. Jayal v. Union of India (2003), [2004] 9 S.C.C. 362 (India S.C.).
Narmada Bachao Andolan v. Union of India (2000), [2000] 10 S.C.C. 664 (India S.C.).
People United for Better Living in Calcutta-Public v. State of West Bengal (1992), [1993] A.I.R. Cal. 215 (Calcutta H.C.).
Research Foundation for Science and Technology and Natural Resource Policy v. Union of India (2007), [2007] S.C. 890 (India S.C.) (indlaw.com).
Research Foundation for Science and Technology and Natural Resource Policy v. Union of India, [2007] Writ Petition (Civil) No. 657 (India S.C.), online: The Judgments Information System <http://www.courtnic.nic.in/courtnicsc.asp>.
Research Foundation for Science and Technology and Natural Resource Policy v. Union of India (2007), [2007] S.C. 1484 (India S.C.) (indlaw.com).
Research Foundation for Science Technology and Natural Resource Policy v. Union of India (2007), [2007] A.I.R. S.C. 3118 (India S.C.).
Research Foundation for Science, Technology and Natural Resources Policy v. Union of India (12 March 2007), Writ Petition (Civil) 657 of 1995 (India S.C.), online: The Judgments Information System <http://www.courtnic.nic.in/courtnicsc.asp>.
Research Foundation for Science, Technology and Natural Resources Policy v. Union of India (5 June 2006), Writ Petition (Civil) 657 of 1995 (India S.C.), online: The Judgments Information System <http://www.courtnic.nic.in/courtnicsc.asp>.
Research Foundation for Science, Technology and Natural Resources Policy v. Union of India, [1995] Writ Petition (Civil) No. 657 (India S.C.), online: The Judgments Information System <http://www.courtnic.nic.in/courtnicsc.asp>.
Research Foundation for Science, Technology and Natural Resources Policy v. Union of India (14 October 2003), Writ Petition (Civil) 657 of 1995 (India S.C.), online: The Judgments Information System <http://www.courtnic.nic.in/courtnicsc.asp>.
Research Foundation for Science, Technology and Natural Resources Policy v. Union of India (16 January 2006), Writ Petition (Civil) 657 of 1995 (India S.C.), online: The Judgments Information System <http://www.courtnic.nic.in/courtnicsc.asp>.
Research Foundation for Science, Technology and Natural Resources Policy v. Union of India (13 February 2006), Writ Petition (Civil) 657 of 1995 (India S.C.), online: The Judgments Information System <http://www.courtnic.nic.in/courtnicsc.asp>.
Rural Litigation and Entitlement Kendra Dehradun v. State of Uttar Pradesh (1985), [1985] (2) S.C.C. 431 (India S.C.).

Rural Litigation and Entitlement Kendra, Dehradun v. State of U.P. (1988), [1988] A.I.R. S.C. 2187 (India S.C.).
S. Jagannath v. Union of India (1986), [1997] 2 S.C.C. 87 (India S.C.).
Samatha v. State of A.P. (1997), [1997] 8 S.C.C. 191 (India S.C.).
Shantistar Builders v. Narayan Khimalal Totame (1990), [1990] 1 S.C.C. 520 (India S.C.).
State of H.P. v. Ganesh Wood Products (1995), [1995] 6 S.C.C. 363 (India S.C.).
State of Tamil Nadu v. M.S. Hind Stone (1981), A.I.R. 1981 S.C. 711 (India S.C.).
State of U.P. v. Jai Bir Singh (2005), [2005] 5 S.C.C. 1 (India S.C.).
Steel Authority of India Ltd. v. National Union Waterfront Workers (2001), [2001] 7 S.C.C. 1 (India S.C.).
Subhash Kumar v. State of Bihar (1991), [1991] A.I.R. S.C. 420 (India S.C.).
Suo Motu v. State of Rajasthan (2004), [2005] A.I.R. Raj. 82 (Rajasthan H.C.).
T.K. Rangarajan v. Government of Tamil Nadu (2003), [2003] A.I.R. S.C. 3032 (India S.C.).
T.N. Godavarman Thirumulkpad v. Union of India (1996), [1997] 2 S.C.C. 267 (India S.C.).
T.N. Godavarman Thirumulpad v. Ashok Khot (2006), [2006] 5 S.C.J. 662 (India S.C.).
T.N. Godavarman Thirumulpad v. Union of India, [1995] Writ Petition (Civil) 202 (India S.C.), online: The Judgments Information System <http://www.courtnic.nic.in/courtnicsc.asp>.
Union of India v. G. Ganayutham (1997), [1997] 7 S.C.C. 463 (India S.C.).
Vellore Citizens' Welfare Forum v. Union of India (1996), [1996] 5 S.C.C. 647 (India S.C.).

Netherlands

Stichting Greenpeace Nederland, a Foundation that has its Seat in Amsterdam v. the State Secretary for Housing, Spatial Planning and the Environment (21 February 2007), Case No. 200606331/1 (Council of State, Administrative Jurisdiction Division), online: Basel Convention: Dismantling of Ships – Relevant Caselaw <http://www.basel.int/ships/relevcaselaw.html>.
Upperton Ltd., with offices in Mauritius v. The Minister of Housing, Spatial Planning and the Environment, (19 June 2002), LJN-No.: AE4310 Case No.: 200105168/2 (Council of State, The Hague), reproduced in Recycling of Ships: Decision Concerning the Export of the "Sandrien" for Ship Recycling, Submitted by Greenpeace International, IMO/MEPC 48/INF.11, 10 July 2002 (KR-CON).

Turkey

Cemsan Ship Dismantling Metal and Steel Industry Trade Ltd. Co. v. Ministry of Environment, Ankara (2003/1184), Case No. 2002/496 (Izmir 2nd Admin. Ct.) (unofficial translation), online: Basel Convention <http://www.basel.int/ships/Turkey300903e.pdf>.

United Kingdom

In *Queen on the Application of Gregan v. Hartlepool BC* (2003), [2003] EWHC 3278 (Admin), [2004] J.P.L. 1088 (Q.B.D. Admin) (WL).
Queen on the Application of Friends of the Earth Ltd. v. Environment Agency (2003), [2003] EWHC 3193 (Admin), [2004] Env. L.R. 31 (Q.B.D. Admin) (WL).

United States

Basel Action Network v. Maritime Administration, 285 F.Supp.2d 58 (D.D.C., 2003).
Basel Action Network v. Maritime Administration, 370 F.Supp.2d 57 (D.D.C., 2005).

International Cases

International Court of Arbitration

Trail Smelter Arbitration (U.S. v. Can.), U.N. Rep. Int'l Arb. Awards 1905 (1949), 3 reprinted in (1941) 35 Am. J. Int'l L. 684.

International Court of Justice

Case Concerning the Gabcikovo – Nagymaros *Project* (*Hungary/Slovakia*) (1997), [1997] I.C.J. 7.

Government Documents

India

India, Government of India Ministry of Labour and Employment, *The Second National Commission on Labour Report* (New Delhi: Ministry of Labour and Employment, 2002), online: Government of India Ministry of Labour <http://www.labour.nic.in/lcomm2/nlc_report.html>.

India, Gujarat Pollution Control Board, *Inspection Report of 'Blue Lady'* (India: Gujarat Pollution Control Board, 2006), online: GPCB Office <http://www.gpcb.gov.in/PDF/bluelady_main_report.pdf>.

India, Ministry of Environment & Forests, *Report of the High Powered Committee on Management of Hazardous Wastes*, vol. 1, at 1–15, online: Ministry of Environment & Forests <http://envfor.nic.in/cpcb/hpcreport/>.

India, *Report of the Committee of Technical Experts on Ship Breaking Activities* (Chairman: Dr. Prodipto Ghosh, New Delhi: Ministry of Environment & Forests, 30 August 2006) [unpublished] [on file with the author].

India, *Report of the Committee to Review the Coastal Regulation Zone Notification 1991* (Chairman: Prof. M.S. Swaminathan, New Delhi: Ministry of Environment & Forests, February 2005).

India, *Report of the Technical Experts Committee on Dismantling Plan of the Ship 'Blue Lady', Beached at Alang (Gujarat)*, (Chairman: Dr. Prodipto Ghosh, New Delhi: Ministry of Environment & Forests, 10 May 2007) [on file with the author].

India, Supreme Court Monitoring Committee on Hazardous Wastes, *First Report of the Monitoring Committee on the Entry of the French Warship "Le Celmenceau" for Breaking at Alang in India with Recommendations* (Chairman: Dr. G. Thyagarajan, New Delhi: Ministry of Environment & Forests, 2006) [unpublished] [on file with the author].

India, Supreme Court Monitoring Committee on Hazardous Wastes, *Final Report on Clemenceau: Minority Report* (Chairman: Dr. G. Thyagarajan, New Delhi: Ministry of Environment & Forests, 2006) [unpublished] [on file with the author].

India, Supreme Court Monitoring Committee on Hazardous Wastes, *Final Report on Clemenceau: Majority Report* (Chairman: Dr. G. Thyagarajan, New Delhi: Ministry of Environment & Forests, 2006) [unpublished] [on file with the author].

India, Supreme Court Monitoring Committee on Management of Hazardous Wastes (For the period: November 2003 to October 2006), *Report of the Supreme Court Monitoring Committee on Management of Hazardous Wastes* Vol. 1 (Chairman: Dr. G. Thyagarajan, New Delhi: Ministry of Environment & Forests, 2006) [unpublished] [on file with the author].

United Kingdom

Department for Environment, Food & Rural Affairs, "UK Ship Recycling Strategy: Consultation Paper, March 2006", online: Basel Convention <http://cop8.basel.int/ships/docs/16e.pdf>.

U.K., Department for Environment, Food and Rural Affairs, *Overview of Ship Recycling in the UK*: *Guidance* (London: Crown Copyright, Department for Environment, Food and Rural Affairs, 2007), online: Defra <http://www.defra.gov.uk/environment/waste/strategy/pdf/shiprecycle-strategy-overview.pdf>.

U.K., H.C., "Environment, Food and Rural Affairs Committee on Dismantling Defunct Ships in the UK", HC 834 (18th Report of Session 2003–2004), online: United Kingdom Parliament <http://www.publications.parliament.uk/pa/cm200304/cmselect/cmenvfru/834/834.pdf>.

United States

U.S., Environment Protection Agency, *A Guide for Ship Scrappers: Tips for Regulatory Compliance* (EPA 315-B-00-001) (Washington, D.C.: Office of Enforcement and Compliance Assurance,

2000), online: U.S. Environment Protection Agency <http://www.epa.gov/oecaerth/resources/publications/civil/federal/shipscrapguide.pdf>.

U.S., Maritime Administration, *National Defense Reserve Fleet (NDRF), Suisun Bay, CA, Vessel Environmental Review*, (Oakland: 2007), online: BayAreaNewsGroup <http://bayareanewsgroup.com/multimedia/cct/multimedia/pdf/suisunpollute.pdf>.

U.S., Maritime Administration, *Report to Congress on the Progress of the Vessel Disposal Programme* (Department of Transportation, 2007) at 16–17, online: U.S. Department of Transportation <http://www.marad.dot.gov/Ship%20Disposal/ReportToCongressFinal03_19_07.pdf>.

U.S., Maritime Administration, *Transfer of National Defense Reserve Fleet Vessels from the James River Reserve Fleet for Disposal at Able UK Facilities, Teesside, UK Environmental Assessment: Final* (Department of Transportation, 2004), online: U.S. Department of Transportation <http://www.marad.dot.gov/Programs/Scrapping/web/02-MARAD%20UK%20FinalEA%20Text_cameraready_062904.pdf>.

Secondary Materials

Articles

Abraham, C.M. & Rosencranz, Armin. "An Evaluation of Pollution Control Legislation in India" (1986) 11 Colum. J. Envtl. L. 101 (HeinOnline).

Abraham, Phillip. "Offshore Oil and Gas Facility Decommissioning in Nova Scotia and Newfoundland" (2001) 24 Dal. L.J. 333 (HeinOnline).

Adams, Rebecca. "The Ecological Decline of the Black Sea," Notes, [1997] Colo. J. Int'l Envtl. L. & Pol'y 209 (WLeC).

Adede, Andronico O. "The Treaty System from Stockholm (1972) to Rio de Janeiro (1992)" (Statement at the Rio Environmental Treaties Colloquium, Pace Law School Center for Environmental Legal Studies, 30 October 1994), (1995) 13 Pace. Envtl. L. Rev. 33.

Agyebeng, William K. "Theory in Search of Practice: The Right of Innocent Passage in the Territorial Sea" (2006) 39 Cornell Int'l L.J. 371 (HeinOnline).

Ahluwalia, Montek S. "Economic Reforms in India since 1991: Has Gradualism Worked?" (2002) 16:3 J. Econ. Persp. 67.

Ahmed, Zubair. "Stay out, India tells toxic ship" *BBC News* (6 January 2006), online: BBC News <http://news.bbc.co.uk/2/hi/south_asia/4588922.stm>.

Allen, Mark E. "Slowing Europe's Hazardous Waste Trade: Implementing the Basel Convention into European Union Law", Comments, (1995) 6 Colo. J. Int'l Envtl. L. & Pol'y 163.

Andersen, Aage Bjorn. "The Clemenceau Case: Potential Hazardous Materials Assessment" *Greenpeace International* (2 February 2006), online: Greenpeace International <http://www.greenpeace.org/raw/content/international/press/reports/the-clemenceau-case-potentia.pdf>.

Anderson, Micheal R. "Litigation and Activism: The Bhopal Case" (1993) Third World Legal Stud. 177 (HeinOnline).

"Anti-fouling Systems" (2002) Discussion Paper, 3, online: International Maritime Organization <http://www.imo.org/includes/blastDataOnly.asp/data_id%3D7986/FOULING2003.pdf>.

Bailey, Paul J. "Is There a Decent Way to Break Up Ships?" (2000) Discussion Paper (ILO, Sectoral Activities Programme), online: International Labour Organization <http://www.ilo.org/public/english/dialogue/sector/papers/shpbreak/index.htm#N_1_>.

"Bangladesh Country Environmental Analysis" (2006) Bangladesh Development Series Paper No: 12 (The World Bank Office, Dhaka), online: World Bank <http://siteresources.worldbank.org/BANGLADESHEXTN/Resources/295759-1173922647418/complete.pdf>.

Basel Action Network, Briefing Paper, "Legal Shipwreck: IMO Convention Legalizes Toxic Ship Dumping Running from Basel to Turn Back the Clock" (May 2009), online: Basel Action Network <http://www.ban.org/Library/BP05_May_2009.pdf>.

Basel Action Network, "Legal Analysis of Letter from Mr. Raja to Ms. Hedegaard regarding the legal application of the Basel Convention to the Kong Frederik IX (aka Riky) (3 May 2005)", online: Basel Action Network <http://www.ban.org/Library/BAN_analysis_Ricky.pdf>.

Basel Action Network, Media Release, "I.M.O. Ship Recycling Convention Denounced as "Legal Shipwreck": Activists Call for Ban on Toxic Ship Beaching" (11 May 2009), online: Basel Action Network <http://www.ban.org/BAN_NEWS/2009/090511_imo_legal_shipwreck.html>.

Basel Action Network, "Needless Risk: The Bush Administration's Scheme to Export Toxic Waste Ships to Europe" (20 October 2003), online: Basel Action Network <http://www.ban.org/Library/Needless%20Risk%20Final.pdf>.

Basel Action Network, Press Release, "Toxic Ghost Fleet Ship Export to UK Stopped" (31 May 2007), online: Basel Action Network <http://www.ban.org/ban_news/2007/070531_toxic_ghost_fleet.html>.

Bavadam, Lyla. "Profits over safety" *Frontline [of India]* 23:2 (28 January 2006), online: Frontline <http://www.flonnet.com/fl2302/stories/20060210007701100.htm>.

——. "Shipload of trouble" *Frontline [of India]* 24:22 (3 November 2007), online: Frontline <http://www.hinduonnet.com/fline/fl2422/stories/20071116505108300.htm>.

"Basel Non-Compliance Notification Report" *Basel Action Network* (31 January 2006), online: Basel Action Network <http://www.ban.org/notifications/BNN2006_1.pdf>.

Bidwai, Praful. "Confused and ambivalent" *Frontline [of India]* 22:11 (21 May 2005), online: Frontline <http://www.hinduonnet.com/fline/fl2211/stories/20050603005811600.htm>.

"Blue Lady's fate uncertain as activists to challenge SC" *The Economic Times [of India]* (18 September 2007) (Prowler).

Brancato, Brian P. "Asbestosis Litigation and Marine Insurance" (1995) 20 Tul. Mar. L.J. 105 (WLeC).

Brautigam, Lisa A. "Control of Aquatic Nuisance Species Introductions Via Ballast Water in the United States: Is the Exemption of Ballast Water Discharges from Clean Water Act Regulation a Valid Exercise of Authority by the Environmental Protection Agency?" (2001) 6 Ocean & Coast. L.J. 33 (WLeC).

Brown, Valerie J. "Warfare: Iraq's Toxic Shipwrecks" *Environmental Health Perspectives* 113:4 (April 2005) at A230 (JSTOR).

"Chirac orders 'toxic' ship home", *BBC News* (16 February 2006), online: BBC News < http://news .bbc.co.uk/2/hi/europe/4716472.stm>.

Campbell, Viola Blayre. "Ghost Ships and Recycling Pollution: Sending America's Trash to Europe" (2004) 12 Tulsa J. Comp. & Int'l. L. 189.

Carfield, Maggi. "Enhancing Poor People's Capabilities through the Rule of Law: Creating an Access to Justice Index", Notes, (2005) 83 Wash. U.L.Q. 339 (HeinOnline).

Cartwright, Christine. "Natural Resource Damage Assessment: The Exxon Valdez Oil Spill and Its Implications," Note, (1991) 17 Rutgers Computer & Tech. L.J. 451 (WLeC).

Cassels, Jamie. "The Uncertain Promise of Law: Lessons from Bhopal" (1991) 29 Osgoode Hall L.J. 1 (HeinOnline).

Cha, Mijin J. "A Critical Examination of the Environmental Jurisprudence of the Courts of India" (2005) 10 Alb. L. Envtl. Outlook 197 (HeinOnline).

Chimni, B.S. "A Just World under Law: A View from the South" (2007) 22 Am. U. Int'l L. Rev. 199 (WLeC).

Chircop, Aldo. "Law of the Sea and International Environmental Law Considerations for Places of Refuge for Ships in Need of Assistance", in Aldo Chircop & Olof Linden, eds., *Places of Refuge for Ships: Emerging Concerns of a Maritime Custom* (Boston: Martinus Nijhoff Publishers, 2006) 231.

Cohen, Matt. "U.S. Shipbreaking Exports: Balancing Safe Disposal with Economic Realities" (2005) 28 Environs Envtl. L. & Pol'y J. 237.

Colburn, Jamison E. "Turbot Wars: Straddling Stocks, Regime Theory, and a New U.N. Agreement" (1997) 6 J. Transnat'l L. & Pol'y 323 (HeinOnline).

Comments on Statement of French Government on Clemenceau by the Basel Action Network on Behalf of the Greater Coalition Demanding Return of Clemenceau to France for Decontamination (24 January 2006), online: Basel Action Network <http://www.ban.org/Library/statement_of_french_government_on_clemenceau.pdf>

Comments on the Draft Technical Guidelines on Environmentally Sound Management for Full and Partial Dismantling of Ships, (10 January 2001), online: Basel Action Network <http://www.ban .org/Library/BANComments%20ShipbreakingGuidelines2.PDF>.

Comments on the Indian Committee Inspection Report on the Hazardous Materials onboard the SS Blue Lady (31 July 2006), online: Basel Action Network <http:www.ban.org/Library/NGO_Platform_Critique_on_TC_Inspection_Report_Final.pdf>.

Conner, Gary. "Scraping the Hull: Ridding your organization of barnacles" (6 April 2004), online: thefabricator.com <http://www.thefabricator.com/CEOs/CEOs_Article.cfm?ID=877>.

Cooper, Jeremy. "Poverty and Constitutional Justice: The Indian Experience" (1993) 44 Mercer L. Rev. 611 (WLeC).

Das, Prafulla. "Broken lives" *Frontline [of India]* 23:2 (28 February 2006), online: Frontline <http://www.flonnet.com/fl2302/stories/20060210007301600.htm>.

Dasgupta, Manas. "Slow death for ship-breaking industry" *The Hindu [of India]* (6 March 2006), online: The Hindu <http://www.thehindu.com/>.

Dias, Ayesha. "Judicial Activism in the Development and Enforcement of Environmental Law: Some Comparative Insights from the Indian Experience" (1994) 6 J. Envtl. L. 243 (HeinOnline).

Dixit, Vinod. "Justice as Fairness – A Restatement" (2005) 1 J. Nat'l Judicial Academy 396.

Dodds, David. "Breaking Up is Hard to Do: Environmental Effects of Shipwrecking and Possible Solutions under India's Environmental Regime", Comment, (2007) 20 Pac. McGeorge Global Bus. & D.L.J. 207.

Dzidzornu, David M. "Coastal State Obligations and Powers Respecting EEZ Environmental Protection under Part XII of the UNCLOS: A Descriptive Analysis" (1997) 8 Colo. J. Int'l Envtl. L. & Pol'y 283 (WLeC).

Earthjustice, Press Release, "Legal Challenge Keeping "Ghost Fleet" Ship Scrapping Jobs in the US" (15 October 2004), online: Earthjustice <http://www.earthjustice.org/news/press/004/legal_challenge_keeping_ghost_fleet_ship_scrapping_jobs_in_the_us.html>.

ECORYS Transport, "The Ship Recycling Fund: Financing Environmentally Sound Scrapping and Recycling of Sea-going Ships, Final, (1 February 2005), 26, online: Greenpeace International <http://www.greenpeace.org/raw/content/india/press/reports/the-ship-breaking-fund.pdf>.

Elkins, David. "Responding to Rawls: Toward a Consistent and Supportable Theory of Distributive Justice" (2007) 21 BYU J. Pub. L. 267.

Evans, S.M. Leksono, T. & McKinnell, P.D. "Tributyltin Pollution: A Diminishing Problem Following Legislation Limiting the Use of TBT-Based Anti-fouling Paints" (1995) 30:1 Mar. Pollution Bull. 14 (ScienceDirect).

Fayette, Louise de La. "New Developments in the Disposal of Offshore Installations" (1999) 14 Int'l J. Mar. & Coast. L. 523 at 525 (HeinOnline).

——. "The London Convention 1972: Preparing for the Future" (1998) 13 Int'l J. Mar. & Coast. L. 515 (HeinOnline).

Firestone, Jeremy & Corbett, James J. "Coastal and Port Environments: International Legal and Policy Responses to Reduce Ballast Water Introductions of Potentially Invasive Species" (2006) 7:1 S.D. L. & Pol'y 45 (WLeC).

Fisher, Emily. "Sustainable Development and Environmental Justice: Same Planet, Different Worlds?" (2003) 26 Environs Envtl. L. & Pol'y J. 201 (WLeC).

Fu, Chung-Te & Wu, Shian-Chee. "Bioaccumulation of Polychlorinated Biphenyls in Mullet Fish in a Former Ship Dismantling Harbour, a Contaminated Estuary, and Nearby Coastal Fish Farms (2005) 51 Mar. Pollution Bull. 932 (ScienceDirect).

Goldsmith, Rt. Hon. Lord QC. "Keynote Address" (Remarks delivered at the Stanford Law Review Symposium in Global Constitutionalism at the Stanford Constitutional Law Center, 16 February 2007), (2007) 59 Stan. L. Rev. 1155 (WLeC).

Greenpeace, "Destination Unknown: European single hull oiltankers: no place to go" *Greenpeace International*, (December 2004), online: Greenpeace International <http://www.greenpeace.org/raw/content/india/press/reports/destination-unknown.pdf>.

Greenpeace, "Special Report, Shipbreaking: Toxic Waste in Disguise the China Connection" at 2, online: Greenpeace International <http://www.greenpeace.org/raw/content/international/press/reports/shipbreaking-toxic-waste-in-d.pdf>.

Gregory, Kimberly K. "The Basel Convention and the International Trade of Hazardous Waste: The Road to the Destruction of Public Health and the Environment is Paved with Good Intentions" (2001) 10:2 Currents 80 (WLeC).

Hassan, Daud. International Conventions Relating to Land-Based Sources of Marine Pollution Control: Applications and Shortcomings (2004) 16 Geo. Int'l Envtl. L. Rev. 657 (WLeC).

Hassan, Parvez & Azim, Azim Azfar. "Securing Environmental Rights through Public Interest Litigation in South Asia", Comment, (2004) 22 Va. Envtl. L.J. 215 (WLeC).

Head, III, Thomas R. "PCBs-The Rise and Fall of an Industrial Miracle" (2005) 19:4 Nat. Resources & Env't. 15.

Hodson, John D. "Nozick, Libertarianism and Rights" (1977) 19 Ari. L. Rev. 212 (HeinOnline).

Hugh J. Marbury, "Hazardous Waste Exportation: The Global Manifestation of Environmental Racism", Note, (1995) 28 Vand. J. Transnat'l L. 251 (WLeC).

"India media ban over 'toxic' ship" *BBC News* (13 February 2006), online: BBC News <http://news.bbc.co.uk/2/hi/south_asia/4708530.stm>.

Interview of Claude Alvares by Lyla Bavadam [n.d. 2006?] in Cover Story, "We get the waste and pay for it" *Frontline [of India]* 23:2 (28 January 2006), online: Frontline < http://www.flonnet.com/fl2302/stories/20060210006601900.htm>.

Iyer, V.R. Krishna. "Bhoposhima: Crime without Punishment: Case for Crisis Management Jurisprudence" *Economic and Political Weekly* 26:47 (23 November 1991) 2705.

J.R., "Paint additive hammers coral" *Science News* 167:13 (26 March 2005) 206 (JSTOR).

Kaiser, Mark J. "A Review of Ship Breaking and Rig Scrapping in the Gulf of Mexico" (2008) 39 Ocean Devel. & Int'l Rev. 178.

Kannan, Kurunthachalam *et. al*, "Butyltins in Muscle and Liver of Fish Collected from Certain Asian and Oceanian Countries" (1995) 90:3 Envtl. Pollution 279 (ScienceDirect).

Karp, James P. "Sustainable Development: Toward a New Vision", Essay, (1994) 13 Va. Envtl. L.J. 239 (WLeC).

Kelly, Petra K. "The Need for Eco-Justice" (1990) 14 Fletcher F. World Aff. 327 (HeinOnline).

Kerckhof, Francis Roose, Patrick & Haelters Jan. "The Tricolor Incident: From Collision to Environmental Disaster" (2004) 6:3 Atlantic Sea Birds 85, online: Royal Netherlands Institute for Sea Research <http://www.nioz.nl/public/mee/birds/0272.pdf>.

Keyuan, Zou. "Implementing Marine Environmental Protection Law in China: Progress, Problems and Prospects" (1999) 23:3 Mar. Pol'y 207 (ScienceDirect).

Kilcoyne, Diane. "The Basel Convention: Will It Curtail Hazardous Waste Exports" (1992) 16:2 Environs Envtl. L. & Pol'y J. 47 (HeinOnline).

Killion, David. "Trans-Atlantic Ghost Busting: The Failed Attempt to Dispose of the Chesapeake Ghost Fleet in the United Kingdom", Comment, (2008) 42 U. Rich. L. Rev. 731 (WLeC).

Koh, Tommy T.B. "A Constitution for the Oceans" (Remarks at the Final Session of the Third United Nations Conference on the Law of the Sea, Montego Bay, 1–11 December 1982), online: United Nations <http://www.un.org/Depts/los/convention_agreements/texts/koh_english.pdf>.

Krishna, Gopal. "Blue Lady seduces SC on Environment Day" *Basel Action Network* (6 June 2006), online: Basel Action Network <http://www.ban.org/Library/Editorials/060606_environment_day.html>.

——. "French apex court rules, Clemenceau recalled" *India Together* (16 July 2008), online: India Together <http://www.indiatogether.org/2006/feb/env-shipretn.htm>.

——. "The scrapping of Riky" *India Together* (16 July 2008), online: India Together <http://www.indiatogether.org/2006/mar/env-riky.htm>.

——. "Will the Blue Lady do a Le Clemenceau?" *India Together* (14 July 2008), online: India Together <http://www.indiatogether.org/2007/feb/env-bluelady.htm>.

Langa, Mahesh. "Blue Lady: End of an era" *Express India* (3 August 2006), online: expressindia <http://cities.expressindia.com/fullstory.php?newsid=195268>.

Lipman, Z. "Transboundary Movement of Hazardous Waste: Environmental Justice Issues for Developing Countries" [1999] Acta Jur. 266 275 (HeinOnline).

Low, Tim. "Ballast Invaders: the Problem and Response" *Invasive Species Council, Australia* (September 2003), online: Invasive Species Council <http://www.invasives.org.au/downloads/tl_ballast.pdf>.

Martin, Rex. "Rawls's New Theory of Justice" (1994) 69 Chicago-Kent L. Rev. 737 (WLeC).

Masood, Alauddin. "Ship-breaking attracting entrepreneurs" *Dawn the Internet ed.* (24 December 2001), online: Dawn: the Internet Edition <http://www.dawn.com/2001/12/24/ebr20.htm>.

Matthews, Sue & Brand, Kobie. "Africa Invaded: The Growing Danger of Invasive Alien Species" *The Global Invasive Species Programme*, 2004, online: GiSP <http://www.gisp.org/publications/invaded/gispAfrica.pdf>.

McCloskey, Michael. "The Emperor has No Clothes: The Conundrum of Sustainable Development" (1999) 9 Duke Envtl. L. & Pol'y F. 153 (WLeC).

McConnell, Moira L. "Ballast and Biosecurity: The Legal, Economic and Safety Implications of the Developing International Regime to Prevent the Spread of Harmful Aquatic Organisms and Pathogens in Ships' Ballast Water" (2003) 17 Ocean Y.B. 213.

——. & Gold, Edgar. "The Modern Law of the Sea: Framework for the Protection and Preservation of the Marine Environment?" (1991) 23 Case W. Res. J. Int'l L. 83 (WLeC).

McGee, Sarah. "Proposals for Ballast Water Regulation: Biosecurity in an Insecure World" [2001] Colo. J. Int'l Envtl. L. & Pol'y 141 (WLeC).

Mehta, M.C. "Growth of Environmental Jurisprudence in India" [1999] Acta Jur. 71 (HeinOnline).

——. "The Accountability Principle: Legal Solutions to Break Corruption's Impact on India's Environment" Book Excerpt of *In the Public Interest* by M.C. Mehta, (2006) 21 J. Envtl. L. & Litig. 141 (WLeC).

Mensink, Berend P. *et al.* "Tributyltin Exposure in Early Life Stages Induces the Development of Male Sexual Characteristics in the Common Whelk, *Buccinum undatum*" (1996) 42:I-4 Mar. Envtl. Research 151 (ScienceDirect).

Mitra, Kushan. "The Slow Death of A Graveyard; Even as Alang awaits the arrival of another vessel laden with asbestos, the Blue Lady, its ship breakers know that their days are numbered" *Business Today [of India]* (16 July 2006) 224.

Morita, Takako. "N.I.M.B.Y. Syndrome and the Ticking Time Bomb: Disputes over the Dismantling of Naval Obsolete Vessels", Note, (2005) 17 Geo. Int'l Envtl. L. Rev. 723 at 726 (WLeC).

Naravane, Vaiju. "A shipload of trouble from France" *The Hindu [of India]* (3 January 2006), online: The Hindu <http://www.thehindu.com/>.

"Nature reserve or scrap yard?" *Greenpeace News* (10 April 2003), online: Greenpeace <http://www.greenpeace.org/international/news/nature-reserve-or-scrap-yard>.

Negre, Andrew P. *et al.* "Understanding Ship-grounding Impacts on a Coral Reef: Potential Effects of Anti-foulant Paint Contamination on Coral Recruitment" (2002) 44 Mar. Pollution Bull. 111 (ScienceDirect).

Neşer, Gökdeniz. "The Shipbreaking Industry in Turkey: Environmental, Safety and Health Issues" (2008) 16 J. Cleaner Production 350 (ScienceDirect).

Nester, Katie. "*Norfolk & Western Railway Company v. Ayers*: Asbestosis-Inflicted Plaintiffs and Fear of Cancer Claims", Notes, (2004) 23 St. Louis U. Pub. L. Rev. 367.

NGO Platform on Shipbreaking, Press Release, "Human Rights and Environmental Groups Condemn India for Callous Disregard for its Poorest Labourers and the Law after Shipbreaking Ruling" (12 September 2007), online: Basel Action Network <http://www.ban.org/ban_news/2007/070912_shipbreaking_ruling.html>.

Noland, L. Christopher. "The Ghouls That Won't Go Away – The Dire Environmental Consequences Posed by the Ghost Fleet in the James River", Note, (2006) 30 Wm. & Mary Envtl. L. & Pol'y Rev. 513 (WLeC).

Nussbaum, Martha. "Human Rights and Human Capabilities" (2007) 20 Harv. Hum. Rts. J. 21 (WLeC).

O'Shea, Sharonne & Cangelosi, Allegra. "Trojan Horses in Our Harbors: Biological Contamination from Ballast Water Discharge" (1996) 27 U. Tol. L. Rev. 381 (WLeC).

Olson, Molly Harriss. "Accepting the Sustainable Development Challenge" (1995) 31 Willamette L. Rev. 253 (WLeC).

Park, Rozelia S. "An Examination of International Environmental Racism through the Lens of Transboundary Movement of Hazardous Wastes", Note, (1998) 5 Ind. J. Global Legal Stud. 659 (WLeC).

Portas, Pierre. "The Basel Convention, Back to the Future", Commentary, (2006) 6:3 S.D.L. & Pol'y 38 (HeinOnline).

Pukkala, Eero & Saarni, Heikki. "Cancer Incidence among Finnish Seafarers, 1967-92" (1996) 7:2 Cancer Causes & Control 231 (JSTOR).

Puthucherril, Tony George & Vijayabalan, Lekshmi. "Judiciary in Quest for Gender Justice to Women" (2006) 4 Sri Lanka L.C.L. Rev. 3.

——., Tony George. "Ballast Waters and Aquatic Invasive Species: A Model for India" (2008) 19 Colo. J. Int'l Envtl. L & Pol'y 381 (WLeC).

——., Tony George. "Harnessing the Atom: Strengthening the Regulatory Board for Nuclear Safety in India Based on the Canadian Experience" (2008) 26:4 J. Energy & Nat'l Res. L. 553.

"Radioactive material aboard Blue Lady" *The Times of India* (30 July 2007) (Prowler).

Rahman, Md. Arifur & Shahin, Muhammed Ali. "Shipbreaking Industry in Bangladesh Needs Modernization and Policy Planning" *The Bangladesh Observer* (9 November 2007) at 4.

Ramesh, M.K. "Environmental Justice: Courts and Beyond" (2002) 3:1 Indian J. Envtl. L. 20.

Rath, Dwarik Nath. "In the face of death" *PUCL Bulletin* (June 2003), online: What's new on the site <http://www.pucl.org/by-month-2003.htm>.

Reddy, M. Srinivasa *et al.* "Description of the Small Plastics Fragments in Marine Sediments Along the Alang-Sosiya Ship-Breaking Yard, India" (2006) 68 Estuarine, Coast. & Shelf Sci. 656 (ScienceDirect).

——. *et al.* "Distribution, Enrichment and Accumulation of Heavy Metals in Coastal Sediments of Alang–Sosiya Ship Scrapping Yard, India" (2004) 48 Mar. Pollution Bull. 1055 (ScienceDirect).

Reznikoff, Paul. "Lead Poisoning", (1942) 42:10 Am. J. Nursing 1123 (JSTOR).

Rosencranz, Armin & Yurchak, Kathleen D. "Progress on the Environmental Front: The Regulation of Industry and Development in India" (1996) 19 Hastings Int'l & Comp. L. Rev. 489 (WLeC).

Røst, Odd Harald. "Shipbreaking: Lives at stake" *Metal World* 4:2 (June 2003) 12, online: International Metalworkers' Federation <http://www.imfmetal.org/main/files/2-2003.pdf>.

Russell, Dawn A. & McConnell, Moira L. "Preface" (1995) 18:1 Dal. L.J. 5 (HeinOnline).

Sainath, P. "17,060 farmer suicides in one year" *The Hindu [of India]* (31 January 2008), online: The Hindu <http://www.thehindu.com/>.

——. "One farmer's suicide every 30 minutes" *The Hindu [of India]* (15 November 2007), online: The Hindu <http://www.thehindu.com/>.

——. "The decade of our discontent" *The Hindu [of India]* (9 August 2007), online: The Hindu <http://www.thehindu.com/>.

Saleh, Heba. "Egypt grants 'toxic' ship passage" *BBC News* (15 January 2006), online: BBC News <http://news.bbc.co.uk/2/hi/south_asia/4615662.stm>.

Sankaran, Kamala. "Protecting the Worker in the Informal Economy: The Role of Labour Law" in Guy Davidov & Brian Langille, eds., *Boundaries and Frontiers of Labour Law* (Portland: Hart Publishing, 2006) 205.

Sanyal, Santanu. "Kolkata Dock to re-build ship-breaking potential" *The Hindu Business Line [of India]* (11 July 2005), online: The Hindu Business Line <http://www.thehindubusinessline.com/2005/07/11/stories/2005071100330600.htm>.

Sawyer, John F. "Shipbreaking and the North-South Debate: Economic Development or Environmental and Labor Catastrophe?", Comment, (2001) 20 Penn St. Int'l L. Rev. 535 (WLeC).

Sharma, Kalpana. "Breaking ships need not break lives" *The Hindu [of India]* (21 August 2006), online: The Hindu <http://www.thehindu.com/>.

Shashikumar, V.K. "Poisoned Lives – IX Alang the Waste Coast" *Tehelka [of India]* (13 August 2005), online: Tehelka – The People's Paper <http://www.tehelka.com/story_main13.asp?filename=Ne081305alang_the.asp>.

Shaw, Bill Winslett Brenda J. & Cross, Frank B. "The Global Environment: A Proposal to Eliminate Marine Oil Pollution" (1987) 27 Nat. Resources J. 157 (HeinOnline).

Silveira, MaryPat Williams. "International Legal Instruments and Sustainable Development: Principles, Requirements, and Restructuring" (1995) 31 Willamette L. Rev. 239 (HeinOnline).

Star, Ronald. "American and Japanese Controls on Polychlorinated Biphenyls (PCBs)" (1976) 1 Harv. Envtl. L. Rev. 561.

Stein, Mark S. "Nozick: A Utilitarian Reformulation", Essay, (1997) 18 N. Ill. U.L. Rev. 339 (WLeC).

"Stifling democratic discussion" *The Hindu [of India]* (16 February 2006), online: The Hindu <http://www.thehindu.com/>.

Stone, Hilary "Flouting the Law – European Shipment of Hazardous Waste" (2007) 21:1 Nat. Resources & Env't 49 (HeinOnline).

Sundram, Muthu S. "Basel Convention on Transboundary Movement of Hazardous Wastes: Total Ban Amendment" (1997) 9 Pace Int'l L. Rev. 1 (WLeC).

Tarak, Pedro. "Introduction" (Seminar on the Law of Sustainable Development, Argentina, 19 April 1995), (1996) 13 Pace. Envtl. L. Rev. 561 (WLeC).

"Tax evasion only hope for industry's survival" *International The News [of Pakistan]* (29 March 2008), online: International The News <http://thenews.jang.com.pk/daily_detail.asp?id=103695>.

Thakker, Nisha. "India's Toxic Landfills: A Dumping Ground for the World's Electronic Waste" (2006) 6 S. D. L. & Pol'y 58 (HeinOnline).

Thant Zin, Cowan. "Single-hull VLCC owners opt for conversion" (6 August 2008), online: tanker-world <http://www.tankerworld.com/news/2008/08/72785>.

"The Sangam of Foreign Investment, Multinational Corporations and Human Rights: An Indian Perspective for a Developing Asia" [2004] Sing. J.L.S. 305 (HeinOnline).

Tsimplis, Michael. "Liability and Compensation in the International Transport of Hazardous Wastes by Sea: The 1999 Protocol to the Basel Convention" (2001) 16 Int'l J. Mar. & Coast. L. 295 (HeinOnline).

VanderZwaag, David. "The Concept and Principles of Sustainable Development: Rio-Formulating Common Law Doctrines and Environmental Laws" (1993) 13 Windsor Y.B. Access Just. 39 (HeinOnline).

Venkatesan, V. "Breaking rules" *Frontline [of India]* 22:25 (3 December 2005), online: Frontline <http://www.flonnet.com/fl2225/stories/20051216002304500.htm>.

——. "Deadly Vessel" *Frontline [of India]* 23:2 (28 January 2006), online: Frontline <http://www.flonnet.com/fl2302/stories/20060210008300400.htm>.

——. "Pressures of monitoring" *Frontline [of India]* 22:25 (3 December 2005), online: Frontline <http://www.flonnet.com/fl2225/stories/20051216002904700.htm>.

Venkateshwarlu, K. "Moves afoot to set up ship-breaking unit near Kakinada" *The Hindu [of India]* (10 October 2004), online: The Hindu <http://www.thehindu.com/>.

Weeramantry, H.E. Judge C.G. "Achieving Sustainable Justice through International Law" in Marie-Claire Cordonier Segger & Judge C.G. Weeramantry, eds., *Sustainable Justice: Reconciling Economic, Social and Environmental Law* (Boston: Martinus Nijhoff Publishers, 2005) 15.

Weinstein, Elaine B. "The Impact of Regulation of Transport of Hazardous Waste on Freedom of Navigation" (1994) 9 Int'l J. Marine & Coast. L. 135 (HeinOnline).

White, Marianne. "Warships sunk for artificial reefs may pose PCB hazard" *The Province [of Canada]* (8 January 2008), online: The Province <http://www.canada.com/theprovince/news/story.html?id=5ef3226a-d6ca-43c4-abeb-275a4b2c20f8>.

Xu, Xiaobing & Wilson, George. "On Conflict of Human Rights" (2006) 5 Pierce L. Rev. 32 (WLeC).

Yee, Amy. "Wreckers in deep water" *Financial Times [London]* (18 April 2007) 4, online: FT.com < http://www.ft.com/home/us>.

Young, Oran R. "Regime Dynamics: The Rise and Fall of International Regimes" in Stephen D. Krasner, ed., *International Regimes* (London: Cornell University Press, 1983) 93.

Zaidi, Annie. "Killer mineral" *Frontline [of India]* 23:20 (7 October 2006), online: Frontline <http://www.hinduonnet.com/fline/fl2320/stories/20061020002610000.htm>.

Zin, Cowan Thant. "Single-hull VLCC owners opt for conversion" (6 August 2008), online: tanker-world <http://www.tankerworld.com/news/2008/08/72785>.

Books

Anand, R.P. *Origin and Development of the Law of the Sea: History of International Law Revisited* (The Hague: Martinus Nijhoff Publishers, 1982).

Bates, John H. & Benson, Charles. *Marine Environment Law*, looseleaf (London: Lloyd's of London Press Ltd., 1993) ¶ 8.1.

Colborn, Theo, Dumanoski, Dianne & Meyers, John Peter. *Our Stolen Future* (New York: Penguin, 1997).

Eyres, D.J. *Ship Construction*, 6th ed. (Boston: Butterworth-Heinemann, 2007) (ScienceDirect).

Gold, Edgar. *Maritime Transport: The Evolution of International Marine Policy and Shipping Law* (Massachusetts: Lexington Books, 1981).

——., Chircop, Aldo & Kindred, Hugh. *Maritime Law* (Toronto: Irwin Law, 2003).

Harrison, Jeffery L. *Law and Economics*, 2d ed., West Nutshell Series (St. Paul, MN: West Group, 2000).

Maritime Labour Conventions and Recommendations including Standards Relating to Fishing, Dock Work and Inland Navigation, 4th ed. (N.p., 1998).

Nozick, Robert. *Anarchy, State and Utopia*, 1st ed. (New York: Basic Books, 1974).

Rawls, John. *Justice as Fairness: A Restatement*, Ering Kelly (Massachusetts: Harvard University Press, 2001).

——., John. *A Theory of Justice*, 1st ed. (Oxford: Oxford University Press, 1999).

Sainath, P. *Everybody Loves a Good Drought: Stories from India's Poorest Districts* (London: Headline Book Publishing, 1998).

Sen, Amartya. *Development as Freedom* (Oxford: Oxford University Press, 1999).

Serageldin, Ismail. & Steer, Andrew. eds. with the collaboration of Michael M. Cernea *et al.*, *Making Development Sustainable: From Concepts to Action*, Environmentally Sustainable Development Occasional Paper Series No. 2 (Washington D.C.: The International Bank for Reconstruction and Development/The World Bank, 1994).

Stopford, Martin. *Maritime Economics* 2d ed. (London: Routledge, 1997).

Van Dyke, Jon M. "Sharing Ocean Resources – In a Time of Scarcity and Selfishness", in Harry N. Scheiber, ed., *Law of the Sea: The Common Heritage and Emerging Challenges* (The Hague: Martinus Nijhoff Publishers) 3.

World Bank. *World Development Report, 1997: The State in a Changing World* (New York: Oxford University Press, 1997).

World Commission on Environment & Development. *Our Common Future* (Oxford: Oxford University Press, 1987).

Electronic Sources

About Ecodock, online: nv ecodock <http://www.ecodock.info/uk/overecodock.php>.

About the Basel Action Network – BAN, online: Basel Action Network <http://www.ban.org/main/about_BAN.html>.

Additional Case Studies: What a Long, Strange Trip it Has Been, online: Principles of Environmental Science <http://highered.mcgraw-hill.com/site/0072919833/student_view0/chapter13/additional_case_studies.html>.

Alang Ship Recycling, online: Gujarat Maritime Board <http://www.gmbports.org/alang_alang.htm>.

Alang Ship Recycling: Compliance, online: Gujarat Maritime Board <http://www.gmbports.org/alang_shecomp.htm>.

Anti-fouling Systems, online: International Maritime Organization <http://www.imo.org/>.

ARSBC Artificial Reefs in British Columbia, online: Artificial Reef Society of British Columbia <http://www.artificialreef.bc.ca/OurReefs/index.htm>.

Ban Amendment to the Basel Convention on the Control of Transboundary Movements of Hazardous Wastes and their Disposal Geneva, 22 September 1995, online: Basel Convention <http://www.basel.int/ratif/ban-alpha.htm>.

Basel Action Network on behalf of Global NGO Platform on Shipbreaking, *Critique of the Draft I.M.O. "International Convention for Safe and Environmentally Sound Recycling of Ships"* (15 March 2006), online: Basel Action Network <http://www.ban.org/Library/IMO_Draft_Convention_CritiqueFINAL.pdf>.

Basel Convention: Dismantling of Ships – Relevant Caselaw <http://www.basel.int/ships/relevcaselaw.html>.

BIMCO <http://www.bimco.org/>.

BIMCO Policy on Ship Recycling, online: BIMCO <http://www.bimco.org/Members%20Area/News/General_News/2007/01/0126-BIMCO%20Policy%20on%20Ship%20Recycling.aspx>.

BIMCO Standard Contract for the Sale of Vessels for Demolition and Recycling Code Name: "DEMOLISHCON", osnline: BIMCO <http://www.bimco.org/~/media/9698E1F48AFE4174B3761F31CA80E361.ashx>.

Breaking News, NGO platform on shipbreaking <http://www.shipbreakingplatform.com/>.

Craig Eason "DNV Predicts Fleet Overcapacity Could Reach 10,000 Vessels" *Lloyd's List* (10 March 2009), online: Lloyd's List <http://www.lloydslist.com/ll/news/viewArticle.htm?articleId=20017626328>.

——., Craig. "Eitzen Cancels a Third of its Newbuild Orders" *Lloyd's List* (8 April 2009), online: Lloyd's List <http://www.lloydslist.com/ll/news/viewArticle.htm?articleId =20017637394>.

——., Craig. "Srab Walks Away from Newbuilding Order" *Lloyd's List* (20 March 2009), online: Lloyd's List <http://www.lloydslist.com/ll/news/viewArticle.htm?articleId=20017630516>.

DEMOLISHCON: SALESCRAP 87 Contract Revised and Re-named DEMOLISHCON, online: BIMCO <http://www.bimco.org/Corporate%20Area/Documents/Document_samples/Sundry_Other_Forms/DEMOLISHCON.aspx>.

EU Comments Received Pursuant to OEWG6 Decisions, "Assessment on Ship Dismantling with Particular Reference to the Levels of Control and Enforcement Established by the Basel Convention and the Expected Level of Control and Enforcement to be provided by the Draft Ship Recycling Convention in their Entirety", online: Basel Convention <http://www.basel.int/ships/commentsOEWG6/oewg6.html>.

Forest Case Update, online: Forest Case Update <http://www.forestcaseindia.org/>.

Gaddani Ship-breaking, online: Lasbela Government <http://www.lasbeladistrictgovt.com/gaddani/ship%20breaking.htm>.

"Global Ballast Water Management Programme" *Ballast Water News* Issue 3 (October-December 2000), online: Ballast Water News <http://globallast.imo.org/BallastWaterNews3.pdf>.

Global Programme for Sustainable Ship Recycling, online: Basel Convention <http://www.basel.int/ships/gpssr/index.html>.

Goal and Objective, online: Project AIDS Alang <http://www.projectaidsalang.org/>.

Government of India, *Ministry of Environment & Forests*, online: Ministry of Environment & Forests <http://envfor.nic.in/>.

Greenpeace International/Basel Action Network (BAN), "The IMO Guidelines on Ship Recycling Annotated", 21 November 2003, online: Basel Action Network <http://www.ban.org/Library/greenpeace-ban-iii-3e.pdf>.

Greenpeace, online: Greenpeace <http://www.greenpeaceweb.org/shipbreak/>.

Greenpeace, *Playing Hide and Seek* (December 2003), online: Greenpeace Nederland <http://www.greenpeaceweb.org/shipbreak/rapport_DEF_versie_2.pdf>.

Greenpeace-FIDH in cooperation with YPSA, End of Life Ships: The Human Cost of Breaking Ships, Jacqueline Schuiling (Netherlands: Greenpeace International, 2005), online: Greenpeace International <http://www.greenpeace.org/raw/content/international/press/reports/end-of-life-the-human-cost-of.pdf>.

Gujarat Maritime Board <http://www.gmbports.org/alangship.htm>.

IMO Meeting Adopts Accelerated Single-hull Tanker Phase-out, New Regulation on Carriage of Heavy Fuel Oil, MEPC 50th Sess. (1 & 4 December 2003), online: International Maritime Organization <http://www.imo.org/About/mainframe.asp?topic_id=758&doc_id=3341>.

IMO, *The London Convention and Protocol: Their Role and Contribution to Protection of the Marine Environment*, online: International Maritime Organization <http://www.imo.org/includes/blastDataOnly.asp/data_id%3D21278/LC-LPbrochure.pdf>.

Industry Code of Practice on Ship Recycling, (August 2001), online: Basel Convention <http://www.basel.int/ships/docs/08e.pdf>.

International Chamber of Shipping, online: International Chamber of Shipping <http://www.marisec.org/ics/ICS%20brochure.pdf>.

International Labour Standards, online: International Labour Organization <http://www.ilo.org/ilolex/english/conventions.pdf>.

International Labour Standards, online: International Labour Organization <http://www.ilo.org/ilolex/english/recommendations.pdf>.

International Maritime Organization Diplomatic Conference on Ship Recycling, online: IMO CONF2009 <http://www.imoconf2009.hk/>.

International Metalworkers' Federation, *Status of Shipbreaking Workers in India: A Survey*, (2006), online: International Metalworkers' Federation <http://www.imfmetal.org/main/files/0604281065779/Shipbreaking_survey.pdf>.

Introduction to IMO, online: International Maritime Organization <http://www.imo.org/>.

ISRA, online: I.S.R.A. <http://www.isra-dis.nl/>.

Jones, Samantha L. "A Toxic Trade: Ship Breaking in China" (A China Environmental Health Project Fact Sheet, 1 February 2007), online: Woodrow Wilson International Center for Scholars <http://www.wilsoncenter.org/topics/docs/shipbreaking_feb1.pdf>.

Justice K.G. Balakrishnan, Chief Justice of India, online: Mediavigil <http://mediavigil.blogspot.com/2007_08_01_archive.html>.

Letter from Ship Recycling Industries Association (India) to H.K. Dash Saheb, Vice Chairman of the Gujarat Maritime Board (18 October 2006), online: Ship Recycling Industries Association (India) <http://www.sriaindia.com/home.html>.

Maharashtra Pollution Control Board, *Second ATR-Part V*, (16 March 2004), online: Maharashtra Pollution Control Board <http://mpcb.mah.nic.in/images/2ndanex5.pdf>.

Mc Farlane, James. Address (Speech on the 3rd International Intercessional Meeting of the IMO Working Group on Ship Recycling in Nantes, France, (21–25 January 2008), online: I.S.R.A <http://www.isra-dis.com/documents/Speech%20by%20board%20member%20James%20Mc%20Farlane.pdf>.

Mikelis, Nikos. "Development of the International Convention for the Safe and Environmentally Sound Recycling of Ships" (PowerPoint presented to the Preparatory Discussions on Promoting Sustainable Ship Recycling through the Global Programme, Dhaka, Bangladesh, 13 January 2008), online: Basel Convention <http://www.basel.int/ships/gpssr/index.html>.

Nairobi International Convention on the Removal of Wrecks, 2007, online: International Maritime Organization <http://194.196.162.45/Legal/mainframe.asp?topic_id=1319>.

Obligations and Opportunities for a Mandatory Alternate or Additional Instrument to the Basel Convention for End-of-Life Ships, Submitted by Greenpeace & the Basel Action Network, 26 May 2005, online: Basel Action Network <http://www.ban.org/Library/mepc_submission_long.pdf>.

Oftedal, Sveinung. *Development of the Draft International Convention for the Safe and Environmentally Sound Recycling of Ships*, online: European Commission Environment <http://ec.europa.eu/environment/greenweek2007/sources/files/day2_session8_2/Oftedal.pdf>.

Ormond, Thomas. "Developments in Europe: The European Commission's Green Paper on Better Ship Recycling" (PowerPoint presented to the Preparatory Discussions on Promoting Sustainable Ship Recycling through the Global Programme, Dhaka, Bangladesh, 13 January 2008), online: Basel Convention <http://www.basel.int/ships/gpssr/index.html>.

Parties to the Basel Convention, online: Basel Convention <http://www.basel.int/ratif/convention.htm>.

P&O Nedlloyd and China Partnership in Ship Recycling, online: Hong Kong Shipowners Association <http://www.hksoa.org/member/Technical%20Papers/Tech/Tom%20Peter%20Blankestijn.pdf>.

"Preparatory Discussions on Promoting Sustainable Ship Recycling through the Global Programme: Discussion Document", Global Programme on Sustainable Ship Recycling, Dhaka, Bangladesh, 13 January 2008, online: Basel Convention <http://www.basel.int/ships/gpssr/index.html>.

Programme on Safety and Health at Work and the Environment (SafeWork,), online: International Labour Organization <http://www.ilo.org/global/lang–en/index.htm>.

Proposed ILO Declaration on Social Justice for a Fair Globalization, online: International Labour Organization <http://www.ilo.org/wcmsp5/groups/public/—ed_norm/—relconf/documents/meeting-document/wcms_094042.pdf>.

Rahman, Ataur & Ullah, AZM Tabarak. "Ship Breaking: A Background Paper Prepared for the ILOs Sectoral Activities Programmes" (1999), online: International Labour Organization <http://www.ilo.org/public/english/protection/safework/sectors/shipbrk/shpbreak.htm>.

Role in Steel Economy, online: Ship Recycling Industries Association (India) <http://www.sriaindia.com/steel_economy.html>.

Roy, Bishwajit. "Health Problems Among the Workers in the Ship Breaking Industry", online: BILS <http://66.102.1.104/scholar?hl=en&lr=&q=cache:LBrRFQaDcUQJ:www.bils-bd.org/Journal_Archive_pages/2004/April_June/English/Health_Problems.pdf+Health+Problems+among+the+Workers+in+the+Ship+Breaking+industry>.

Rugarabamu, Donata. "Inaugural Discussions on the Global Programme on Sustainable Ship Recycling" (PowerPoint presented to the Inaugural Discussions on the Global Programme on Sustainable Ship Recycling, Dhaka, Bangladesh, 12 January 2008), online: Basel Convention <http://www.basel.int/ships/gpssr/index.html>.

Safe Work Shipbreaking, online: International Labour Organization <http://www.ilo.org/public/english/protection/safework/sectors/shipbrk/index.htm>.

Safety and Health in Shipbreaking: Guidelines for Asian Countries and Turkey, ILO, Interregional Tripartite Meeting of Experts on Safety and Health in Shipbreaking for Selected Asian Countries and Turkey, Bangkok, 7–14 October 2003, MESHS/2003/1, online: International Labour Organization <http://www.ilo.org/public/english/standards/relm/gb/docs/gb289/pdf/meshs-1.pdf>.

Sector Meetings, online: International Labour Organization <http://www.ilo.org/public/english/dialogue/sector/techmeet/iloimobc08/index.htm>.

Shahin, Md. Ali. "Shipping Hazardous Waste" (PowerPoint presented to the Preparatory Discussions on Promoting Sustainable Ship Recycling through the Global Programme, Dhaka, Bangladesh, 13 January 2008), online: Basel Convention <http://www.basel.int/ships/gpssr/index.html>.

Shahjahan, Md. "Ship Recycling in Bangladesh", (PowerPoint presented to the Inaugural Discussions on the Global Programme on Sustainable Ship Recycling, Dhaka, Bangladesh, 12 January 2008), online: Basel Convention <http://www.basel.int/ships/gpssr/index.html>.

Sharma, Atul. "Implementation of Recent Directives of the Hon'ble Supreme Court of India Regarding Sustainable and Environment Friendly Ship Recycling in India" (PowerPoint presented to the National Workshop on International Convention for Safe and Environmentally Sound Recycling of Ships, Maritime Training Institute, Mumbai, India, 9 January 2008) [unpublished], online: Gujarat Maritime Board <http://www.gmbports.org/international_workshop.htm>.

Ship Breaking Unit (Kozhikode) Port Based, online: Steel Industrials Kerala Limited <http://www.steelinkerala.com/html/s0220div.htm#div>.

Ship Breaking Unit at Kakinada Port, online: Department of Ports <http://www.andhraports.com/kkdshipbreakingunit.htm>.

Ship Recyclers' Association of Turkey, online: Ship Recyclers' Association of Turkey <http://www.gemisander.org/index.php?lang=en>.

Ship Recycling Industries Association (India), online: Ship Recycling Industries Association (India) <http://www.sriaindia.com/>.

Ship Recycling, online: EMSA <http://www.emsa.eu.int/end185d012d002.html>.

Ship Recycling, online: U.S. Department of State <http://www.state.gov/r/pa/ei/pix/b/sat/82342.htm>.

South Asia Environment and Social Development Unit, *Pakistan Strategic Country Environmental Assessment*, World Bank Report No. 36946-PK, Vol. 1: Main Report (21 August 2006), online: World Bank <http://siteresources.worldbank.org/SOUTHASIAEXT/Resources/Publications/448813-1188777211460/pakceavolume1.pdf>.

Summary of Status of Convention: As at 31 August 2009, online: International Maritime Organization <http://www.imo.org/conventions/mainframe.asp?topic_id=247>.

The 1998 Pulitzer Prize Winners Investigative Reporting, online: The Pulitzer Prizes< http://www.pulitzer.org/works/1998,Investigative+Reporting>.

The Conference of the Parties, online: Basel Convention <http://www.basel.int/meetings/cop/cop-0.html>.

The Problem, online: GloBallast Partnerships <http://globallast.im-o.org/index.asp?page=problem.htm&menu=true>.

U.S. Department of Transportation Maritime Administration, *The National Defense Reserve Fleet, Ready Reserve Force*, online: U.S. Department of Transportation <http://www.marad.dot.gov/Ship%20Disposal/PRESS_NDRF_RRF_4qtr05.pdf>.

UNEP Global Programme of Action for the Protection of the Marine Environment from Land-Based Activities, online: UNEP <http://www.gpa.unep.org/index.html?ln=6>.

Wingfield, Susan. "Introduction to the Global Programme Concept" (PowerPoint presented to the Preparatory Discussions on Promoting Sustainable Ship Recycling through the Global Programme, Dhaka, Bangladesh, 13 January 2008), online: Basel Convention <http://www.basel.int/ships/gpssr/index.html>.

Young Power in Social Action, *Workers in Shipbreaking Industries: A Base Line Survey of Chittagong (Bangladesh)* (Bangladesh, 2005), online: Young Power in Social Action <http://www.ypsa.org/publications/shipbreaking_baseline_survey.pdf>.

1996 Protocol to the London Convention 1972: Overview of Contracting Parties, online: International Maritime Organization <http://www.imo.org/includes/blastDataOnly.asp/data_id%3D25138/PROTPART.1st.doc>.

2003 Amendments to MARPOL, 4 December 2003, entry into force April 2005, online: IMO <http://www.imo.org/Conventions/contents.asp?doc_id=678&topic_id=258#2001>.

Encyclopedias

Halsbury's Laws of India, vol. 17 (India: Butterworths, 2003) "Employment-I (Unorganised Labour)" at 1 ¶ 130.001.

The Oxford Encyclopedia of Economic History, e-reference ed., Joel Mokyr ed. (USA: *Oxford University Press, 2005)* "India".

The Oxford Encyclopedia of Economic History, e-reference ed., Joel Mokyr ed. (USA: *Oxford University Press, 2005)* "Pakistan".

The Oxford Encyclopedia of Economic History, e-reference ed., Joel Mokyr ed. (USA: *Oxford University Press, 2005)* "Bangladesh".

The Oxford Encyclopedia of Maritime History, e-reference ed., John J. Hattendorf ed. (USA: Oxford University Press, 2007) "Fouling".

Monographs

Andersen, Aage Bjørn. "An Issues Paper: Worker Safety in the Ship-breaking Industries" (2001) (ILO, Sectoral Activities Programme), online: International Labour Organization <http://www.ilo.org/public/english/dialogue/sector/papers/shpbreak/wp-167.pdf>.

Bangladesh, Water Resources Planning Organisation. *Institutional Aspects of Ship Breaking Industry in Bangladesh*, WP044 (Dhaka: Ministry of Water Resources, 2005) (Prepared by Khurshed Alam), online: Integrated Coastal Zone Management, Bangladesh <http://www.iczmpbangladesh.org/rep/wp044.pdf>.

Denmark, Danish Environmental Protection Agency. *Shipbreaking in OECD*, Working Report No. 18, Danish Environmental Protection Agency, 2003 (by Frank Stuer-Lauridsen, Nikolai Kristensen & Jesper Skaarup Cowi A/S), online: Miljøministeriet <http://www2.mst.dk/Udgiv/publications/2003/87-7972-588-0/pdf/87-7972-589-9.pdf>.

Greenpeace-FIDH in cooperation with YPSA. *End of Life Ships: The Human Cost of Breaking Ships*, Jacqueline Schuiling (Netherlands: Greenpeace International, 2005), online: Greenpeace International <http://www.greenpeace.org/raw/content/international/press/reports/end-of-life-the-human-cost-of.pdf>.

Hess, Ron *et al. Disposal Options for Ships*, RAND Monograph Report (RAND Corporation, 2001), online: RAND Corporation <http://www.rand.org/pubs/monograph_reports/MR1377/>.

Hossain, Md. M. Maruf & Islam, Mohammad Mahmudul. *Ship Breaking Activities and Its Impact on the Coastal Zone of Chittagong, Bangladesh: Towards Sustainable Development*, (Chittagong: Advocacy & Publication Unit, YPSA, 2006), online: Young Power in Social Action <http://www.ypsa.org/publications/Impact.pdf>.

Kanthak Judit & Jayaraman, Nityanand. *Ships for Scrap III: Findings of a Greenpeace Study on Workplace and Environmental Contamination in Alang-Sosiya Ship Breaking Yards, Gujarat, India*, Steel and Toxic Wastes for Asia (The Netherlands: Greenpeace International, 2001), online: Greenpeace Nederland <http://www.greenpeaceweb.org/shipbreak/shipsforscrap3.pdf>.

Kummer, Katharina. *The Basel Convention and Related Legal Rules*, Professor Ian Brownlie, International Management of Hazardous Wastes, Oxford Monographs in International Law (New York: Oxford University Press, 1995).

Matser, Eco, Liu, Howard & Harjono, Marietta. *Ships for Scrap IV: Findings of a Greenpeace Visit to Four Shipbreaking Yards in China*, Steel and Toxic Wastes for Asia (The Netherlands: Stichting Greenpeace Nederland, 2001), online: Greenpeace Nederland <http://www.greenpeaceweb.org/shipbreak/shipsforscrap4.pdf>.

UNESCO. *Impacts and Challenges of Large Coastal Industry: Alang-Sosiya Ship-Breaking Yard, Gujarat, India*, Coastal Region and Small Island Papers 17 (Paris: UNESCO, 2004), online: UNESCO <http://www.unesco.org/csi/pub/papers3/assby.pdf>.

Vardar, Erdem & Harjono, Marietta. *Ships for Scrap V: Greenpeace Report on Environmental, Health and Safety Conditions in Alia˘ga Shipbreaking Yards, Izmir, Turkey*, Steel and Toxic Wastes for Asia (Greenpeace Mediterranean, 2002), online: Greenpeace Nederland <http://www.greenpeaceweb.org/shipbreak/shipsforscrap5.pdf>.

Unpublished Theses

Chircop, Aldo. *Cooperative Regimes in Ocean Management: A Study in Mediterranean Regionalism* (J.S.D. Thesis, Dalhousie Law School, Vol. 1, 1998) [unpublished].

Kaye, Stuart B. *International Fisheries Management: A Comparative Analysis of Legal Approaches to Management in the Context of Polar Fisheries Regimes* (J.S.D. Thesis, Dalhousie Law School, 1999) [unpublished].

Krogstrup, Ea & Arleth, Karen King Nash. *Ship Scrapping – a Floating Scenario* (International Development Studies Final Thesis, University of Roskilde, Denmark, 2006), online: RUDAR <http://rudar.ruc.dk/bitstream/1800/1521/1/Thefinalmaster.pdf>.

Puthucherril, Tony George. *Water Resources Management Law: A Case Study with Reference to the State of Kerala* (M.Phil. (Law) Thesis, W.B. National University of Juridical Sciences, Kolkata, India, 2003) [unpublished].

Vedeler, Katrine Vetaas. *From Cradle to Grave – Value Chain Responsibility in the Ship Scrapping Industry* (Master Thesis, Norwegian School of Economics and Business Administration, 2006), online: BORA-NHH <http://bora.nhh.no/>.

Other Materials

Brown, Julie McElroy. "Shipbreaking at Alang, India "What is the Right Thing for this Place?"" (2006) [unpublished, archived at University of Washington Libraries Digital Initiatives Unit], online: University of Washington Libraries <https://digital.lib.washington.edu/dspace/bitstream/1773/2630/1/McElroyBrown_project.pdf>.

Gold, Edgar. *International Maritime Law: Basic Principles Lecture Notes and Materials* (Faculty of Law, World Maritime University, Sweden, 30 June 1986).

International Metalworkers' Federation, *Status of Shipbreaking Workers in India: A Survey*, online: International Metalworkers' Federation <http://www.imfmetal.org/m ain/files/06042810465779/Shipbreaking_survey.pdf>.

Kazan-Allen, Laurie. *Le Clemenceau: Action and Reaction*, online: International Ban Asbestos Secretariat <http://www.btinternet.com/~ibas/lka_le_clemenceau.htm>.

Letter from Connie Hedegaard, Danish Minister for the Environment to A. Raja, Minister for Environment and Forests, India (15 April 2005), online: Greenpeace Nederland <http://www.greenpeaceweb.org/shipbreak/frederik.pdf>.

Letter from Gopalakrishna Thyagarajan to Bhanujan (24 May 2005), online: Greenpeace Nederland <http://www.greenpeaceweb.org/shipbreak/drthya.pdf>.

Manji, Firoze. Book Review of *Development as Freedom* by Amartya Sen, online: Fahamu <http://www.fahamu.org/downloads/DevelopmentasFreedom.pdf>.

Menon, Professor N.R. Madhava. "Role of the State and Judiciary in Advancing Equity and Equality in a Market Economy" (Address at the National Seminar on Equity & Equality in a Market Economy during the Golden Jubilee Celebrations of the Indian Law Institute, New Delhi, 5 August 2006) [unpublished].

Mikelis, Nikos. "Developments and Issues on Recycling of Ships" (Paper presented to The East Asian Seas Congress, Haikou City, China, 12 December 2006), online: International Maritime Organization <http://www.imo.org/includes/blastDataOnly.asp/data_id%3D17980/Developments.pdf>.

National Workshop on International Convention for Safe and Environmentally Sound Recycling of Ships (7–10 January 2008), online: Gujarat Maritime Board <http://www.gmbports.org/international_workshop.htm>.

Proceedings of Third Meeting of the Committee of Technical Experts on Management of Hazardous Waste Relating to Ship Breaking (20 May 2006) at Annex A-2, page 1, in the Affidavit dated 31 August 2006 filed by Dr. V. Rajagopalan, Member Convenor, Committee of Technical Experts on Ship Breaking in *Research Foundation* [on file with the author].

Puthucherril, Tony George. *One Step Forward, Two Steps Back: Constitutionalism and Environmental Jurisprudence in India,* online: OSGOODE Graduate Law Students' Association <http://www.osgoode.yorku.ca/glsa/2008conference/documents/Tony%20George.pdf>.

Raaymakers, Steve. "The Ballast Water Problem: Global Ecological, Economic and Human Health Impacts" (Paper Presented to the RECSO /IMO Joint Seminar on Tanker Ballast Water Management & Technologies, Dubai, UAE, 16–18 December 2002), online: International

Maritime Organization <http://www.imo.org/includes/blastDataOnly.asp/data_id%3D8595/RaaymakersGlobalImpactsPaper.pdf>.

Swift, Peter M. "How Can Ship Breaking Become a Sound Industry In its Own Right?" (PowerPoint presented in Shanghai, China, 3 March 2005), online: Intertanko <http://www.intertanko.com/templates/DocumentList.aspx?id=20528>.

Index

Legal Aspects of Sustainable Development

General Editor: David Freestone

1. K. Bastmeijer and T. Koivurova (eds), *Theory and Practice of Transboundary Environmental Impact Assessment.* 2008
ISBN 978 90 04 16479 6

2. C. Voigt, *Sustainable Development as a Principle of International Law: Resolving Conflicts between Climate Measures and WTO Law.* 2008
ISBN 978 90 04 16697 4

3. R. Warner, *Protecting the Oceans Beyond National Jurisdiction.* Strengthening the International Law Framework. 2009
ISBN 978 90 04 17262 3

4. S.M.A. Salman, *The World Bank Policy for Projects on International Waterways.* An Historical and Legal Analysis. 2009
ISBN 978 90 04 17837 3

5. T.G. Puthucherril, *From Shipbreaking to Sustainable Ship Recycling.* Evolution of a Legal Regime. 2010
ISBN 978 90 04 17491 7

6. M.A. Palma, M. Tsamenyi and W. Edeson, *Promoting Sustainable Fisheries.* The International Legal and Policy Framework to Combat Illegal, Unreported and Unregulated Fishing. 2010
ISBN 978 90 04 17575 4

www.ingramcontent.com/pod-product-compliance
Lightning Source LLC
LaVergne TN
LVHW011116300926
849160LV00008B/400
* 9 7 8 9 0 0 4 1 7 4 9 1 7 *